VLSI-SoC: RESEARCH TRENDS IN VLSI AND SYSTEMS ON CHIP

IFIP – The International Federation for Information Processing

IFIP was founded in 1960 under the auspices of UNESCO, following the First World Computer Congress held in Paris the previous year. An umbrella organization for societies working in information processing, IFIP's aim is two-fold: to support information processing within its member countries and to encourage technology transfer to developing nations. As its mission statement clearly states,

> *IFIP's mission is to be the leading, truly international, apolitical organization which encourages and assists in the development, exploitation and application of information technology for the benefit of all people.*

IFIP is a non-profitmaking organization, run almost solely by 2500 volunteers. It operates through a number of technical committees, which organize events and publications. IFIP's events range from an international congress to local seminars, but the most important are:

• The IFIP World Computer Congress, held every second year;
• Open conferences;
• Working conferences.

The flagship event is the IFIP World Computer Congress, at which both invited and contributed papers are presented. Contributed papers are rigorously refereed and the rejection rate is high.

As with the Congress, participation in the open conferences is open to all and papers may be invited or submitted. Again, submitted papers are stringently refereed.

The working conferences are structured differently. They are usually run by a working group and attendance is small and by invitation only. Their purpose is to create an atmosphere conducive to innovation and development. Refereeing is less rigorous and papers are subjected to extensive group discussion.

Publications arising from IFIP events vary. The papers presented at the IFIP World Computer Congress and at open conferences are published as conference proceedings, while the results of the working conferences are often published as collections of selected and edited papers.

Any national society whose primary activity is in information may apply to become a full member of IFIP, although full membership is restricted to one society per country. Full members are entitled to vote at the annual General Assembly, National societies preferring a less committed involvement may apply for associate or corresponding membership. Associate members enjoy the same benefits as full members, but without voting rights. Corresponding members are not represented in IFIP bodies. Affiliated membership is open to non-national societies, and individual and honorary membership schemes are also offered.

VLSI-SoC: RESEARCH TRENDS IN VLSI AND SYSTEMS ON CHIP

Fourteenth International Conference on Very Large Scale Integration of System on Chip (VLSI-SoC2006), October 16-18, 2006, Nice, France

Edited by

Giovanni De Micheli
Ecole Polytechnique Fédérale de Lausanne, CH

Salvador Mir
TIMA- Institut National Polytechnique de Grenoble, FR

Ricardo Reis
Universidade Federal do Rio Grande do Sul, BR

 Springer

Library of Congress Control Number: 2007934346

VLSI-SoC Research Trends in VLSI and Systems on Chip

Edited by G. De Micheli, S. Mir, and R. Reis

    p.   cm. (IFIP International Federation for Information Processing, a Springer Series in Computer Science)

    ISSN: 1571-5736 / 1861-2288 (Internet)
    ISBN: 978-0-387-74908-2
    eISBN: 978-0-387-74909-9

Printed on acid-free paper

9 8 7 6 5 4 3 2 1
springer.com

CONTENTS

PREFACE

This book contains extended and revised versions of the best papers that were presented during the fourteenth edition of the IFIP TC10/WG10.5 International Conference on Very Large Scale Integration, a global System-on-a-Chip Design & CAD conference. The 14th conference was held at the Hotel Boscolo, Nice, France (October 16-18, 2006). Previous conferences have taken place in Edinburgh, Trondheim, Vancouver, Munich, Grenoble, Tokyo, Gramado, Lisbon, Montpellier, Darmstadt and Perth.

The purpose of this conference, sponsored by IFIP TC 10 Working Group 10.5 and by the IEEE Council on Electronic Design Automation (CEDA), is to provide a forum to exchange ideas and show industrial and academic research results in the field of microelectronics design. The current trend toward increasing chip integration and technology process advancements brings about stimulating new challenges both at the physical and system-design levels, as well in the test of these systems. VLSI-SOC conferences aim to address these exciting new issues.

The 2006 edition of VLSI-SoC maintained the traditional structure, which has been successful at the previous VLSI-SOC conferences. The quality of submissions (154 papers from 26 countries) made the selection process difficult, but finally 54 papers and 16 posters were accepted for presentation in VLSI-SoC 2006. Out of the 54 full papers presented at the conference, 13 regular papers and 6 papers from special sessions were chosen by a selection committee to have an extended and revised version included in this book. These selected papers came from Brazil, Belgium, France, Italy, Japan, The Netherlands, Portugal, Singapore, Spain, Switzerland, the U.K. and the United States of America.

Furthermore, this book includes two papers related to invited talks presented at the conference as part of a panel, by Shekhar Borkar from Intel Corporation and Khrisna Palem, from Georgia Institute of Technology.

VLSI-SoC 2006 was the culmination of many dedicated volunteers: paper authors, reviewers, session chairs, invited speakers and various committee chairs, especially the local arrangements organizers. We thank them all for their contribution.

This book is intended for the VLSI community mainly to whom that did not have the chance to take part in the VLSI-SOC 2006 Conference. The papers were selected to cover a wide variety of excellence in VLSI technology and the advanced research they describe. We hope you will enjoy reading this book and find it useful in your professional life and to the development of the VLSI community as a whole.

The editors

Architectures for High Dynamic Range, High Speed Image Sensor Readout Circuits

Sam Kavusi, Kunal Ghosh, and Abbas El Gamal

Department of Electrical Engineering, Stanford University, Stanford CA 94305, USA

Abstract. The stringent performance requirements of many infrared imaging applications warrant the development of precision high dynamic range, high speed focal plane arrays. In addition to achieving high dynamic range, the readout circuits for these image sensors must achieve high linearity and SNR at low power consumption. We first review four high dynamic range image sensor schemes that have been developed for visible range imaging and discuss why they cannot meet the stringent performance demands of infrared imaging. We then describe a new dynamic range extension scheme, Folded Multiple Capture, that can meet these performance requirements. Dynamic range is extended using synchronous self-reset while high SNR is maintained using few non-uniformly spaced captures and least-squares fit to estimate pixel photocurrent. We conclude with a description of a prototype of this architecture targeted for 3D-IC IR focal plane arrays.

1 Introduction

Precision high dynamic range (HDR), high speed imaging is finding growing applications in the automotive, surveillance, tactical, industrial, and medical and diagnostic instrumentation (e.g., fluorescence detection and spectroscopy) arenas. These applications can be broadly segmented into those operating in the visible range (typically 400nm $< \lambda <$ 800nm) and those operating in the infrared (IR) range (typically 4μm $< \lambda < 12\mu$m). Precision HDR, high speed IR imaging applications, specifically, are fraught with challenges. In addition to the ability to capture scenes with large variations in irradiance due to object temperatures, the imaging system must be able to deal with undesirable scene disturbances, due to, for example, sun reflection or laser jamming. The imaging system must also have highly linear, shot noise limited readout in order to achieve the stringent sensitivity requirements. In [1], it is argued that low power IR focal plane arrays (FPAs) with > 120dB dynamic range operating at 1000 frames/sec are needed for such applications. These performance requirements are far more aggressive than is achievable with present-day IR FPAs.

The advent of 3D-IC technology, which has increased the effective pixel area available, enables implementation of recently-developed HDR schemes with high pixel count. In this chapter, we discuss the main design and implementation challenges that limit the performance of these architectures. We find that these

Please use the following format when citing this chapter:

Kavusi, S., Ghosh, K. and El Gamal, A., 2007, in IFIP International Federation for Information Processing, Volume 249, VLSI-SoC: Research Trends in VLSI and Systems on Chip, eds. De Micheli, G., Mir, S., Reis, R., (Boston: Springer), pp. 1–23

architectures cannot achieve the high speed, HDR requirements without a significant penalty in pixel area and power consumption, and therefore do not lend themselves well to the aforementioned IR imaging applications. We then describe an HDR extension scheme, called Folded Multiple Capture (FMC) [7], that can achieve all the requirements stated in [1].

The rest of this chapter is organized as follows. In Section 2, we first briefly review the fundamentals of image sensors and introducing some needed terminology. In Sections 3-6, we discuss several of the recently-developed HDR schemes. In Section 7, we discuss the architecture and operation of FMC, implementation of a prototype, and experimental results obtained.

2   Background

An image sensor consists of an array of photodetectors followed by circuits for readout. Sensor performance is therefore a function of both the photodetector used and the readout circuits. Each photodetector in a conventional image sensor, e.g., CCD, CMOS APS, or IR FPA, converts incident photon flux into photocurrent i_{ph}. In visible range imaging, the incident photon flux corresponds to light reflected off of objects in the scene, while in IR imaging, the incident photon flux corresponds to object thermal radiation. A simplified Signal-to-Noise ratio (SNR) of the integrated photocurrent is given by

$$\text{SNR}(i_{ph}) = \frac{(i_{ph}t_{\text{int}})^2}{qi_{ph}t_{\text{int}} + q^2\sigma_{\text{Readout}}^2}, \quad \text{for } i_{ph} \leq \frac{qQ_{\text{max}}}{t_{\text{int}}},$$

where t_{int} is the integration time, q is the charge of an electron, Q_{max} is the saturation charge or well capacity, and σ_{Readout} is the readout noise. Note that this simplified SNR only considers integrated shot noise and readout noise and assumes that correlated-double-sampling (CDS) is performed, thus eliminating the reset noise and offset contributions. We also assume that gain FPN and dark current are either negligible or calibrated for, as is usually the case for state-of-the-art visible and IR sensors.

Image sensor dynamic range (DR) is defined as the ratio of the largest nonsaturating photocurrent to the minimum detectable photocurrent, typically defined as the standard deviation of the noise under dark conditions. In visible range imaging, this corresponds to the range of intrascene illumination levels that can be imaged, while in IR imaging, this corresponds to the range of intrascene temperatures that can be imaged. Assuming the above sensor model, $i_{\text{max}} = qQ_{\text{max}}/t_{\text{int}}$ and $i_{\text{min}} = q\sigma_{\text{Readout}}/t_{\text{int}}$ and dynamic range is given by

$$\text{DR} = \frac{i_{\text{max}}}{i_{\text{min}}} = \frac{Q_{\text{max}}}{\sigma_{\text{Readout}}}.$$

Since the dynamic range of image sensors is generally limited by the readout circuitry, HDR extension schemes modify a conventional sensor's readout circuits to improve its DR. Extending DR at the low end requires reducing i_{min}, which can be achieved by either reducing σ_{Readout} or increasing t_{int}. DR extension at the low end obtained by decreasing the diode or sense node capacitance to reduce σ_{Readout}, as is usually done in visible range image sensors, reduces Q_{max} which is not desirable for IR imaging. Extending dynamic range at the high end requires increasing i_{max}, which can be achieved by adapting the integration times to photocurrent or increasing the effective well capacity.

In the following sections we review some of the recently developed HDR architectures.

3   Time-to-Saturation

The time-to-saturation scheme [9] attempts to achieve high dynamic range with high SNR by converting each photocurrent i_{ph} into its time-to-saturation t_{sat} $(i_{ph}) = qQ_{\text{max}}/i_{ph}$. A block diagram of the scheme and plot of the integrator output as a function of time are given in Figure 1. After the photodiode and the time reference capacitor $C_{\text{T-Ref}}$ are reset, the output of the integrator is read out for CDS. Photocurrent is then integrated and converted to voltage, which is compared to a reference V_{max}. Concurrently, $C_{\text{T-Ref}}$ follows the time-ramp. When the integrator output reaches V_{max}, the comparator flips and $t_{\text{sat}}(i_{ph})$ is stored on $C_{\text{T-Ref}}$. At the end of integration, $v(t_{\text{int}})$ and t_{sat} are read out. If $t_{\text{sat}} < t_{\text{int}}$, the signal is estimated as $qQ_{\text{max}}/t_{\text{sat}}$, otherwise the signal is estimated using $v(t_{\text{int}})$ only.

The filter is defined by

$$\hat{i_{ph}} = \frac{Cv(t)}{t_{\text{sat}}}.$$

Note that the minimum detectable signal is given by $i_{\text{min}} = q\sigma_{\text{Readout}}/t_{\text{int}}$, which has the same form as that of the conventional sensor. The maximum nonsaturating signal depends on the comparator delay and offset as well as the noise associated with $t_{\text{sat}}(i_{ph})$ due to time-ramp noise, kTC of $C_{\text{T-Ref}}$, and the readout noise. Let σ_{sat} be the total rms of the noise added to $t_{\text{sat}}(i_{ph})$, then the maximum detectable signal is given by $i_{\text{max}} = qQ_{\text{max}}/\sigma_{\text{sat}}$. Therefore, the maximum achievable dynamic range for a given t_{int} is given by

$$\text{DR} = \frac{Q_{\text{max}}t_{\text{int}}}{\sigma_{\text{Readout}}\sigma_{\text{sat}}}.$$

To calculate SNR, note that qQ_{max} corresponds to the expected value of the integrator charge at t_{sat}. Assuming CDS the integrator charge is given by

$$Q(t_{\text{sat}}) = \frac{1}{q}\left(i_{ph}t_{\text{sat}} + Q_{\text{Shot}} + Q_{\text{Readout}}\right).$$

Note that t_{sat} has an Inverse Gaussian distribution [8]. Linear approximation of $\hat{i}_{ph}$ can be used to calculate the error term caused by T_{Error}, readout noise of

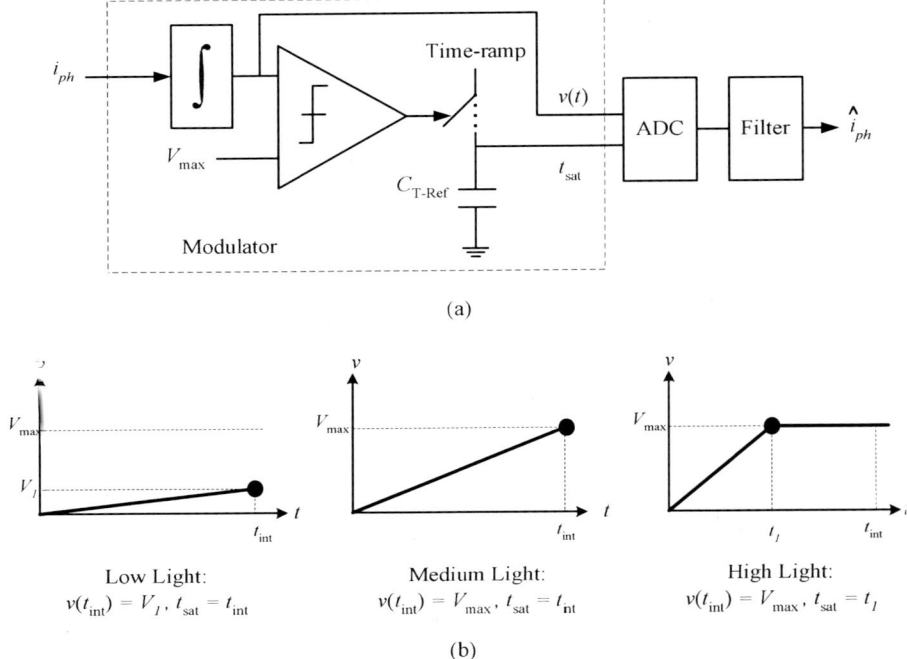

(a)

(b)

Fig. 1. Time-to-saturation block diagram.

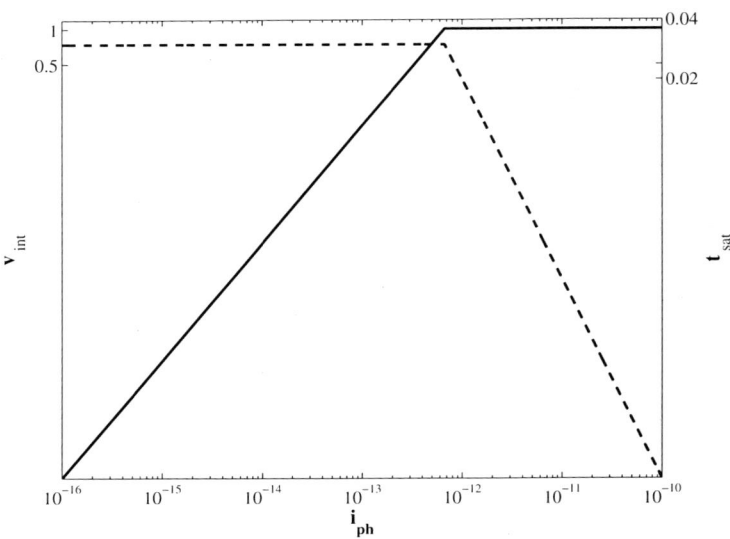

Fig. 2. Time-to-saturation signals.

time-to-saturation:

$$
\begin{aligned}
\hat{i}_{ph} &= \frac{qQ_{\max}}{t_{\text{sat}} + T_{\text{Error}}} \\
&= \frac{(i_{ph} t_{\text{sat}} + Q_{\text{Shot}} + Q_{\text{Readout}})}{t_{\text{sat}} + T_{\text{Error}}} \\
&\approx i_{ph} + \frac{Q_{\text{Shot}}}{t_{\text{sat}}} + \frac{Q_{\text{Readout}}}{t_{\text{sat}}} - i_{ph} \frac{T_{\text{Error}}}{t_{\text{sat}}}
\end{aligned}
$$

Therefore, the total output noise power is given by

$$
\sigma_i^2 =
\begin{cases}
\dfrac{q i_{ph}}{t_{\text{int}}} + \dfrac{q^2 \sigma_{\text{Readout}}^2}{t_{\text{int}}^2} & \text{if } i_{ph} \le \dfrac{qQ_{\max}}{t_{\text{int}}} \\[2ex]
\dfrac{q i_{ph}}{t_{\text{sat}}(i_{ph})} + \dfrac{q^2 \sigma_{\text{Readout}}^2}{t_{\text{sat}}(i_{ph})^2} + \dfrac{(qQ_{\max})^2 \sigma_{\text{sat}}^2}{t_{\text{sat}}(i_{ph})^4} & \text{if } i_{ph} > \dfrac{qQ_{\max}}{t_{\text{int}}},
\end{cases}
$$

and SNR is thus given by

$$
\text{SNR}(i_{ph}) =
\begin{cases}
\dfrac{(i_{ph} t_{\text{int}})^2}{q i_{ph} t_{\text{int}} + q^2 \sigma_{\text{Readout}}^2} & \text{if } i_{ph} \le \dfrac{qQ_{\max}}{t_{\text{int}}} \\[2ex]
\dfrac{(qQ_{\max})^2}{q^2 Q_{\max} + (i_{ph} \sigma_{\text{sat}})^2 + q^2 \sigma_{\text{Readout}}^2} & \text{if } i_{ph} > \dfrac{qQ_{\max}}{t_{\text{int}}}.
\end{cases}
$$

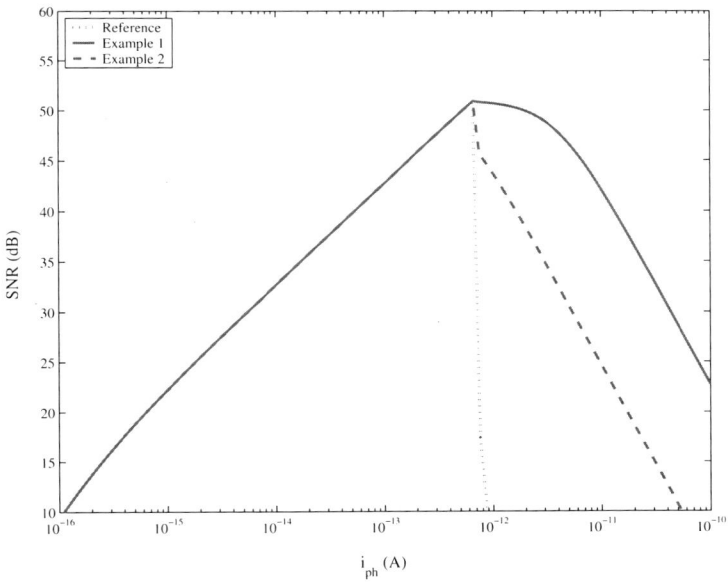

Fig. 3. SNR versus i_{ph} for time-to-saturation for two different examples.

Figure 3 plots SNR versus i_{ph} assuming $Q_{\text{sat}} = 125,000e-$, $\sigma_{\text{Readout}} = 5e-$, $t_{\text{int}} = 30msec$. Example 1 assumes $\sigma_{\text{sat}} = 0.0005t_{\text{int}}$ and achieves DR= 156dB. Example 2 assumes $\sigma_{\text{sat}} = 0.004t_{\text{int}}$ and achieves DR= 136dB.

Note that SNR is identical to that of the reference sensor within the reference DR. For larger i_{ph}, SNR drops monotonically as $1/i_{ph}^2$ (-20dB per decade) due to the effect of σ_{sat}. Thus, DR is increased but at the expense of reduction in SNR. In Example 2 σ_{sat} is the dominant term degrading SNR at high end.

In the original implementation [10] only the time-to-saturation of the pixel is readout and therefore at low end cannot be differentiated producing $t_{\text{sat}} = t_{\text{int}}$ (see Figure 2). By adding the residue readout in [9,11] the time-to-saturation is combined with a conventional readout and achieve very high dynamic range. The digital samples read out include the integrator voltage $v(t_{\text{int}})$ and the time to saturation t_{sat} shown in Figure 2. Using this readout, the integrating capacitor dynamic range is multiplied by the dynamic range of $C_{\text{T-Ref}}$ and thus a much wider dynamic range is achieved in a small area.

More details concerning implementation and correction for several nonidealities can be found in [9,11]. Note that the implementation in [9] actually uses two time-ramps and two capacitors to reduce σ_{sat}. We accounted for this indirectly by using small σ_{sat} in the examples.

4   Multiple-Capture

The multiple-capture scheme [12,13] increases dynamic range by sampling the signal nondestructively multiple times during integration. The HDR image can be constructed using the last-sample-before-saturation algorithm [12] as illustrated in Figure 4. Note that with this algorithm DR is only increased at the high end. DR at the low end can be increased by a combination of image blur prevention and weighted averaging [15], which requires significant on-chip memory and DSP capability.

To define DR and SNR, we assume uniform sampling time t_{capt} and that the filter only performs last-sample-before-saturation and digital CDS. The maximum nonsaturating signal is given by $i_{\max} = qQ_{\max}/t_{\text{capt}}$ and the minimum detectable signal is given by $i_{\min} = q\sigma_{\text{Readout}}/t_{\text{int}}$. Thus

$$\text{DR} = \frac{Q_{\max}t_{\text{int}}}{\sigma_{\text{Readout}}t_{\text{capt}}}.$$

To define SNR note that it follows that of a conventional sensor for $t_{\text{sat}}(i_{ph}) \geq t_{\text{int}}$. In the extended range we use

$$\text{SNR} \approx \phi(i_{ph})Q_{\max}$$

approximation, where ϕ is the normalized integrator output voltage. Note that $\phi(i_{ph}) = t_{\text{last-sample}}(i_{ph})/(t_{\text{sat}}(i_{ph}))$, where $t_{\text{sat}}(i_{ph}) = qQ_{\max}/i_{ph}$.

Note that when $t_{\text{sat}}(i_{ph}) < t_{\text{int}}$

$$t_{\text{last-sample}}(i_{ph}) = \left\lfloor \frac{t_{\text{sat}}(i_{ph})}{t_{\text{capt}}} \right\rfloor t_{\text{capt}},$$

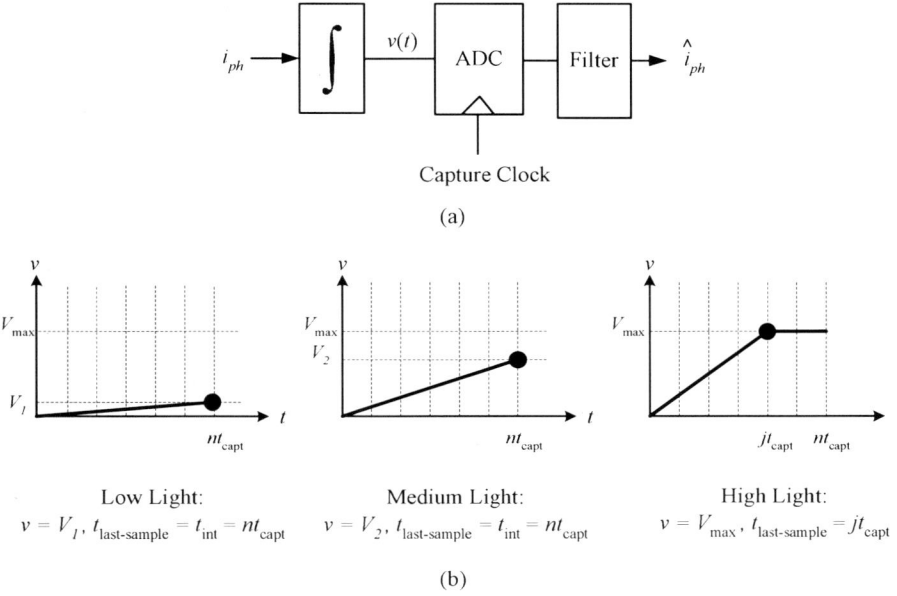

Fig. 4. Multiple-capture block diagram.

and therefore $(1 - t_{\text{capt}}/t_{\text{sat}}(i_{ph})) Q_{\max} < \text{SNR}(i_{ph}) < Q_{\max}$.

Figure 5 plots SNR versus i_{ph} assuming $Q_{\text{sat}} = 125,000e-$, $t_{\text{int}} = 30msec$. Example 1 assumes 10 bit ADC, $\sigma_{\text{Readout}} = 35e-$, $t_{\text{capt}} = 150\mu sec$ and achieves DR= 117dB. Example 2 assumes 9 bit ADC, $\sigma_{\text{Readout}} = 70e-$, $t_{\text{capt}} = 100\mu sec$ and achieves DR= 114dB. The parameters t_{capt} and σ_{Readout} assume comparison time of 100ns and readout time per row per bit of 10nsec and 512×512 pixel array.

Note that unlike time-to-saturation, SNR does not degrade in the extended range, since $t_{\text{last-sample}}$ has the same accuracy as the multiple capture clock. However, DR suffers at the low end due to the large quantization noise of the per pixel ADC.

Moderate dynamic range increase has been also achieved using dual capture technique for CCD and CMOS APS sensors [18]. In dual capture, a scene is imaged only twice, once using a short integration time and another using a much longer integration time, and the two images are combined into a high dynamic range image. In [14] the authors cleverly take a second capture with a short integration time while the rest of the image is read out using dual outputs. There is a trade-off between the extension achieved by dual capture and the minimum SNR over the extended range. To find the SNR dip over its peak, minimum ϕ can be calculated. For example in [14] $\phi = 1/64$ and minimum SNR = 30dB with $Q_{\max} = 60,000e$.

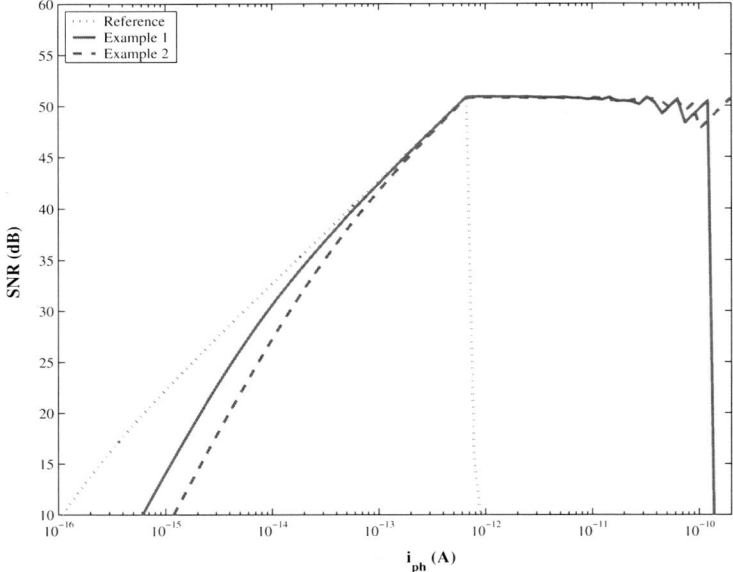

Fig. 5. SNR versus i_{ph} for multiple-capture for two different examples.

5   Extended Counting $\Sigma\Delta$

We first discuss the first order $\Sigma\Delta$ readout scheme and its variations. In Subsection 5.2, we analyze the extended counting scheme.

5.1   First Order Incremental $\Sigma\Delta$

The block diagram of the first order single-bit $\Sigma\Delta$ [19] is shown in Figure 6. At each clock cycle, the integrator output $v(t)$ is compared to the threshold value $V_{max}/2$. If the comparator flips, $V_{max}/2$ is subtracted off $v(t)$, thus preventing saturation of the integrator. The subtraction is typically implemented using a switched capacitor circuit. Note that the number of resets during the exposure time is proportional to the photocurrent value. A filter, which can be as simple as a counter, is used to estimate the photocurrent from the binary comparator output sequence. In *incremental* $\Sigma\Delta$ [17], the integrator is reset at the beginning of each frame. Such resetting improves SNR [19], because, unlike the free-running case, the integrator value at the beginning of each frame is known [5]. We, therefore, focus on incremental $\Sigma\Delta$.

To quantify the SNR and DR achieved by incremental $\Sigma\Delta$, we use the equivalent integrator output ramp shown in Figure 7. Note that the output sequence from the $\Sigma\Delta$ modulator is identical to the sequence generated by comparing the

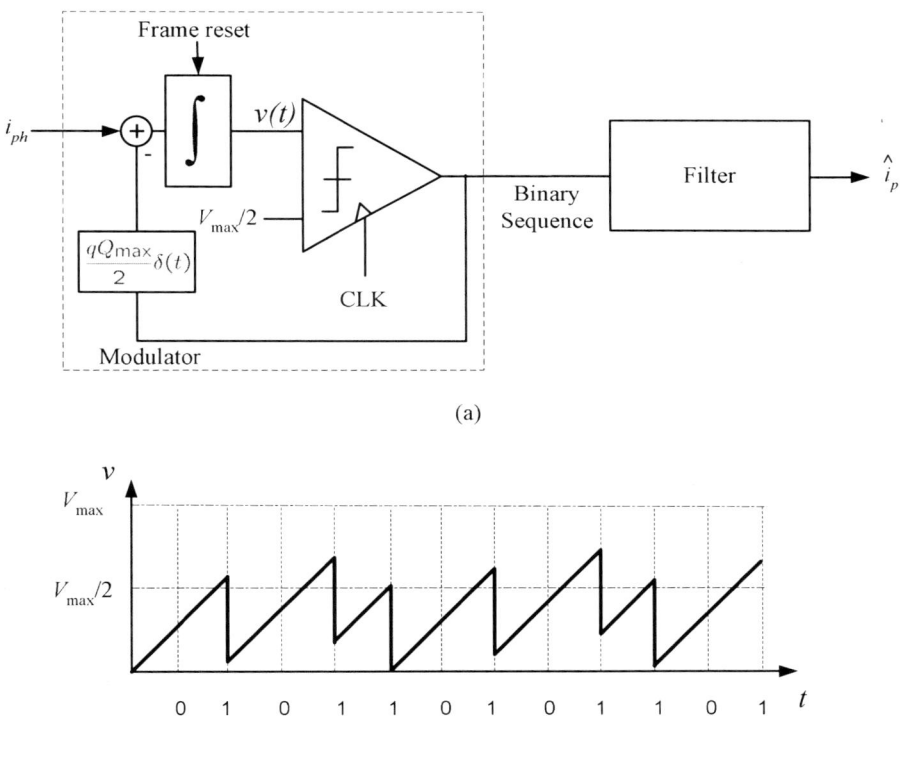

(a)

(b)

Fig. 6. Single-bit incremental $\Sigma\Delta$ block diagram.

equivalent ramp to the cumulative sum of the sequence, scaled by and biased by $V_{max}/2$.

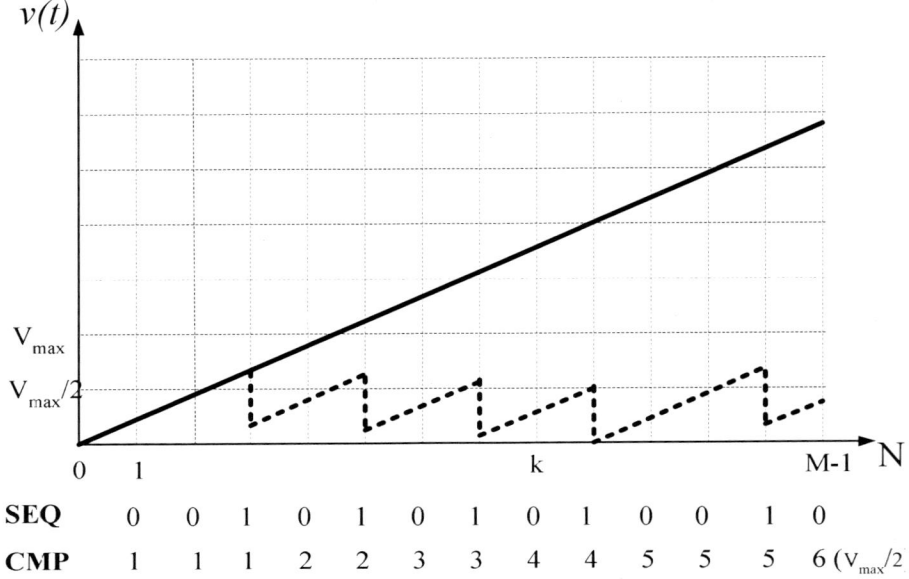

Fig. 7. Equivalence of $\Sigma\Delta$ output sequence (SEQ) to the sequence obtained by comparing the equivalent ramp (solid line) to the cumulative sum (CMP) of the sequence scaled and biased by $V_{max}/2$.

Assuming that a counter is used for decimation, then at the end of integration time, the counter value is

$$n_{counter}(i_{ph}) = \lfloor 2i_{ph}t_{int}/CV_{max} \rfloor.$$

Now, assuming that the $V_{max}/4$ bias in the counter readout is compensated for, and that quantization noise is signal independent and uncorrelated with other noise sources, the standard deviation of the effective readout noise is approximately given by

$$\sigma_{Readout-eff} = \sqrt{\frac{(CV_{max})^2}{48q^2} + n_{counter}(i_{ph})\sigma_{Switch}^2 + \sigma_{Reset}^2},$$

where, σ_{Switch} is the noise due to charge subtraction and σ_{Reset} is the reset noise. The first term in $\sigma_{Readout-eff}$ corresponds to quantization noise $\Delta^2/12$ with $\Delta = V_{max}/2$. Therefore, the minimum detectable signal is given by

$$i_{min} = q\sigma_{Readout-eff}/t_{int} \approx CV_{max}/4\sqrt{3}t_{int}.$$

From Figure 7, the maximum non-saturating signal is given by

$$i_{max} = CV_{max}/2t_{clk}.$$

Therefore, the maximum achievable dynamic range for a given t_{int} is given by

$$DR = \frac{2\sqrt{3}t_{int}}{t_{clk}}.$$

In order to derive SNR, we need to consider the variation in charge subtraction, which translates into gain FPN. Denoting the standard deviation of charge subtraction by σ_{Offset}, we obtain

$$SNR(i_{ph}) = \frac{(i_{ph}t_{int})^2}{qi_{ph}t_{int} + (q\sigma_{Readout-eff})^2 + (n_{counter}q\sigma_{Offset})^2}.$$

Figure 8 plots SNR versus i_{ph} and compares it with SNR of the reference sensor. Both examples assume $Q_{max} = 125,000e-$, $t_{int} = 30msec$, $t_{clk} = 10\mu sec$, $\sigma_{Switch} = 57e-$ and achieve DR= 80dB. Example 1 assumes $\sigma_{Offset} = 0.0005Q_{max}$. Example 2 assumes $\sigma_{Offset} = 0.01Q_{max}$.

Note that with the same Q_{max} and t_{int} the DR of this scheme is shifted to the right with respect to the reference sensor DR, that is, this scheme has very poor low signal performance. Also note that SNR at the low end is quantization limited, whereas at the high end, it becomes gain FPN limited. The reason for the SNR degradation at the low end is the coarseness of the single-bit quantization and the filter used.

Reducing the size of the integrating capacitor or lowering V_{max} may improve low end performance. However, these solutions increase σ_{Offset}, which would degrade SNR at the high end. SNR at the low end can also be improved by using more sophisticated filters such as triangular, zoomer, recursive, etc. To demonstrate the extent of possible SNR improvement, in Figure 9 we compare the performance using a counter to that using the optimal filter [5]. Examples assume $Q_{max} = 125,000e-$, $t_{int} = 30msec$, $t_{clk} = 10\mu sec$, $\sigma_{Switch} = 57e-$, $\sigma_{Offset} = 0.0005Q_{max}$ and achieve DR= 80dB. Note that substantial improvement in SNR is possible, but at the expense of higher digital circuit complexity and increased power consumption, required to achieve thermal noise level below quantization noise. As discussed, SNR at the high end is limited by the gain FPN due to variation in charge subtraction.

The extended counting scheme we discuss in the following section solves the coarse quantization problem of the single-bit $\Sigma\Delta$ schemes by quantizing the residue at the end of integration, $v(t_{int})$, using a multi-bit ADC.

5.2 Extended Counting

A block diagram of the extended counting scheme [16] is shown in Figure 10. Except for the additional residue ADC step, the architecture is identical to the single-bit $\Sigma\Delta$ architecture with a counter, discussed in the previous section.

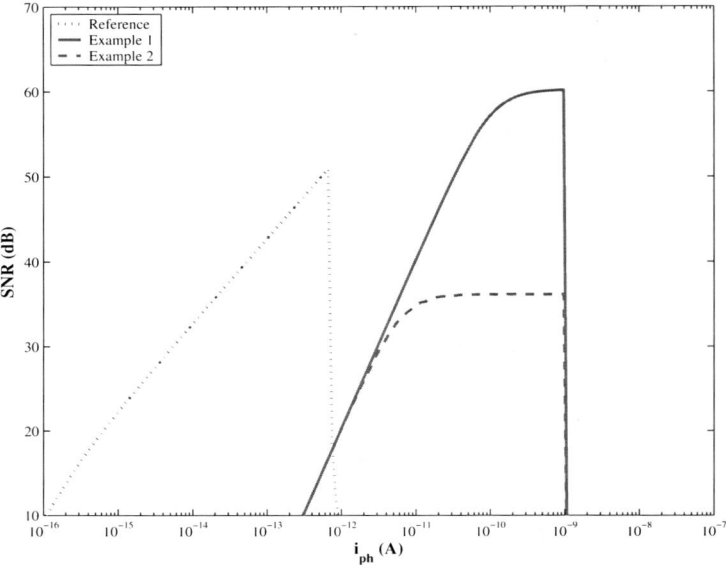

Fig. 8. SNR versus i_{ph} for single-bit incremental $\Sigma\Delta$ ADC for two different examples.

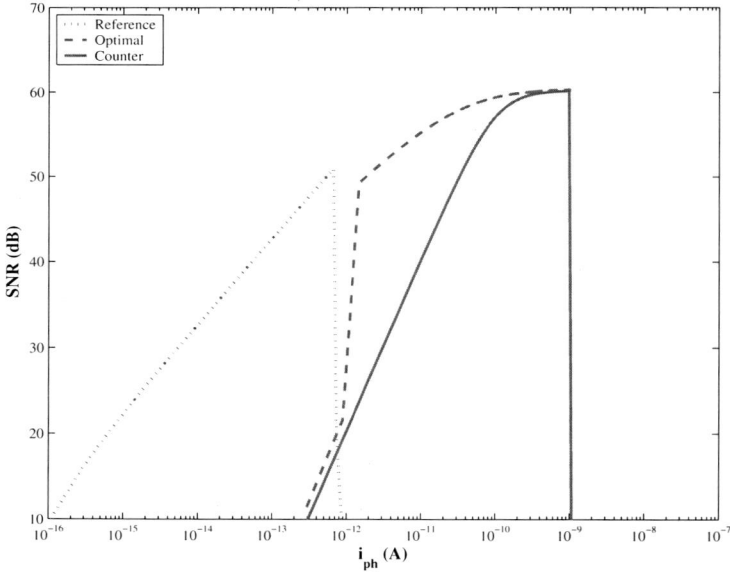

Fig. 9. SNR versus i_{ph} for single-bit $\Sigma\Delta$ with counter and optimal filter.

The counter value at the end of the integration time and the digitized residue are combined to estimate the photocurrent as

$$\hat{i}_{ph} = \frac{qQ_{max}}{t_{int}}\left(\frac{1}{2}n_{counter} + \frac{v(t_{int})}{V_{max}}\right).$$

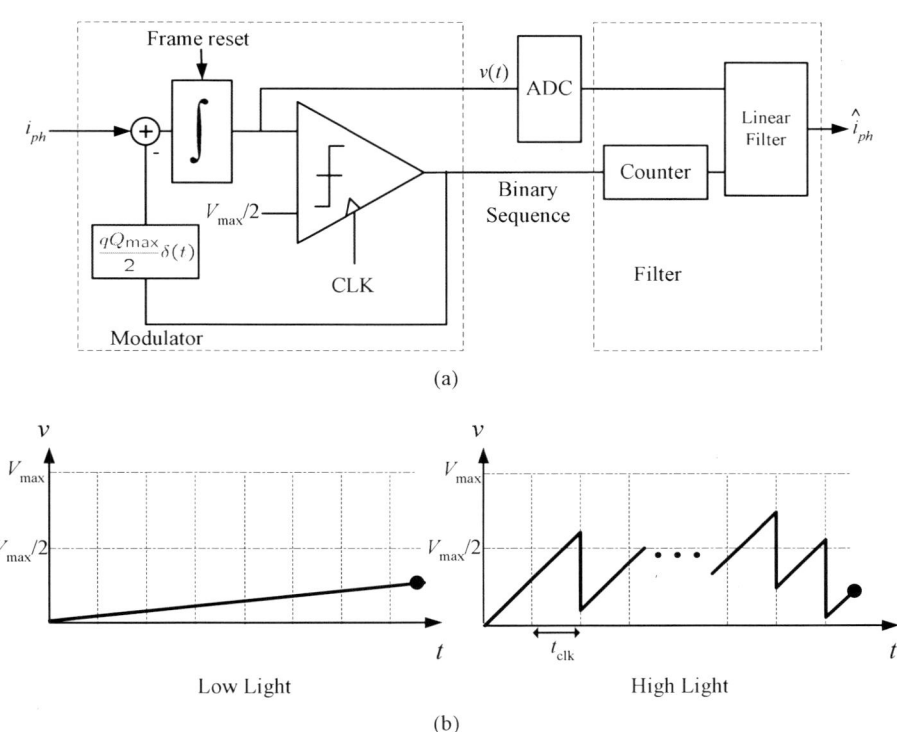

Fig. 10. Extended counting block diagram.

In order to calculate DR and SNR, we note that the standard deviation of the effective readout noise is given by

$$\sigma_{Readout-eff} = \sqrt{\sigma_{ADC-Readout}^2 + n_{counter}(i_{ph})\sigma_{Switch}^2 + \sigma_{Reset}^2},$$

where, $\sigma_{ADC-Readout}$ is the quantization noise, σ_{Switch} is the switched capacitor noise due to charge subtraction, σ_{Reset} is the reset noise and $n_{counter}(i_{ph})$ is the counter output at the end of t_{int}. Thus, the minimum detectable and maximum non-saturating signals are

$$i_{min} = q\sigma_{Readout-eff}/t_{int} = q\sqrt{\sigma_{ADC-Readout}^2 + \sigma_{Reset}^2}/t_{int}, \text{ and}$$

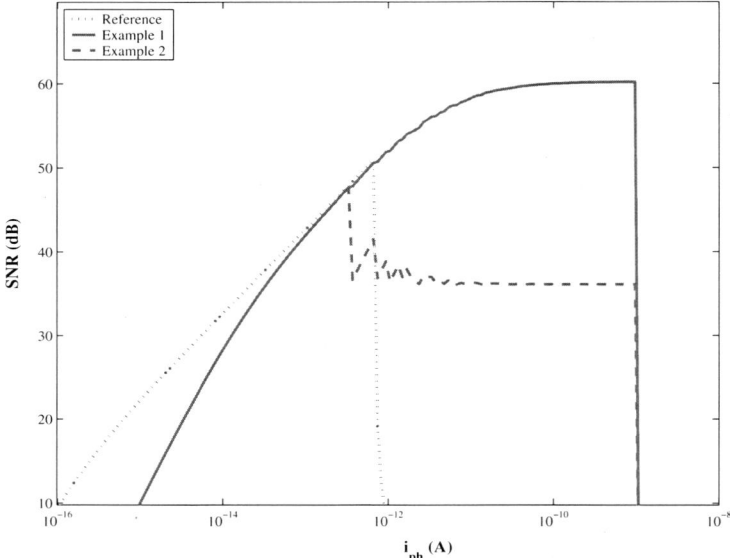

Fig. 11. SNR versus i_{ph} for extended-counting for two different examples.

$$i_{\max} = CV_{\max}/2t_{\mathrm{clk}}.$$

Therefore, the maximum achievable dynamic range for a given t_{int} is given by

$$\mathrm{DR} = \frac{Q_{\max}t_{\mathrm{int}}}{2\sqrt{\sigma_{\mathrm{ADC-Readout}}^2 + \sigma_{\mathrm{Reset}}^2 t_{\mathrm{clk}}}}.$$

In order to derive SNR, note that any variation of charge subtraction, σ_{Offset} will translate to gain fixed pattern noise. Thus SNR is given by

$$\mathrm{SNR}(i_{ph}) = \frac{(i_{ph}t_{\mathrm{int}})^2}{qi_{ph}t_{\mathrm{int}} + (q\sigma_{\mathrm{Readout}})^2 + (n_{\mathrm{counter}}q\sigma_{\mathrm{Offset}})^2}.$$

The SNR is plotted versus signal in Figure 11 assuming $\sigma_{\mathrm{Switch}} = 57e-$, $\sigma_{\mathrm{ADC-Readout}} = 9e-$ and achieving DR= 130dB. Example 1 assumes $\sigma_{\mathrm{Offset}} = 0.0005Q_{\max}$. Example 2 assumes $\sigma_{\mathrm{Offset}} = 0.01Q_{\max}$.

6   Synchronous Self-reset with Residue Readout

In this section we discuss the synchronous self-reset with residue readout scheme proposed in [21]. The scheme is described in Figure 12. The photocurrent is integrated and converted into voltage $v(t)$, which is periodically compared to a reference voltage $V_{\max}$. If $v(t) \geq V_{\max}$, the comparator switches, the integrator is reset, and the counter is incremented. At the end of integration, the digitized value of $v(t_{\mathrm{int}})$ and the reset count are combined to estimate the photocurrent. Let n_{Reset} be the number of resets, then

$$\hat{i}_{ph} = \frac{qQ_{\max}}{t_{\mathrm{int}}} \left(n_{\mathrm{Reset}} + \frac{v(t_{\mathrm{int}})}{V_{\max}} \right).$$

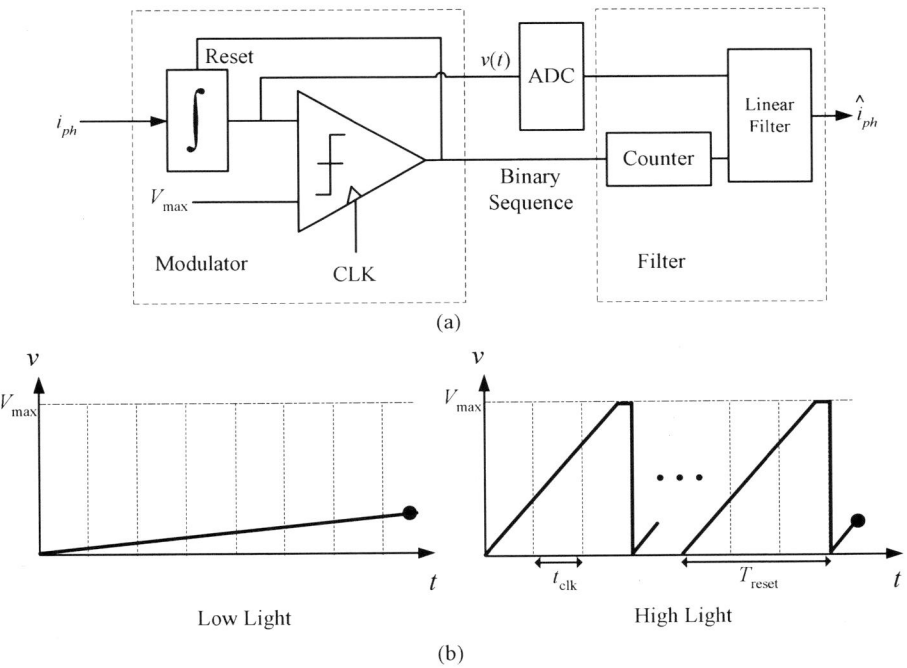

Fig. 12. Synchronous self-reset block diagram.

To compute DR and SNR, we first compute the distortion due to the underestimation of charge resulting from saturation before synchronous resetting takes place (see the waveform in Figure 8(b)). At the high end, assuming no noise $T_{\mathrm{reset}} = \lceil qQ_{\max}/(i_{ph}t_{\mathrm{clk}}) \rceil t_{\mathrm{clk}}$, and the counter output is given by $n_{\mathrm{Reset}} = \lfloor t_{\mathrm{int}}/T_{\mathrm{reset}} \rfloor$. Therefore, we can write

$$i_{ph} - \frac{i_{ph}t_{\mathrm{clk}}}{t_{\mathrm{int}}} \times \left\lfloor \frac{t_{\mathrm{int}}}{\lceil qQ_{\max}/(i_{ph}t_{\mathrm{clk}}) \rceil t_{\mathrm{clk}}} \right\rfloor < \hat{i}_{ph} < i_{ph}.$$

The total average distortion power is therefore given by

$$\sigma_{\text{Distortion}}^2 = \frac{1}{3} \left(\frac{i_{ph} t_{\text{clk}}}{t_{\text{int}}} \times \left\lfloor \frac{t_{\text{int}}}{\lceil qQ_{\max}/(i_{ph} t_{\text{clk}}) \rceil t_{\text{clk}}} \right\rfloor \right)^2 .$$

To find the total noise power we need to add contributions from shot noise, reset noise, residue readout noise, and gain FPN. To estimate the average noise power due to shot noise, we approximate the total integration time for shot noise by t_{int}. Gain FPN in the extended DR is due in part to the comparator and reset offsets that result in offset variation, σ_{Offset}, in $Q_{\max}$. Combining these noise terms with the distortion, we obtain the total noise power

$$\sigma_i^2 = \sigma_{\text{Distortion}}^2 + \frac{q i_{ph}}{t_{\text{int}}} + (n_{\text{Reset}} + 1) \left(\frac{q \sigma_{\text{Reset}}}{t_{\text{int}}} \right)^2 + \left(\frac{q \sigma_{\text{Readout}}}{t_{\text{int}}} \right)^2 +$$

$$+ \left(\frac{q \sigma_{\text{Offset}} n_{\text{Reset}}}{t_{\text{int}}} \right)^2 + (\sigma_H i_{ph})^2 .$$

Therefore SNR is given by

$$\text{SNR}(i_{ph}) = \frac{(i_{ph} t_{\text{int}})^2}{\sigma_i^2} .$$

To compute DR for the scheme, note that $i_{\min}$ is given by $i_{\min} = q \sqrt{\sigma_{\text{Readout}}^2 + \sigma_{\text{Reset}}^2} / t_{\text{int}}$ and $i_{\max} = \sqrt{3} q Q_{\max} / t_{\text{clk}}$. Therefore,

$$\text{DR} = \frac{\sqrt{3} Q_{\max} t_{\text{int}}}{\sqrt{\sigma_{\text{Readout}}^2 + \sigma_{\text{Reset}}^2} t_{\text{clk}}} .$$

Figure 13 plots SNR versus i_{ph} for two examples, assuming $Q_{\text{sat}} = 125{,}000e-$, $\sigma_{\text{Readout}} = 35e-$, $t_{\text{int}} = 30msec$, $t_{\text{clk}} = 1\mu sec$, Example 1: $\sigma_{\text{Offset}} = 0.001Q_{\max}$, Example 2: $\sigma_{\text{Offset}} = 0.01Q_{\max}$, both achieve DR= 161dB. Note that SNR in the extended DR first increases as i_{ph} (10dB per decade) then drops as $1/i_{ph}^2$ (-20dB per decade). In particular note the sudden decrease in SNR for the example with high σ_{Offset}.

This technique suffers from poor quantization at the high end. Figure 14 shows the transfer function assuming no noise. Note that the plot is logarithmic in both axes and the quantization regions are growing. Constant SNR can be achieved if the number of quantization regions were the same in all decades; however, as shown in Figure 14 this number is decreasing with i_{ph}.

In the following section, we discuss the new Folded Multiple Capture HDR scheme [7], which by combining features of the synchronous self-reset and multiple capture schemes discussed above, can satisfy the precision imaging requirements in IR with low power consumption and robust circuits. We first discuss the architecture and operation of FMC. We then describe a prototype of the architecture and experimental results obtained.

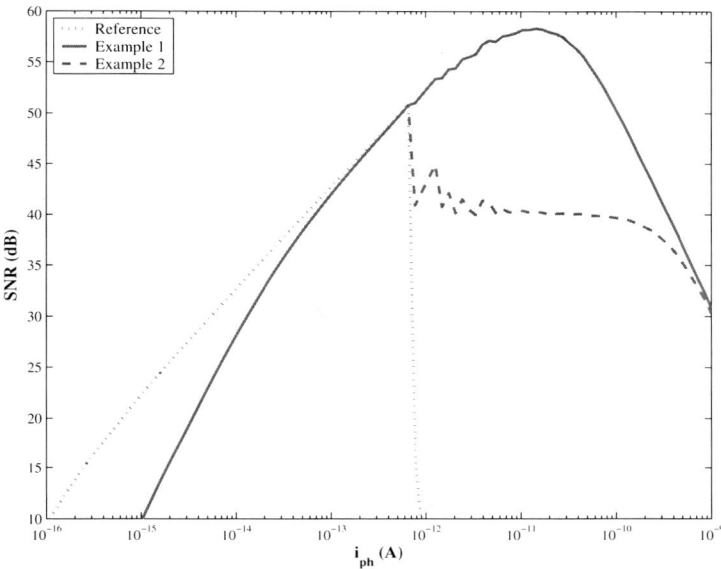

Fig. 13. SNR versus i_{ph} for synchronous self-reset for two different examples.

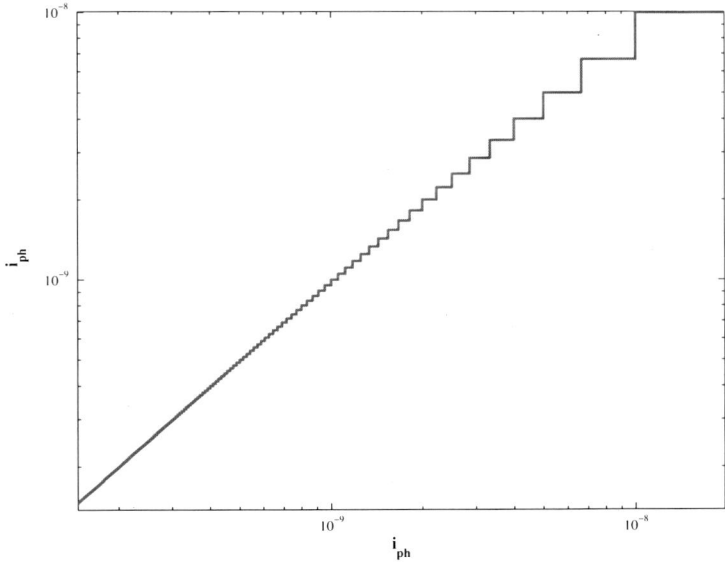

Fig. 14. Transfer Function of the synchronous self-reset scheme assuming no noise.

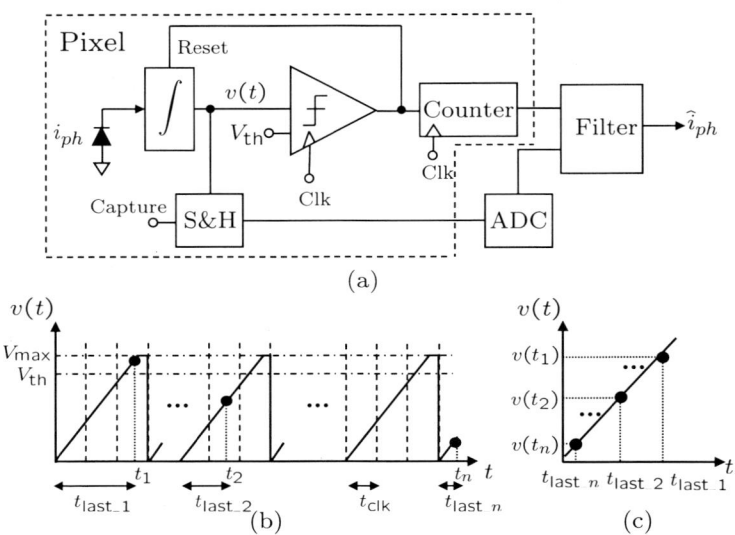

Fig. 15. Above: FMC architecture. Below: FMC operation.

7   Folded Multiple Capture

A block diagram of the FMC architecture is shown in Fig. 15(a). Each pixel consists of an integrator, with reset that is controlled by a comparator, a counter, and a sample-and-hold (S&H). The S&H output is digitized by a fine ADC, whose output along with the counter values are fed to a filter that generates the photocurrent estimate. At each clock cycle, the integrated photocurrent, $v(t)$, is compared to a threshold voltage V_{th}. The integrator is reset when the comparator output flips creating the folded waveform shown in Fig. 15(b). Meanwhile, the integrator output is sampled and digitized at predefined sampling or capture times $t_1, t_2, \ldots, t_n$. The capture times are synchronized with Clk, shifted by $t_{\text{C k}}/2$ to avoid simultaneous reset and capture. The counter is incremented by the clock and reset by the comparator output signal. Its value, which corresponds to the *effective integration time* $t_{\text{last\_i}}$ (the time from the last reset), is read out at each capture time. The slope of the linear least-squares fit of the digitized capture values and their corresponding integration times is used to estimate the photocurrent (see Fig. 15(c)). In effect, FMC performs n regular captures during an exposure time and combines them to achieve a high fidelity estimate of the photocurrent. Dynamic range is extended by $2t_{\text{int}}/t_{\text{Clk}}$ over the integrating capacitor dynamic range. For example, for $t_{\text{int}}/t_{\text{Clk}} = 1000$, DR increases by 66 dB. Fig. 16 shows example waveforms for $t_{\text{int}}/t_{\text{Clk}} = 8$ and four capture times. A low input photocurrent (see Fig. 16(a)) results in no reset and the scheme reduces to a conventional FPA with Fowler readout [20]. A high photocurrent (see Fig. 16(c, d)) results in periodic reset. Unlike other self-reset schemes discussed earlier, however, the number of resets is not used to estimate the signal.

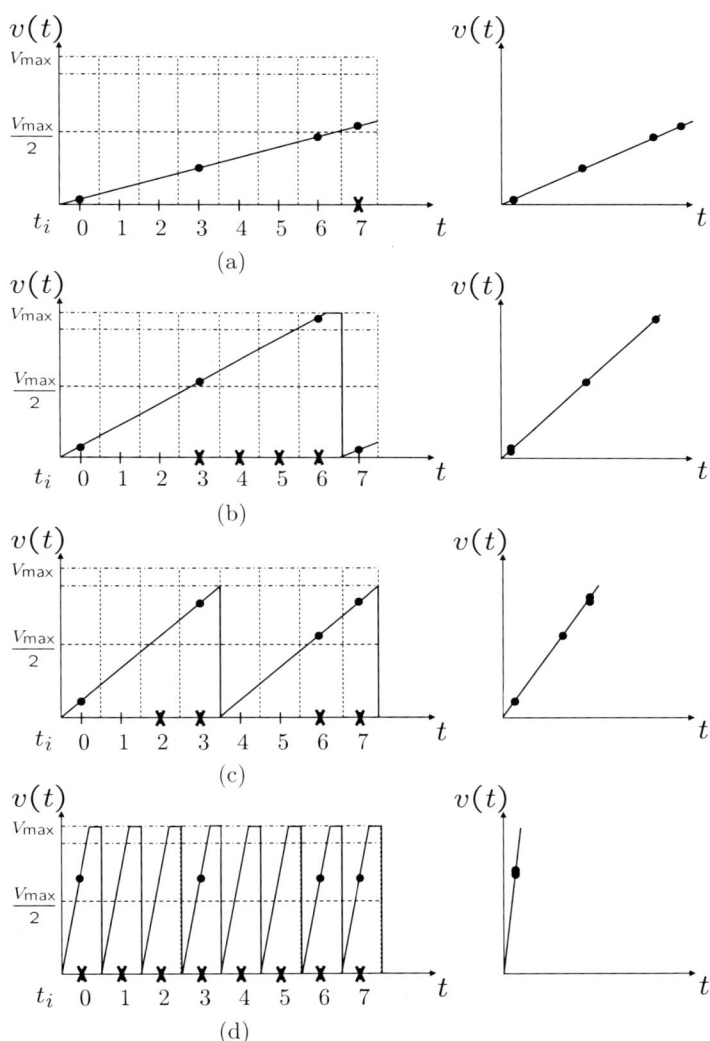

Fig. 16. Integrator output (left) and corresponding least-squares estimate (right) for four photocurrent values with capture times, $(t_1, t_2, t_3, t_4) = (0, 3, 6, 7)t_{Clk}$. The $\times$ indicate capture times that satisfy $Q_{int} > Q_{max}/2$.

For low power, the number of captures used in achieving the high fidelity estimate of the photocurrent must be small. A surprising fact about FMC is that only 3 to 4 scene-independent globally set captures are needed to achieve uniformly high SNR. We wish to select capture times to guarantee a minimum SNR of $Q_{\max}/2$, for photocurrents $\geq qQ_{\max}/t_{\text{int}}$. Note that a single capture only guarantees this requirement for a certain range of photocurrents. To illustrate this point, consider the example in Fig. 16 again. A capture at $t_4 = 7t_{\text{Clk}}$ satisfies the above SNR condition for the examples in Fig. 16(a),(c),(d). However, using only this capture results in SNR ≈ 0 for example (b). The problem is solved by using another capture, e.g., between $3t_{\text{Clk}} \leq t_3 \leq 6t_{\text{Clk}}$. It can be shown that capture times $(t_2, t_3, t_4) = (3, 6, 7)t_{\text{Clk}}$ for $t_{\text{int}}/t_{\text{Clk}} = 8$ ensure that for all photocurrent values, at least one of the capture values has a value higher than $Q_{\max}/2$. To perform offset cancelation, a low value capture, e.g., at $t_1 = 0$, is also required.

While the above algorithm guarantees minimum SNR of $Q_{\max}/2$, least-squares fit of the captures and corresponding effective integration times to estimate photocurrent further improves SNR by canceling offsets, e.g., due to integrator and readout, and reducing the read, shot, and 1/f noise (see [20]). Note that since all signals in FMC are synchronized with a low jitter clock, SNR is not affected by timing inaccuracies. Further area and power reductions are achieved by relaxing the comparator specifications. As discussed earlier, the variation of the reset period with comparator offsets results in fixed pattern noise (FPN) that typically degrades SNR of HDR schemes. Since in FMC reset periods are not used to estimate photocurrent, the associated FPN is avoided and a simple regenerative architecture can be used for the comparator, obviating the need for a larger, power consuming gain stage. A relaxed comparator design also means that the highest clock frequency is not limited by the comparator speed, but by the settling time of the S&H circuit.

7.1   Implementation

A prototype of the FMC architecture has been implemented in a $0.18\mu m$ CMOS double-poly, five metal-layer process. The chip micrograph is shown in Fig. 17. Four columns have pixels with NWELL/PSUB diodes and the fifth has pixels driven by external current sources. Provisions have been made for bump-bonding IR detectors adjacent to the diodes. The analog and digital periphery circuits are placed at opposite ends of the pixel array. The Timing Control block generates all control signals. The clock rate (and thus dynamic range) and capture times are programmable via a scan chain. Each pixel occupies an area of $30\mu m \times 150\mu m$ (40% analog, 60% digital). In a 3D-IC implementation of the fully integrated imaging system [1], each pixel is estimated to be $30\mu m \times 30\mu m$ with 2 analog and 1 digital circuit layers.

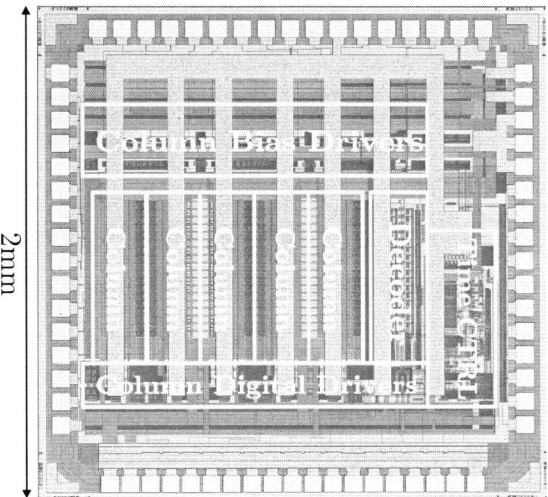

Fig. 17. FMC chip micrograph.

7.2   Experimental Results

A uniform LED illuminator is used as the light source for characterization. The chip analog column outputs, digitized using an on-board ADC, and the chip digital column outputs are transferred to a PC via an FPGA-based data acquisition board. Least-squares fit of the digitized capture values and corresponding effective integration times to estimate photocurrent is then performed in software.

The linearity and SNR are characterized locally at multiple random intervals. Experimental SNR versus i_{ph} results are shown in Figure 18. Read noise is expected to be lower with test setup improvements.

The power consumption per pixel is 25.5μW and dominated by the analog front-end. This corresponds to energy consumption of 25.5nJ for each pixel readout with DR = 138dB and SNR = 60dB. Note that this power consumption can be significantly reduced, e.g., using switched biasing, with knowledge of the detector parameters.

8   Conclusion

This chapter discusses the need for precision high dynamic range, high speed focal plane arrays for IR imaging applications. High dynamic range schemes targeted for visible range imaging are reviewed. A new HDR scheme, Folded Multiple Capture, that meets the stringent performance requirements of IR imaging applications is discussed. A prototype of the architecture targeted towards 3D-IC IR FPAs is described.

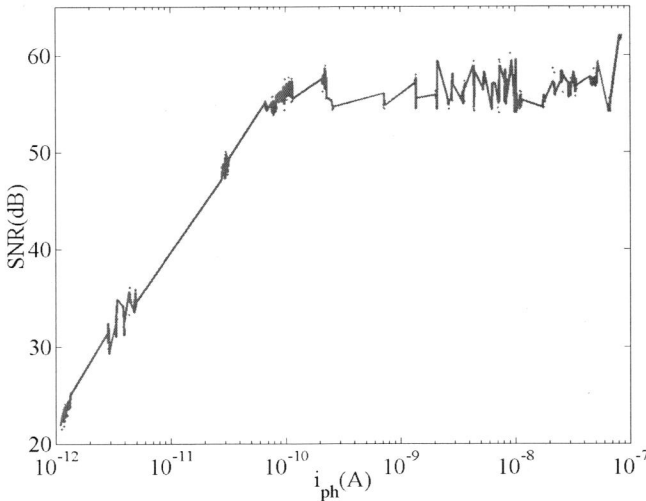

Fig. 18. Experimental scatter plot of SNR vs. i_{ph} for FMC.

References

1. R. Balcerak,"Vertically Integrated Sensor Arrays (VISA)," DARPA/MTO Photonics Symposium, Dec 2004.
2. S. Kavusi and A. El Gamal, "Quantitative Study of High-Dynamic-Range Image Sensor Architectures," Proceedings of the SPIE, vol. 5301, pp. 264-275, Jun 2004.
3. S. Kavusi, H. Kakavand, and A. El Gamal, "Quantitative Study of High Dynamic Range $\Sigma\Delta$-based Focal Plane Array architectures," Proceedings of the SPIE, vol. 5406, pp. 341-350, Aug 2004.
4. S. Kavusi, K. Ghosh, A. El Gamal, "Architectures for High Dynamic Range, High Speed Image Sensor Readout Circuits," *invited*, IFIP VLSI-SoC, pp. 36-41, Oct 2006.
5. S. Kavusi, H. Kakavand, A. El Gamal, "On Incremental Sigma-Delta Modulation with Optimal Filtering," IEEE Transactions on Circuits and Systems-I: Regular papers, vol. 53, no. 5, pp. 1004-1015, May 2006.
6. S. Kavusi and A. El Gamal, "Per-Pixel Analog Front End Architecture for High Dynamic Range Disturbance-Tolerant IR Imaging," Proceedings of the SPIE, vol. 5406, pp. 351-360, Aug 2004.
7. S. Kavusi, K. Ghosh, K. Fife and A. El Gamal, "A 0.18μm CMOS 1000 frames/sec, 138dB Dynamic Range Readout Circuit for 3D-IC IR Focal Plane Arrays, IEEE Custom Integrated Circuits Conference, pp. 229-232, Sep 2006.
8. V. Seshadri, *The Inverse Gaussian Distribution: A Case Study in Exponential Families*, Oxford University Press, 1994.
9. D. Stoppa et al., "Novel CMOS Image Sensor with a 132-dB Dynamic Range," IEEE Journal of Solid-State Circuits, vol. 37, no. 12, pp. 1846 -1852, Dec 2002.
10. V. Brajovic, T. Kanade, "A Sorting Image Sensor: an Example of Massively Parallel Intensity-to-time Processing for Low-latency Computational Sensor," Proceedings

of the 1996 IEEE International Conference on Robotics and Automation, vol.2, pp. 1638-1643, Apr 1996.

11. T. Lulé, B. Schneider, and M. Bohm, "Design and Fabrication of a High Dynamic Range Image Sensor in TFA Technology," IEEE Journal of Solid-State Circuits, vol. 34, no. 5, pp. 704-711, May 1999.

12. D. Yang, B. Fowler, A. El Gamal, and H. Tian, "Image Sensor with Ultrawide Dynamic Range Floating-Point Pixel-Level ADC," IEEE Journal of Solid-State Circuits, vol. 34, no. 12,pp. 1821-1834, Dec 1999.

13. W. Bidermann et al., "A 0.18μm High Dynamic Range NTSC/PAL Imaging System-on-Chip with Embedded DRAM Frame Buffer," IEEE International Solid-State Circuits Conference, pp. 212-213, Feb 2003.

14. O. Yadid-Pecht, E. R. Fossum, "Wide Intrascene Dynamic Range CMOS APS Using Dual Sampling," IEEE Transactions on Electron Devices, vol. 44, no. 10, pp. 721 - 1723, Oct 1997.

15. X. Q. Liu, A. El Gamal, "Synthesis of High Dynamic Range Motion Blur Free Image From Multiple Captures," IEEE Transactions on Circuits and Systems I: Fundamental Theory and Applications, vol. 50, no. 4, pp. 530-539, Apr 2003.

16. C. Jansson, "A High-Resolution, Compact, and Low-Power ADC Suitable for Array Implementation in Standard CMOS," IEEE Transactions on Circuits and Systems-I, vol. 42, no. 11, pp. 904-912, Nov 1995.

17. J. Markus, J. Silva and G.C. Temes, "Theory and applications of incremental $\Delta\Sigma$ converters,"IEEE Transactions on Circuits and Systems I: Fundamental Theory and Applications, volume 51, number 4, pp. 678-690, April 2004.

18. O. Yadid-Pecht "Wide Dynamic Range Sensors," Optical Engineering, vol. 38, no. 10, pp. 1650-1660, Oct 1999.

19. B. Fowler, A. El Gamal, and D. Yang, "A CMOS Area Image Sensor with Pixel-Level A/D Conversion," IEEE International Solid-State Circuits Conference, pp. 226-227, Feb 1994.

20. A. M. Fowler and I. Gatley, "Noise Reduction Strategy for Hybrid IR Focal Plane Arrays," Proceedings of the SPIE, vol. 1541, pp. 127-133, Jul 1991.

21. A. Bermak, A. Bouzerdoum, and K. Eshraghian, "A Vision Sensor with On-Pixel ADC and Built-in Light Adaptation Mechanism," Microelectronics Journal, vol. 33, no. 12, pp 1091-1096, 2002.

Oversampled Time Estimation Techniques for Precision Photonic Detectors

Robert Henderson, Bruce Rae, David Renshaw

School of Engineering and Electronics
University of Edinburgh
Edinburgh, Scotland, UK
Robert.Henderson@ed.ac.uk

Edoardo Charbon

Ecole Polytechnique Fédérale de Lausanne (EPFL)
CH-1015 Lausanne, Switzerland
Edoardo.Charbon@epfl.ch

Abstract. The use of oversampling to reduce I/O requirements of time-to-digital converters for arrays of high precision photonic detectors is considered. Simulation results show that the high linearity offered by oversampled converters can be applied to time estimation. The averaging and lowpass filtering inherent in these techniques reduce jitter and enhance estimates of mean time delay. The effect of background illumination on the accuracy of time-of-flight estimates for Lidar range-finding is modeled using a first order sigma-delta modulator. Novel event-driven techniques are proposed for the reduction of sensitivity to background light level.

1   Introduction

Accurate time measurement is commonly required for space science, high energy physics, range finding and fluorescence lifetime sensing. The key component of such systems, integrated Time-to-Digital Converters (TDCs), Time to Analogue Converters (TACs) or gated counters have achieved single-shot resolutions of 10's of picoseconds [1]. Often however, the quantity that must be estimated accurately is an average time delay between a cyclical stimulus and response. The conventional approach is to take many repeated single-shot time measurements and to construct event histograms. The average delay is then extracted from the mean of the event histogram. Averaging has two favourable effects; to reduce time uncertainty due to jitter and to increase SNR from spuriously generated events due to background noise. Particular examples of this are time-of-flight (TOF) measurement or time-correlated single-photon counting techniques for fluorescence imaging [2,3].

Detectors with both high time precision and sensitivity include photomultiplier tubes and avalanche photodiodes. Single-photon Avalanche Photodiodes (SPADs) have recently been realised in deep submicron CMOS processes [4]. Such detectors promise massively-parallel, single-photon detection with extremely high timing accuracy and

---

Please use the following format when citing this chapter:

Henderson, R., Rae, B., Renshaw, D. and Charbon, E., 2007, in IFIP International Federation for Information Processing, Volume 249, VLSI-SoC: Research Trends in VLSI and Systems on Chip, eds. De Micheli, G., Mir, S., Reis, R., (Boston: Springer), pp. 25–35.

low dark count. The ability to integrate arrays of SPADs with on-chip TDCs or counters is expected to yield imagers with unprecedented sensitivity and dynamic range. However, the high data bandwidths required to transmit photon arrival times or counts to off-chip memories for histogram construction are likely to have serious implications for power consumption, thermal effects and pin-count.

In this paper, we will apply oversampled techniques to improve the accuracy of average photon arrival time estimation and greatly reduce I/O data bandwidth. This is of particular interest for arrays of photonic detectors such as SPADs which can be integrated together with the other readout and processing circuitry in a single chip.

2   Background

2.1   Time-to-digital Conversion

Time-to-digital conversion is the process of converting time delay linearly into a numeric digital representation. Various architectures have been proposed with time resolutions down to a few picoseconds. However, the linearity of these converters has conventionally been limited to around 10-bits by matching [1]. Jitter and temperature stabilization are other key performance criteria.

Sigma-delta converters have been employed very successfully to achieve very high resolutions and linearity at the cost of reduced bandwidth [5]. To the best of our knowledge, the first use of a sigma-delta modulator within a TDC is described in [6] for the estimation of on-chip clock jitter. The authors construct a cascade of a mixer and lowpass filter with a sigma-delta modulator in order to achieve femtosecond time resolutions.

2.2   Optoelectronic System

The optoelectronic system which will be studied in this paper consists of an illumination source (usually a laser or laser diode) producing very short light pulses (femtosecond or picosecond) at 10-100MHz. The illumination is reflected from a target and returns to a detector and TDC system synchronized to the laser by a clock. This system has been used in the past to perform ranging by the time-of-flight method [7,8]. A similar system may be employed for fluorescence lifetime imaging and various other applications [2,3].

3   Sigma-Delta TDC

Fig. 1 shows a circuit diagram of a simple first order sigma-delta modulator with a time to voltage conversion input. A MASH (Multi-stAge noise Shaping) architecture has been chosen for simplicity and inherent stability although there are many others to which the same concepts may be applied [5]. The operation of the modulator is controlled by the timing waveforms of Fig. 2 and is appropriate to any system with a repetitive pulsed illumination source.

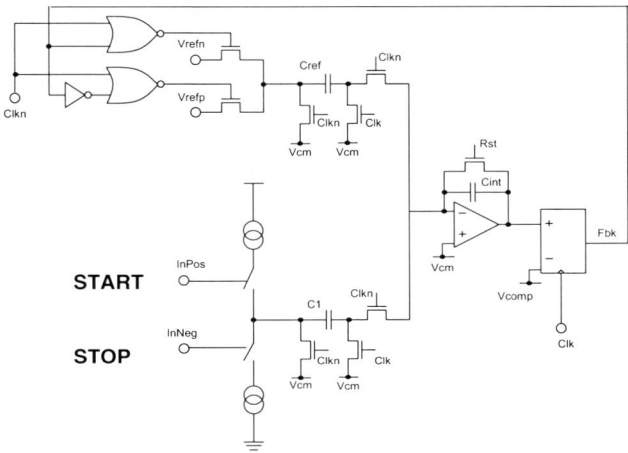

Fig. 1. Sigma-delta Time to Digital Converter based on a first order MASH architecture.

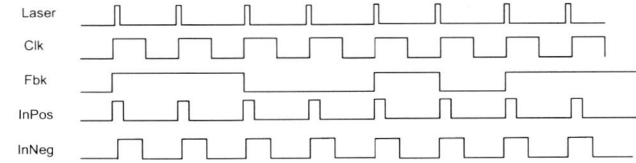

Fig. 2. Sigma-delta TDC timing.

3.1   Operation

The two-phase switched-capacitor implementation of a first order modulator produces an output bit-stream Fbk which will be passed to a lowpass decimation filter (not shown). The clock clk can operate at 10's of MHz synchronized with the pulse repetition rate of the pulsed light source. Fast triggering events from the SPAD or other optical detector generate the waveforms InPos and InNeg. In particular the falling edge of InNeg is related to the detection of the first photon after the laser pulse. Thus the time delay or time-of flight is represented by the delay time between the falling edges of InPos and InNeg.

The current source charge the capacitor C1 to a produce a voltage proportional to the time delay in a similar way to a time-to-analogue converter (TAC) or the charge pump of a PLL. The overlap time between the on-state of the positive and negative current source is based on a technique is used in PLL charge pumps to extend the linear range of conversion and eliminate dead band. Since the current sources are connected passively to the capacitor C1 and common mode voltage Vcm during clk, fast open-loop settling to the nanosecond time intervals of the photonic detector can be achieved. The settling and current requirements of the integrating OTA are determined during the next phase clkn and have a full half clock period. At the end of the period clk, capacitor C1 has been charged to a voltage linearly related to the delay time interval from laser pulse to the first photonic detected. A feedback decision has also been made by the comparator fbk to select either of the reference voltage Vrefn or Vrefp. During the next phase clkn, the selected reference voltage and the voltage or C1 are integrated on the capacitor Cint. This process is repeated over many repetitions of the laser and clock waveform.

3.2   Modeling

To investigate the properties of the system a software model of the modulator and signal source has been developed. We take the particular example of a time-of-flight system where the return signal from the emitted femtosecond pulsed light source is consider to be a Gaussian distributed photon detection peak. This represents the aggregate jitter in the detection system [7] and may originate from a number of sources. The distribution has an adjustable offset representing the TOF and standard deviation representing the jitter.

We also consider a background signal from ambient light or detector dark signal as a Poisson random process parameterized by a mean arrival rate in photons/sec. The output from the detector is considered to be a sequence of delay times of the first photon arrival after the repeated laser pulse. This event may either be triggered by the reflected laser pulse or by a background event photon (internal noise, the dark count rate DCR), whichever occurs first. The detector is considered to generate only one event per clock cycle.

Fig. 3 shows a sample histogram of photon detections for a clock frequency of 25MHz, jitter of 300ps and background arrival rate of 10Mphotons/sec or around 100Lux at 500nm without filtering. We consider that photons detected from the target can be modeled as a set of independent probabilities with different averages as a function of reflectivity. Thus on some clock cycles no photon is returned from the target and events are generated by background illumination, dark count or forced to occur by gating. Note that the jitter and TOF are normalized to the clock period in the forthcoming treatment.

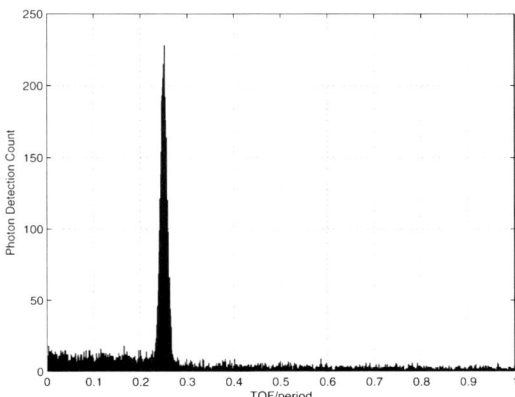

Fig. 3. Sample histogram of the first detected photon for a 25MHz laser repetition rate, 10ns time of flight and 2Mphotons/sec background.

4 Simulation Results

A second order MASH sigma delta modulator with oversampling rate N=256 and 3rd order comb filter has been simulated. A sweep of TOF measurements has been performed and a least mean squares fitting algorithm applied to the decimated modulator output in order to estimate linearity. A number of clock periods (10000) are used before analysis to avoid any transient effects. The noise level at the comb filter output is estimated from the standard deviation of the code over 10000 clock cycles.

Fig. 4 shows the output of the modulator and comb filter with no background noise and a 300ps jitter input with a 25MHz laser and system clock. As expected, the noise has been reduced by sqrt(N) or a factor of 16 from 300ps to 18.75ps. The linearity of the modulator is estimated to be around 10bits, limited only by the RMS noise of the input. As the oversampling factor is increased both jitter and linearity are improved. Note that the jitter on the input signal also acts as a dither and reduces the build up of tones which are known to reduce modulator resolution [5].

Fig. 5 shows the analysis repeated in the presence of 2Mphotons/sec Poisson arrival rate of background illumination. As the TOF is increased there is a greater probability of a background photon triggering the detector rather than the TOF signal. Below 0.1 TOF/period we obtain the same improvement in resolution as in the case without background. Above this level the noise level and distance accuracy is steadily degraded. In Fig. 6 the level of background illumination is varied whilst keeping a fixed TOF input. The minimum and maximum errors from a least mean squares linearity fit on the data is shown in Fig. 7. Below 0.1 TOF/Period the data has good linearity and above 0.1 TOF/Period we see a nonlinear departure and increasing uncertainty.

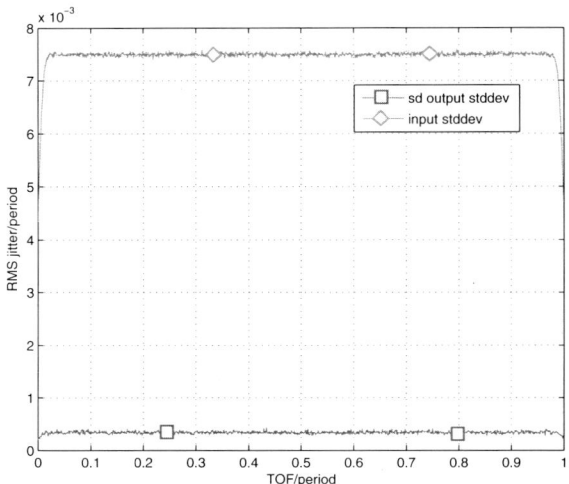

Fig. 4. Relative jitter of a 2nd order modulator/comb filter versus TOF DC level for oversampling ratio 256, period 40ns, input jitter 300ps. A 16x reduction in jitter at the output has been achieved.

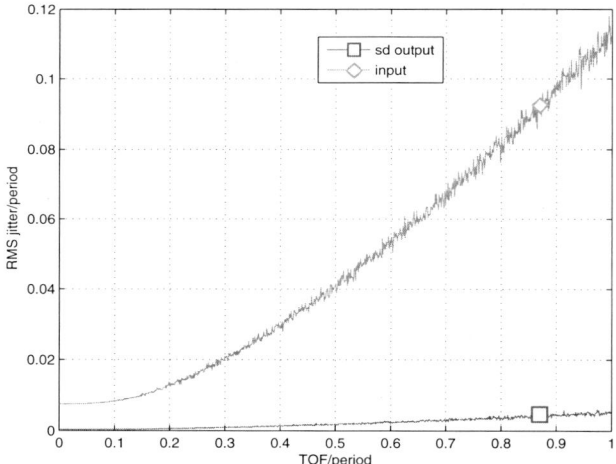

Fig. 5. Relative jitter of a 2nd order modulator/comb filter versus TOF DC level for oversampling ratio 256, period 40ns, input jitter 300ps in the presence of 2Mphotons/sec background illumination.

5   Departure from Linearity

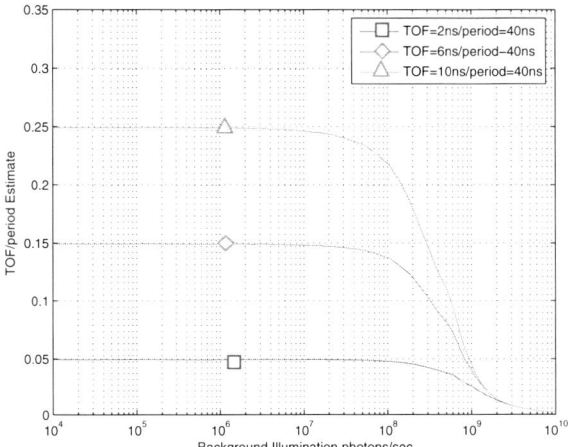

Fig. 6. Modulator output estimate of TOF versus background illumination level for three different TOFs. Shorter TOFs are more resistant to background illumination level.

The simulation results indicate that improvements in linearity, jitter and data rate can be obtained over histogram construction from single-shot TDCs. However, background light will cause departure from linearity when the TOF delay is comparable to the mean Poisson photon inter-arrival time (Fig. 6). Linearity above 10-bit matching level is achievable dependent on correct choice of oversampling rate and input TAC stage.

Fig. 7 shows how the maximum linearity error varies with relative time of flight. The uncertainty is roughly proportional to the relative time of flight. The histogram in Fig. 3 makes an assumption that a photon is received either from the pulsed emission source or the background within every half clock period interval. Two circumstances invalidate this assumption; 1) in a low light environment the probability of receiving a photon will be greatly diminished 2) a low reflectivity target will also greatly reduce the number of detected photons returning from the target. The former case is likely to be encountered in low signal environments such as in fluorescent imaging [9] and the latter in laser ranging with black surfaces or distant targets.

A standard sigma-delta modulator expects an input sample on every clock cycle. In the absence of a trigger from the detector, a full-scale input would be generated to the sigma-delta modulator resulting in a secondary peak of counts at exactly T/2 as shown in Fig. 8. These spurious integrations will skew the average integrated pulse delay [4].

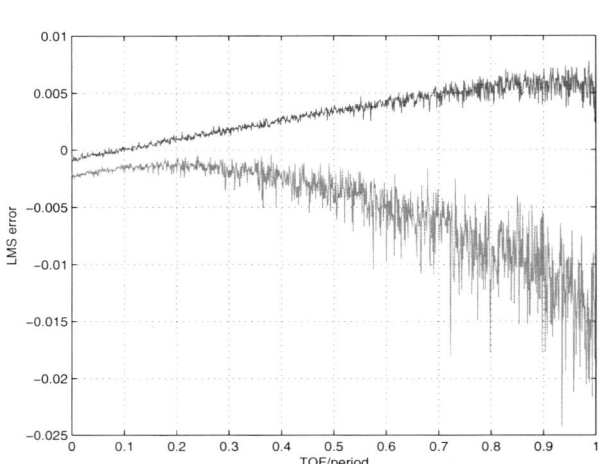

Fig. 7. Maximum and minimum linearity error of a 2nd order sigma-delta modulator versus TOF level for oversampling ratio 256, period 40ns and input jitter 300ps in the presence of 2Mphotons/sec background illumination level.

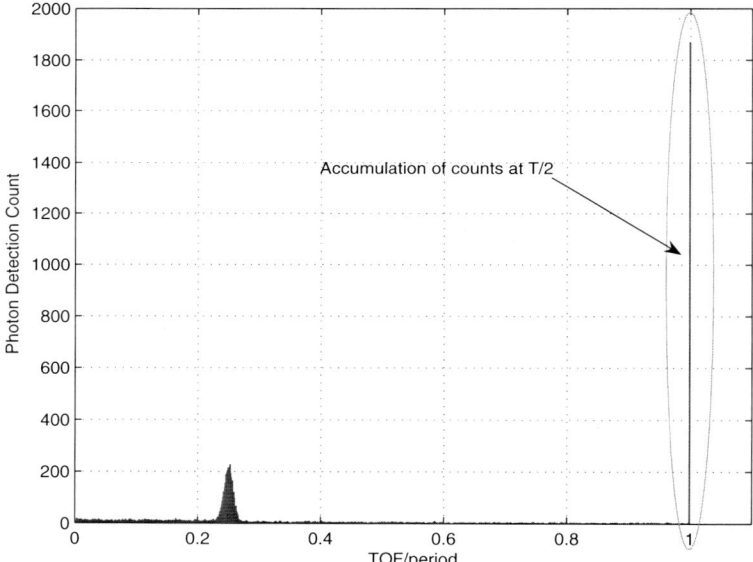

Fig. 8. A sample histogram of the first detected photon for a 25MHz laser repetition rate, 10ns time of flight and 2Mphotons/sec background illumination level. The accumulation of photons at T/2 is due to low probability of photon return from target.

6   Event-driven or Time-windowed Operation

The effects of high background illumination and low photon detection rate can be mitigated by a simple modification of the system operation: event driven clocking of the modulator triggered by a time windowed detector input.

Let us define a lower and upper time bound T_{lo} and T_{hi} where

$$0 < T_{lo} < T_{hi} < T/2 \tag{1}$$

Consider $T_{event}(i)$ to be the time of the first detector event after the ith rising edge of the clock Clk

$$0 < T_{event}(i) < T/2 \tag{2}$$

We generate a new integrating clock signal Clkint(i) such that if
$T_{lo} < T_{event}(i) < T_{hi}$: Clkint(i) = 1
$T_{event}(i) > T_{hi}$: Clkint(i) = 0
$T_{event}(i) < T_{lo}$: Clkint(i) = 0

The integrating clock causes the modulator to integrate the analogue time estimation charge from Cin only if the detector event occurs within the bounds T_{lo} and T_{hi}.

In low light environments, setting $T_{lo}=0$ and $T_{hi}=T/2$ will suppress integration of spurious full scale signals due to the absence of an event in the half period time window.

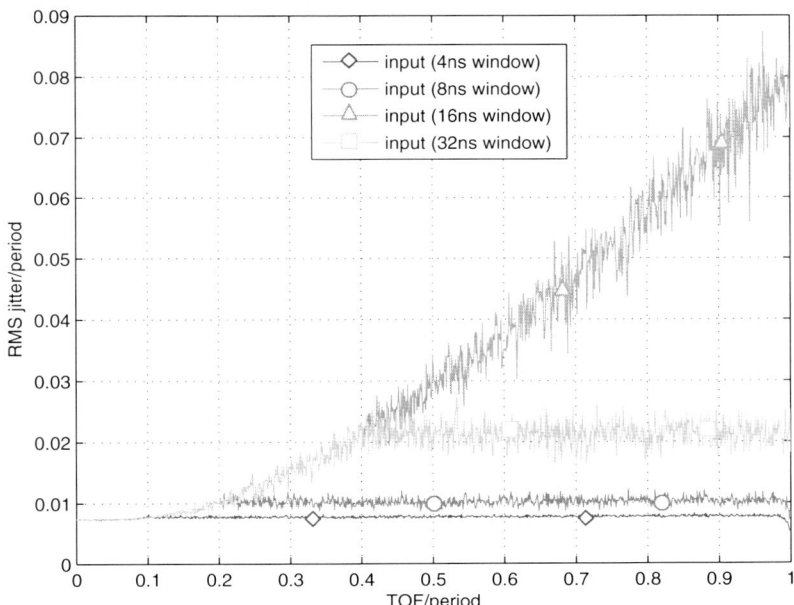

Fig. 9. Event driven operation of a sigma-delta TDC with various time window intervals. Relative jitter of a 2nd order modulator/comb filter versus TOF DC level for oversampling ratio 256, period 40ns, input jitter 300ps in the presence of 2Mphotons/sec background illumination.

In high background light environments T_{lo} and T_{hi} should be narrowed around the mean value of T_{event}. Triggers from the detector are inhibited until the falling edge of Clk. This pulse can be scanned by a variable delay to a position close to the mean TOF. Thus spurious integrations due to background can be minimized.

Fig. 9 shows how the residual jitter on the relative time of flight can be restored to the value in darkness by reducing the time window interval. In this case, the RMS jitter can be restored to the value without ambient light level with a relatively coarse sampling window of 4ns which would be relatively easy to implement in an integrated circuit. Wider time windows cause a ceiling on the maximum jitter following the un-windowed jitter versus relative TOF curve. In Fig. 10, we see that a 4ns time window is also sufficient to restore the linearity of the converter. The choice of time window duration is related to the maximum tolerable background light level and desired linearity and output accuracy of the converter.

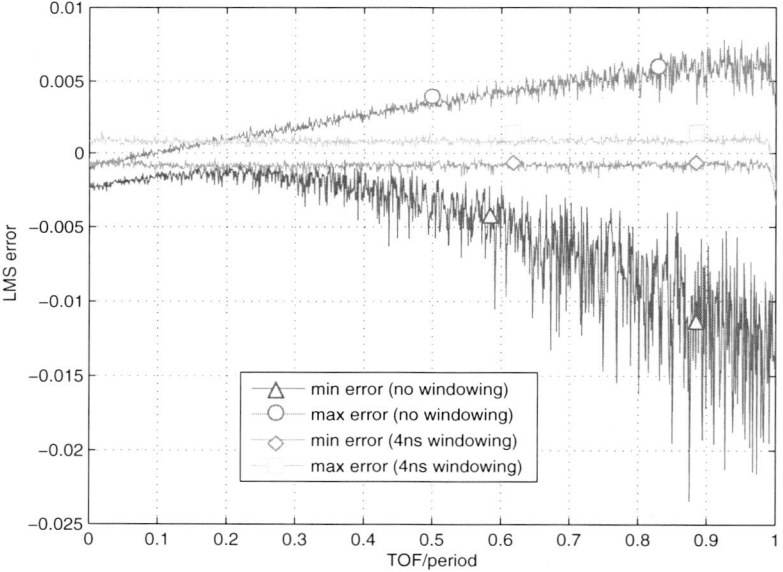

Fig. 10. Comparison of maximum and minimum linearity error of a 2nd order sigma-delta modulator versus TOF level for oversampling ratio 256, period 40ns, input jitter 300ps in the presence of 2Mphotons/sec background illumination level for windowed and unwindowed systems.

A 1st order sigma-delta TDC with a SPAD detector has recently been designed and sent for manufacture in a 0.35μm CMOS technology. The circuit occupies an area of

100um x 200um. This is relatively compact for a time-to-digital converter and promises miniaturized arrays of highly accurate time-estimators for time-resolved imagers.

7   Conclusions

Estimation of mean time-of-flight or decay time has been identified as an oversampled system. We have shown that sigma-delta converters provide a compact, efficient solution to achieve high time resolution, low jitter and reduced system IO bandwidth. A development of traditional sigma-delta converters has been proposed for low-light conditions or for suppression of high ambient light whereby the conversion cycles are event-driven. Simulations show that implementation of a narrow time window for event triggering is sufficient to reject ambient light and restore converter linearity and jitter. Circuit implementations of the trigger and event driven mechanism will be proposed in future work.

8   References

[1]   R. Staszewski, S. Vemulapalli, P. Vallur, J. Wallberg, and P.T. Balsara, "1.3 V 20 ps Time-to-digital converter for frequency synthesis in 90-nm CMOS", IEEE Transactions on Circuits and Systems—II, Vol. 53, No. 3, pp. 220-224, March 2006.

[2]   J.C. Jackson et al., "Characterization of geiger mode avalanche photodiodes for fluorescence decay measurements", Proc. of SPIE, Vol. 4650-07, Photonics West, San Jose, CA, Jan. 2002.

[3]   C. Niclass, A. Rochas, P.A. Besse, and E. Charbon, "Design and Characterization of a CMOS 3-D image sensor based on single photon avalanche diodes", IEEE Journal of Solid-State Circuits, Vol. 40, No. 9, Sep. 2005.

[4]   C. Niclass, M. Sergio, and E. Charbon, "A single photon avalanche diode array fabricated in deep submicron technology", Design Automation and Test Europe Conference, Munich 2006.

[5]   S. Norsworthy, R. Schreier, and G. Temes, Eds., "Delta–sigma data converters, theory, design, and simulation". New York: IEEE Press, 1997.

[6]   M. Collins, B.M. Al-Hashimi, and P.R. Wilson, "On-chip timing measurement architecture with femtosecond resolution", Electronics Letters, Volume 42, Issue 9, 27, pp. 39-40, April 2006.

[7]   S. Pellegrini, G.S. Buller, J.M. Smith, A.M. Wallace, and S. Cova, "Laser-based distance measurement using picosecond resolution time-correlated single-photon counting", Meas. Sci. Technology, 11, pp. 712-716, 2000.

[8]   S.B. Gokturk, H. Yalcin, and C. Bamji, "A Time-Of-Flight Depth Sensor - System Description", Issues and Solutions", Computer Vision and Pattern Recognition Workshop, pp. 35-39, June 2004.

[9]   W. Becker, "Advanced Time-Correlated Single Photon Counting Techniques", Springer-Verlag, Berlin 2005.

Innovative Optoelectronic Approaches to Biomolecular Analysis with Arrays of Silicon Devices

C. Guiducci[1], C. Stagni[1], M. Brocchi[1], M. Lanzoni[1], B. Riccò[1], A. Nascetti[2], D. Caputo[2], and A. De Cesare[2]

[1] DEIS University of Bologna, Viale Risorgimento 2, 40136 Bologna-ITALY,
cguiducci@deis.unibo.it
[2] DIE University of Rome "La Sapienza", Via Eudossiana 18, 00184 Roma-ITALY
decesare@die.uniroma1.it

Abstract. This paper intends to provide a brief survey of the most advanced implementations of molecular analysis tools based on microtechnologies and to discuss a new approach for the development of integrated molecular analysis. The technique presented hereafter combine a label-free method for molecular characterization based on UV absorbance and two technologies of silicon photodetectors addressed to fulfill the requirements of different applications.

1 Introduction

Microtechnologies play a crucial role in the development of innovative low-cost and mass-produced assays for bio-sensing and bio-interactive purposes. In some areas, their use in the development of bio-interactive systems is already well-established. For instance, microsystems for in-body drug delivery and passive silica chip (or slides) for high-parallel molecular analysis are among the most important innovations.

The advantages derived by the use of advanced devices can be summarized as follows:

- The miniaturization of the reaction sites and cells allows the use of a reduced amount of sample and reagents.
- The miniaturization of the measurement system leads to portability and a higher signal-to-noise ratio.
- The possibility to implement high-parallel analysis tools increases of orders of magnitude the speed of analysis.
- The integration of mechanical and fluidic functions for sample handling, delivery, mixing, purification, separation, and amplification leads to stand-alone and easy-to-use devices. In case of miniaturized handling, volumes are often in the nanoliter to picoliter range rather than the microliter range needed for conventional experiments.
- The possibility to provide low-cost and mass-produced assays by batch production.

Please use the following format when citing this chapter:

Guiducci, C., Stagni, C., Brocchi, M., Lanzoni, M., Riccò, B., Nascetti, A., Caputo, D. and De Cesare, A., 2007, in IFIP International Federation for Information Processing, Volume 249, VLSI-SoC: Research Trends in VLSI and Systems on Chip, eds. De Micheli, G., Mir, S., Reis, R., (Boston: Springer), pp. 37–53

– The generation of electrical read-out by the integration of electronic sensors allows the use of microelectronic circuits available for electrical signal conditioning, amplification, filtering, modulation, and transmission.
– The sensing performance can be increased by using high-sensitive electron devices.

The following section provide a survey of some of the most interesting implementations of molecular analysis assays based on advanced technologies, from the point of view of sample handling functions (Sect. 2.1), of electronic devices for parallel molecular detection (Sect. 2.2) and of micromachining implementations for molecular analysis (Sect. 2.3). In Sect. 3 we present the state-of-the-art of integrated optical sensing of molecular reactions and we describe the new approaches we tested based on UV absorbance.

2   Microtechnologies for Biomolecular Analysis

2.1   Total Analysis Systems on a Chip

Integrated microfluidics may be made of plastic, glass, quartz, or silicon. Bulk and surface micromachining performed with sophisticated etching, patterning and deposition techniques are at the basis of channels implementation.

One of the most relevant microfabricated implementation on chip is the Polymerase Chain Reaction (PCR) molecular amplification. This approach has been widely investigated exploiting the good properties of thermal conductivity of silicon and its ability to easily integrate thermal resistances [1],[2],[3].

A large number of basic fluidic components have been assembled in different ways to perform various other chemical process. Many of these are based on electrokinetic transport principles, and include valves, mixing structures, chemical reactors, and chemical separation channels. In addition, chemical separation mechanisms have been miniaturized, including gel electrophoresis, solvent programmed chromatography, isoelectric focusing, isotachophoresis and two-dimensional separations based on liquid chromatography and free-solution electrophoresis. Surface interactions have been exploited for solid-phase extraction to process samples for hybridization of target DNA molecules, and nanoliter-scale reactors have been demonstrated for continuous flow, stopped flow, and thermal cycling reactions [4],[5],[6].

2.2   Electronic Circuits for Bio-sensing Transduction and Processing

Printed Circuit Technology for DNA Detection An electrical-based DNA analysis system has been developed by Motorola (Clinical Microsensors Division). It is based on an electronic instrumentation and on disposable chips implemented with printed circuit board technology. Few dozens of gold electrodes of $250\,\mu m$ are defined on the board where DNA capture probes are immobilized by a self-assembled monolayer technology. Target DNA molecules are detected by specific probes in a first stage. Then, a second probe, which hosts an electroactive label, is used to generate a current signal on the corresponding site [7]. The electronic instrumentation performs alternating current voltammetry on the disposable chip and identifies the site where specific DNA hybridization has occurred.

An Active Chip for Direct DNA Detection Recently, Infineon Technologies has developed two generations of fully-electronic sensor arrays for DNA detection based on CMOS technology.

The chips are implemented in $0.5\,\mu m$ CMOS process extended with additional process steps meant to form gold electrodes on the top. The electrodes are connected to the integrated circuits by means of vias of composite structures of different metals. A single sensing site array is made of two interdigitated gold electrodes arranged within a circular surface of down to $100\,\mu m$ diameter. The spacing between the fingers and their width is $1\,\mu m$. Single-stranded DNA probe molecules are deposited by microspotting technology and immobilized on the gold surface by covalent AU-S bonds. After immobilization, a liquid sample containing the target molecules to be detected is applied to the surface of the whole chip and, in case of matching sequences, an hybridization reaction occurs capturing permanently the targets.

In the first chip – a 16×8 sensor array – an enzyme label (alkaline phosphatase) is beforehand bound to the targets molecules in order to generate an electrochemical signal after the hybridization reaction. The electrochemical activity of the enzyme is detected by applying a suitable chemical substance (p-aminophenyl-phosphate) and performing redox-cycling measurements. The magnitude of the redox current between the finger electrodes depends on the amount of detected targets. The circuits within each position allow to detect sensor currents in a range between $10^{-12}\,A$ and $10^{-7}\,A$ [8].

In a further chip realization, a label-free detection technique has been successfully applied. The detection principle relies on the interface capacitance change led by the hybridization reaction. Indeed, the interface capacitance depends on the configuration of the layer of ions in the vicinity of the electrodes which is affected by the molecules immobilized on the gold surface. The electrodes which captured target molecules ad thus host double-stranded molecules, exhibit a capacitance 20% smaller than the one of the other electrodes. Input/output signals are both analog and digital but the output is fully digital, as CMOS circuits placed below each of the 128 interdigitated electrode perform internally capacitance measurement and analog-to-digital conversion by mixed signal. The conversion is performed in parallel and results are multiplexed on the output using the address signals [9].

2.3   Micromachining for High-sensitive Structures

Microcantilevers for Molecular Mass Sensing Surface and bulk micromachining techniques may help with providing innovative solutions for molecular sensing at the microscale level. Silicon nitride cantilevers have been tested for sensing of molecular binding events. The detection principle is based on the detection of mass changes or on the surface stress change at the cantilever by measuring the change of the resonant frequency or of the bending, respectively. Both label-free and label-mediated techniques have been successfully employed. The recent discovery of the origin of nanomechanical motion generated by DNA hybridization and proteinligand binding provided some insight into the specificity of the technique. DNA hybridization detection, including accurate positive/negative detection of one base pair mismatches have been reported by [10] and [11].

Porous Materials Porous materials are suitable structures for surface sensing techniques as they provide a high surface/volume ratio. Moreover, this technology is meant to create inexpensive devices. High surface areas provide a mechanism to achieve detection sensitivities that are in the range of parts per billion on a short time scale.

DNA hybridization has been detected in a porous silicon media by the interferometric Fabry-Perot technique [12]. Other powerful optical techniques, like the super-prism phenomenon are also under investigation [13].

3   Optical Sensing of Molecular Reactions

Glass arrays for molecular analysis are widely employed in fields like drug discovery, genetic research and medical diagnostics (at the research level). Each site can recognize and capture specifically an identifying part of a gene, a RNA strand or a protein.

At present, these detection systems need a preparation step of optical functionalization of target biomolecules (labeling) and the use of high-cost scanning fluorescence detectors. They relies on the quantification and/or the localization of target biomolecules by means of fluorescent labels. The slides host two-dimensional arrays of small sites (from $20\,\mu m^2$ to $400\,\mu m^2$), on which different molecular probes are immobilized.

Recent works witness the effort of developing arrays of solid state optical devices able to detect on-site the emitted light of the fluorescent labels. The interest in integrated detectors is led by the demand for low-cost devices, to be employed even outside specialized laboratories and addressed to mass-production. These photodiodes arrays are meant both to be coupled with glass structures [14],[15],[16],[17], and to become active sensing substrates for the molecular-spotted arrays [18],[19],[20]. In the latter approach, probes may be spotted directly on the top of the chip by chemically modifying the silicon dioxide passivation. Nevertheless, the integration of optoelectronic detection on existing fluorescence microarrays is still far to be achieved [21], being the reason the non trivial design issues due to the need for integrated filters to screen the excitation light. Moreover, fluorescent labels have been demonstrated to be quite unreliable employed in common scanning procedures [22].

We recently considered a different sensing approach which relies on the intrinsic optical absorbance of biomolecules in the far-UV range. The absorption properties of polynucleotide molecules (DNA and RNA) are highly specific, well-characterized and able to provide quantitative information as well as molecular composition indications [23]. UV absorbance is widely employed to characterize volume samples and even molecular layers [24],[25],[26].

Here, we present some innovative approaches based on the measurement of UV absorbance for DNA strands quantification and detection, suitable for fully-integrated analysis systems.

A system integration perspective must start from the analysis of the existing array surfaces and their needs in terms of densities according to the different applications. At present, existing arrays differs for both dimensions and density of molecular spots per square centimeters. The use of a single chip of mono-crystalline silicon seems to be the natural electronic evolution of quartz high-throughput, high-density microarrays, like the one created by Affymetrix (surface $1.2\,cm^2$). These passive chips, implemented with photolithographic techniques are able to test a whole genome in parallel, with densities of one million sites per square centimeter [27].

Nevertheless, low or medium density array (from ten sites [28],[29] to forty thousand sites [30]) - suitable for analysis targeted to a restricted number of genes - are usually spotted on very large areas (several square centimeters). As a result, to provide on site optoelectronic detection a different technology should be considered.

In what follows, the paper presents different technologies for UV-based molecular detection that are suitable to different applications. In specific, Sect. 3.1 concerns the use of amorphous silicon p-i-n junctions in the measurements of extremely low concentrations of molecules and the suitability of these devices for surface detection of long DNA molecules. Section 3.2 presents the use of UV high-sensitive floating gate memories implemented in CMOS technology for the measurements of molecular absorbance. This technology allows very dense implementations and the integration of additional circuitry for processing of large amount of data. In Sect. 4 are drawn some conclusions.

3.1 Amorphous Silicon P-I-N Junctions

Amorphous silicon technology is particularly suitable for microsystem implementations of analysis tools. It involves low-cost and low-temperature processes, thus, it is suitable to be integrated directly on glass or plastic microfluidics. Amorphous silicon photodetectors are well-known to be high-sensitive devices which can be tuned to be specific to a narrow range of wavelengths [31]. Recently, they have been employed in integrated system for fluorescence detection on molecular assays [15],[20].

Here, we present experimental results of molecular characterization obtained by high-sensitive p-i-n structures of composite amorphous silicon. We describe two employments of such devices, namely the characterization of low-molecular concentrations (see Sect. 3.1 and the use of photodetectors coupled with molecular sensing spots (see Sect. 3.1).

Device We describe the design choices which aim at providing very high-sensitive UV detectors. In the attempt to make inexpensive and compact analysis systems there the use of low-molecular concentrations and of low UV radiation intensity levels are needed.

These devices, already presented in [32], are n-type amorphous silicon/intrinsic amorphous silicon/p-type amorphous silicon carbide (a-SiC:H) stacked structures (n-i-p) grown by Plasma Enhanced Chemical Vapor Deposition (PECVD) on a glass substrate covered by Cr/Al/Cr metal layers (top and front view are shown in Fig. 1). The top contact is a Al/Cr metal grid, whose spacing is optimized for charge collection by taking into account the conductivity of the underlying p-layer. The p-layer is the active zone of the device. In order to ensure good collection efficiency, thus sensitivity, the p-layer thickness has to be less than the electron diffusion length in the p-doped layer and more than the distance needed to generate the built-in potential. A good trade-off between these two competitive requirements is obtained by setting the thickness to $5\,nm$. The i-layer composition and thickness are determined to keep the dark current as low as possible. As a first step, this specification can be met by using hydrogenated amorphous silicon for the intrinsic layer; this allows to reduce the defect density in the i-region, hence the contribution to the inverse saturation current due to thermal generation. The optimum value of the thickness is found to be $150\,nm$, which allows to achieve the minimum of the inverse saturation current around $5 \times 10^{-11}\,A/cm^2$ at small reverse bias. The part of the device under the grid is $2 \times 2\,mm^2$. The metal grid of the device has a pitch of $200\,\mu m$ and the width of the fingers is $50\,\mu m$. The responsivity measured is $45\,mA/W$. The quantum yield is around 0.15 for wavelengths below $300\,nm$.

Method The UV optical absorbance A_M of a molecular sample is given by the Lambert-Beer law :

$$A_M = -Log(Int_M/Int_0) \tag{1}$$

where Int_0 is the intensity of light transmitted to the detectors through a reference sample, (i.e. in absence of absorbing molecules) and Int_M the intensity transmitted by an equivalent sample containing molecules. The current flowing through the p-i-n junction and the intensity incident on the detectors being proportional, it is possible to substitute the ratio between the intensities with the ratio between the detector currents I_0 and I_M:

$$A_M = -Log(I_M/I_0) \tag{2}$$

In order to evaluate the characteristics of an optimized device, it is very important to point out the role of the absorbance of the reference sample. As it will be described in Sect. 3.1, when dealing with microarrays of sensing sites

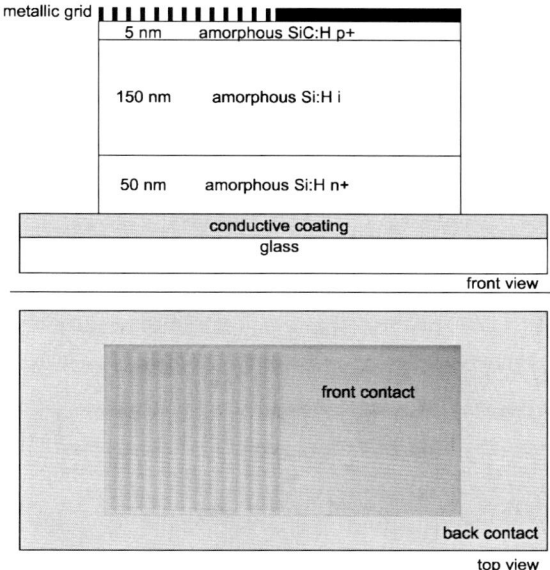

Fig. 1. Front view and top view of the amorphous silicon photodetector. The device surface is $2 \times 4\,mm^2$

the reference absorbance is the one of the sensing site prior to target molecules binding (it is due to the quartz substrate, the linkers and the probes). The resolution of the system (or the minimum detectable absorbance of the captured target molecules), depends on the current working point, defined by the absorbance of the sensing site before molecular recognition reaction. In fact, being eI the absolute error on the photocurrent, a variation of absorbance can be detected if the corresponding variation of the photocurrent is twice the noise current ($6dB\ SNR$). The minimum variation of the current corresponding to the sensing site I_0 that can be detected is $2 \cdot eI$. It follows that:

$$minA_M(I_0) = -Log((I_0 - 2 \cdot eI)/I_0) \qquad (3)$$

where $minA_M$ is the minimum detectable absorbance of captured layer on this sensing site. As I_0 is a function of the absorbance of the sensing site A_0:

$$A_0 = -Log(I_0/I_{void}) \qquad (4)$$

the function $minA_M(A_0)$ can be derived. I_{void} is the photocurrent sensor in absence of the quartz slide and corresponds to $I_{void} = P_0 \cdot R$, where P_0 is the power of the light source and R is the responsivity.

Setup The sensitivity of UV photodetectors has been evaluated by the optical setup drawn in Fig. 2. The light, generated by a mercury lamp, has been filtered

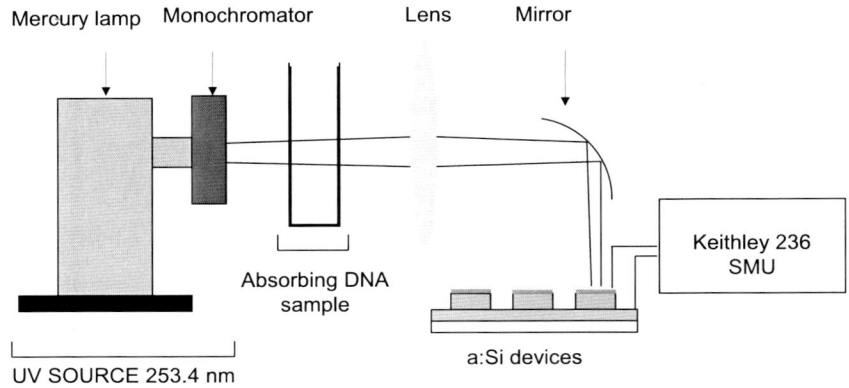

Fig. 2. Optical setup employed in the measurement of DNA molecular samples.

by a *Yobin-Yvon SPEXH10* monochromator. Devices has been tested at a wavelength of $253.4\,nm$ which is located in the absorption peak region of nucleotides (maximum: $260\,nm$) [23]. The incident power on the sensor is $0.35\,\mu W$. UV optics (a lens and a mirror) have been used for beam collimation and focusing while a *Keithley 236 Source Measure Unit* has been employed for the measurement of the a-Si:H sensor current. To validate the UV approach to DNA microarrays, we have compared the absorbance of sensing sites where target DNA binding has and has not taken place, respectively. Bio-functionalized quartz slides have been aligned with the photodetectors (Fig. 3).

High-sensitive molecular sample characterization The sensitivity of the detectors has been investigated by measuring the absorbance of molecules in solution down to very low concentrations. Short sequences of the single-stranded form of DNA have been selected: $30 - mer$ $5'$−gat cat cta cgc cgg acc cgg gca tcg tgg−$3'$ (MW $7669\,g/mole$, extinction coefficient $260700\,L/mole\cdot cm$). The molecules have been diluted to different concentrations into a TAE Mg^{++} buffer and placed in a quartz container.

The DNA absorbance (A_{DNA}) has been calculated by using 2 where the reference sample was a buffer solution of TAE Mg^{++} at room temperature. The short term variability of the mercury lamp intensity has been monitored by measurements of the white field. In Fig. 4 the calculated absorbances for different concentrations (following from 2) have been compared to their nominal value. A good linearity can be observed down to 3×10^{-4} absorbance. Figure 2 shows the setup corresponding to the characterization of molecular samples in solution Sec. 3.1.

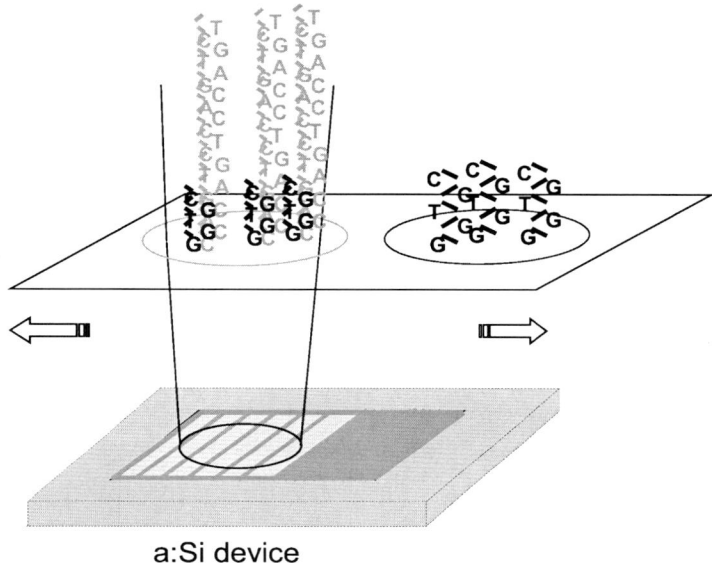

a:Si device

Fig. 3. A site with probes and captured long target sequences has been aligned with the exposed part of the detector. The other site has been aligned on the sensor by a shift of the slide.

Target Molecules Detection on Surface Sites The ability given by a:Si technology to deposit large area arrays allows the implementation of detectors slides meant to integrate most existing DNA microarrays the way they are (for example low/medium density array spotted on standard microscope slides). To demonstrate the viability of the UV approach to DNA microarrays, we detected a layer of captured target molecules on sensing sites with UV a:Si detectors. Sensing sites hosted short probe strands which could capture specifically long strands of target molecules thanks to the complementarity between their sequence and a part of the target.

Device The target molecule is pBR322 linearized with PvuII, a commonly used plasmid cloning vector of 4361 base length (extinction coefficient at $260\,nm$: $4 \times 10^7\,l/mole \cdot cm$). The sites have been exposed to unlabeled target molecules at a $10\,nM$ concentration in a TAE Mg^{++} buffer. The reaction has been realized in a humid chamber at 90°C for fifteen minutes to allow the separation of pBR322 strands and then cooled down to room temperature for two hours. Unbound molecules have been removed by successive rinsing procedures (SSC $2\times$ and $0.2\times$ solutions, 5 minutes each) which ensure high specificity for surface affinity reactions. Capturing oligonucleotides ($5' - SH - (CH_2)_6$gag ctc gga agc cca gta gta ggt$-3'$; MW $8570\,g/mole$; extinction coefficient: $269800\,l/mole\,cm$) have been provided to bind pBR322 strands by the use of of a 21-base long

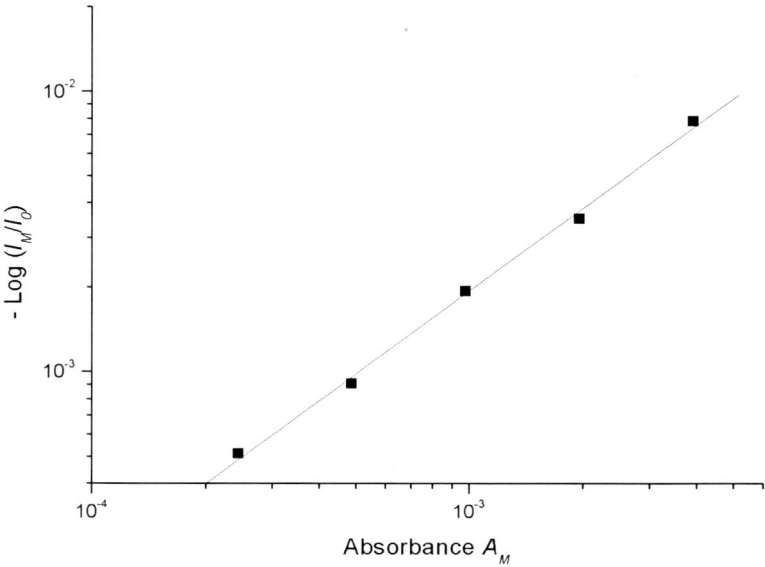

Fig. 4. Plot of the absorbance of molecular samples of low concentrations measured with the photodetector vs. their nominal value.

complementary sequence (acc tac tac tgg gct gct tcc in pBR322) and have been immobilized on sensing site by means of a standard procedure [33].

When target molecules were captured the sites exhibited higher absorbance and, consequently, a lower current was measured from the photodetectors (see in Fig. 5).

3.2   CMOS-Compatible Floating Gate Cells

This section presents an original approach based on the use of floating gate cells aiming at providing high-density arrays for integrated molecular detection and high flexibility in the choice of signal conditioning and processing strategies.

High-UV-sensitive floating gate cells have been exploited as devices to measure DNA molecular absorbance. In these memories the electrons can be injected into the floating gate by tunnel effect and can be completely or partially removed by a certain dose (intensity × time-interval) of UV light. A particular kind of cells, single-poly memories, have been selected because the UV sensitivity of their floating gate was enhanced by specific design choices. A schematic representation of single-poly cells is sketched in Fig. 6, where the main features of the design can be observed:

- the floating gate is the top conductive layer of the device and, in our test-chip, it was entirely exposed to the light (under the passivation oxide);
- the floating gate extends for a large portion of the surface of the cell (which measures $20\,\mu m^2$ for this generation). An advanced modeling and experimental results on the characteristics of these wafers have been already presented in [34].

The test-chips have been fabricated by STMicroelectronics–Italia in $0.25\,\mu m$ CMOS technology. It should be noticed that, although these memory cells are far from being state of the art, their dimensions are comparable to the minimum surface that microfabrication of molecular spots can achieve with the existing technology.

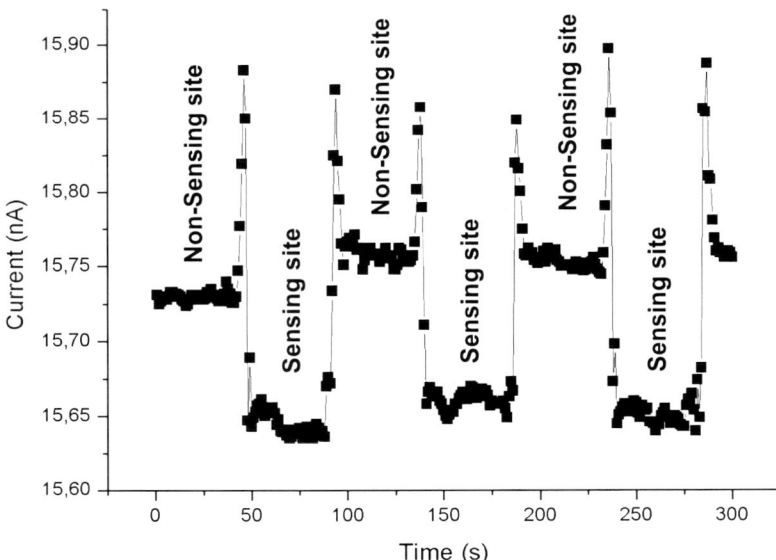

Fig. 5. Sites with only sensing probes (Non-Sensing sites) and sites with sensing probes and captured targets (Sensing Sites) have been aligned successively with a photodetector. The lamp drift was monitored with measurements of the white field and it causes variations in the order of $25\,pA$.

Method The erase-characteristic of the memory cell (the way the threshold voltage of the cell decreases with time during UV irradiation) depends on the intensity of the UV light with an exponential law [35]. The incident photons impart energy to the electrons which have been stored in the floating gate during the programming process and excite them over the oxide energy barrier. The current due to the photo-excitation (I_{UV}) is a linear function of the light intensity at a given wavelength (Int_{UV}) and of the electric field E_{el}. When the control gate is at the same voltage than the substrate, the electric field across the oxide can be written as:

$$E_{el} = \frac{Q_{FG}}{C_{TOT}t_{ox}} \tag{5}$$

where Q_{FG} is the negative charge stored in the floating gate, C_{TOT} the total capacitance of the floating gate and t_{ox} the effective oxide thickness in the photo-excitation area.

The charge excited over the oxide has the following rate:

$$\frac{dQ_{FG}}{dt} = A_{eff}Int_{UV}\left(const1\frac{Q_{FG}}{C_{TOT}t_{ox}} + const2\right) \tag{6}$$

where A_{eff} is the effective area on the floating gate surface. $Const1$ and $const2$ are empirical constants depending on the light wavelength and on the silicon dioxide characteristics. Their ratio $const2/const1$ has the dimension of a field and is negligible with respect to E_{el}. Thus,

$$Q_{FG}(t) = Q_{FG}(0)e^{-\frac{Int_{UV}t}{\vartheta_0}} \tag{7}$$

where $Q_{FG}(0)$ is the charge at the beginning of the erase process and ϑ_0 equals $(A_{eff}const1)/(C_{TOT}t_{ox})$.

The erase characteristic can be written as follows:

$$Vth_{PR} - Vth_E(t) = (Vth_{PR} - Vth_{TOT\_E}) \cdot (1 - e^{-\frac{Int_{UV}t}{\vartheta_0}}) \tag{8}$$

where Vth_{PR} is the threshold voltage at the end of the programming process (beginning of the erasing process), Vth_{TOT_E} the threshold voltage corresponding to the floating gate trapping no electrons and $Vth_E(t)$ the threshold voltage after t seconds of UV light exposure.

For our purposes, let's consider that the UV light intensity can be attenuated by the presence of absorbing molecules on its path, thus affecting the erase characteristic. We demonstrated that floating gate cells may be used to analyze molecular samples (*i.e.* to provide a quantification or to detect the type of molecule) by observing their threshold after UV irradiation.

More particularly, the cell is programmed at a well-defined threshold voltage Vth_{PR} by injecting a corresponding amount of electrons in the floating gate. Then, the cell and the molecular sample are exposed to UV radiation during a certain time t. In the end, a different threshold voltage (Vth_E) – lower than Vth_{PR} – is reached. This threshold voltage depends on the container, solvent, absorbing biomolecules placed as filters on the UV-light path. The erase-characteristic has the following form:

$$Vth_{PR} - Vth_E(t) = (Vth_{PR} - Vth_{TOT\_E}) \cdot (1 - e^{-\frac{tInt_{UV}T}{\vartheta_0}}) \qquad (9)$$

where T is the transmittance due to the absorbing effect of the molecular sample and of the container.

If Vth_{E1} and Vth_{E2} are the threshold voltages reached in case of two samples, their difference is related to the transmittances as follows:

$$Vth_{E1}(t) - Vth_{E2}(t) = (Vth_{PR} - Vth_{TOT\_E}) \cdot (e^{-\frac{tInt_{UV}T_2}{\vartheta_0}} - e^{-\frac{tInt_{UV}T_1}{\vartheta_0}}) \qquad (10)$$

The analysis of the characteristics of a molecular sample is usually done with respect to a reference solution, which corresponds to the same solution in which molecules are diluted during measurements. The reference sample has a defined transmittance T_{REF} for which we can take into account at the denominator inside the constant τ_0. Correspondingly, the threshold voltage after a certain time of irradiation reaches $Vth_{REF}(t)$

For an certain molecular sample having a transmittance T_M:

$$Vth_E(t) - Vth_{REF}(t) = (Vth_{PR} - Vth_{TOT\_E}) \cdot (e^{-\frac{tInt_{UV}}{\tau_0}} - e^{-\frac{tInt_{UV}T_M}{\tau_0}}) \qquad (11)$$

which directly links the threshold voltage of a sample to its transmittance and to its absorbance, as $A_M = -Log(T_M)$.

The difference between the exponentials will be maximum at a certain dose which should be evaluated for each setup.

Setup The source of UV radiation is a Xenon lamp, the distribution of which presents high emission values in the range of interest 250 to $270 \, nm$. Though the lamp is controlled by a special circuit maintaining the supplied power constant $(100W)$, radiation stability over long time periods is not guaranteed. That is why the radiation needs to be measured to compute the effective dose. This system is based on a photodiode the output current of which is amplified by an I/V converter and sampled by the use of an acquisition board equipped with 16-bit converters. The photodiode is placed directly after the UV filter needed to select the radiation range of interest and before the DNA sample container. Then a LabVIEW program controls the shutter and switches off the radiation once the dose has reached the desired value.

The cell characteristic is traced once the light source is masked by means of a HP4156 Semiconductor Parameter Analyzer (SPA) driven by the control PC. The threshold voltage is conventionally defined as the control gate polarization needed to obtain a drain current of $2 \, \mu A$ when the drain is driven to $1 \, V$ and source and bulk are grounded. Resolution in threshold voltage is $1 \, mV$.

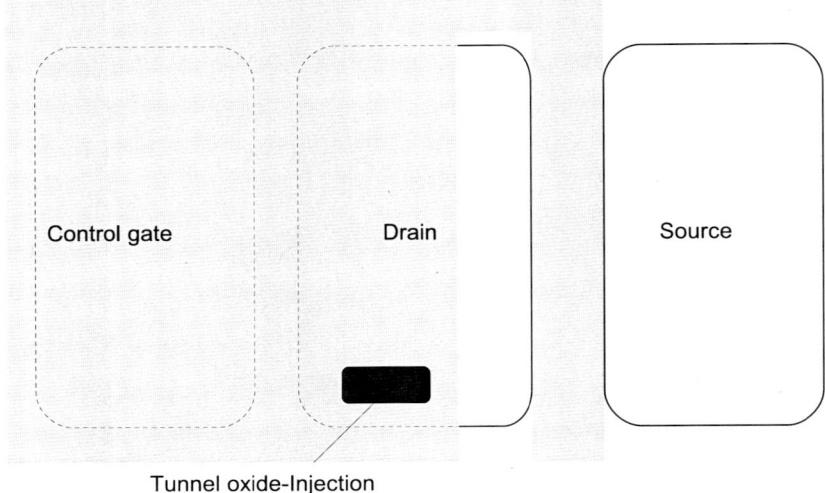

Fig. 6. Schematic top view of single-poly memory cells. The polysilicon floating gate extends for a large portion of the whole cell and it is exposed directly to the light under the silicon dioxide top passivation. The control gate is realized by means of a diffusion as source and drain and not by a second polysilicon layer.

Measurement of DNA hypochromic effect In our experiments, the reference sample (see Sect. 3.2) was a quartz container filled with $1.5\,ml$ buffer solution (TE $1\times$). So the optical path through the liquid sample was $1\,cm$ long. Two different forms of DNA molecule have been considered: single-stranded and double-stranded (two single strands zipped in a double helix). The total absorbance of two separate single strands is 30% bigger than that of the double helix form. This is known as hypochromic effect of DNA which is due to the fact that the interaction of the nucleotide bases with light changes when the strands are bound together.

The strands were 25 or 30 bases long. DNA molecular samples in the two forms have been analyzed in different concentrations by means of floating gate cells erase-characteristics (Fig. 7). On the x axis we reported the concentration of the nucleotide bases for each sample and on the y axis the threshold voltage at the time t of the erasing process. Each measurement is repeated 5 times and the standard deviation is indicated by the errors bars for each point (the latter cannot be easily seen on the plot as they measure $1\,mV$).

The two forms of the molecules are clearly distinguishable even for $1\,\mu M$ bases concentration, which means a resolution equaling the 30% on a 15×10^{-3} absorbance scale corresponding to the single-stranded sample. Such a behavior is comparable to that obtained with standard spectrophotometers [36]. Nevertheless, signal to noise ratio would be remarkably improved by an integrated setup.

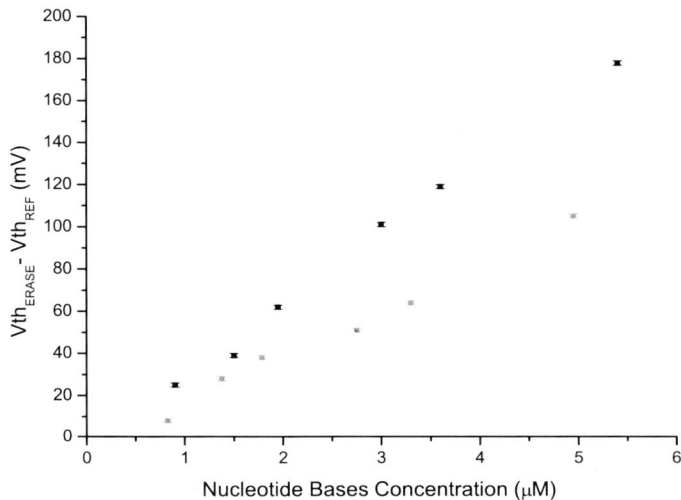

Fig. 7. Experimental plot and linear fit of the measurements of single-stranded (squares) and double-stranded (circles) DNA molecules for different concentrations.

4   Conclusions

Two silicon technologies aimed at the implementation of optical detectors for integrated molecular analysis tools have been selected and tested. The amorphous silicon devices have demonstrated high UV absorbance sensitivity down to 3×10^{-4}. They have also been employed to measure the presence of captured target molecules on a sensing site.

CMOS-compatible floating gate cells, suitable for high-parallel assays have been employed to detect molecular samples of different forms and concentrations. The erase-characteristics have been shown to follow a reproducible exponential law. Besides the cells have been able to detect the hypochromic effect of DNA down to 15×10^{-3} absorbance or at a concentration of $300 \, nM$ short oligonucleotide strands.

References

1. Liu R H, Yang J, Lenigk R, Bonanno J, Grodzinski P, (2004) Anal. Chem., 76:1824 – 1831
2. Lagally E T, Medintz I, Mathies R A, (2001) Anal. Chem., 73:565 – 570
3. Lagally E T, Scherer J R, Blazej R G, Toriello N M, Diep B A, Ramchandani M, Sensabaugh G F, Riley L W, Mathies R A, (2004) Anal. Chem. 76:3162 – 3170
4. Kelly Ryan T, Woolley A T, (2005) Anal. Chem., 77(5):97A – 102A

5. Woolley A T, Sensabaugh G F, Mathies R A, (1997) Anal. Chem., 69:2181 – 2186
6. Shi Y, Simpson P C, Scherer J R, Wexler D, Skibola C, Smith M T, Mathies R A, (1999) Anal. Chem., 71:5354 – 5361
7. Terbrueggen R H, Chen Y-P, Duong H, Millan K M, Mucic R C, Olsen G T, Singhal P, Swami N, Wang H, Welch T W, Yowanto H, Yu C J, Blackburn G F, Kayyem J F, (2001) European Solid-State Devices and Research Conference, 1:119 – 121
8. Schienle M, Frey A, Hofmann F, Holzapfl B, Paulus C, Schindler-Bauer P, Thewes R, (2004) Solid-State Circuits Conference, Digest of Technical Papers, 1:220 – 524
9. Stagni Degli Esposti C, Guiducci C, Benini L, Riccò B, Carrara S, Samorì B, Paulus C, Schienle, Augustyniak M, Thewes R, (2006) Journal of Solid-State Circuits, 41(12):2956 – 2964
10. Wu G, Ram D, Hansen K M, Thundat T, Cote R J, Majumdar A, (2001) Nature Biotechnology, 19:856 – 860
11. Fritz J, Baller M K, Lang H P, Rothuizen H, Vettiger P, Meyer E, Guntherodt H J, Gerber C, Gimzewski J K, (2000) Science, 288:316 – 318
12. Lin V S-Y, Motesharei K, Dancil K, Sailor M J, (1997) Science, 278(5339):840 – 843
13. Prasad T, Colvin V, Mittleman D, (2003) Physical Review B, 67:165103-1 – 165103-7
14. Thrush E, Levi O, Cook L J, Deich J, Kurtz A, Smith S J, Moerner W E, Harris J S Jr, (2005) Sens. and Act. B, 105(2):393 - 399
15. Kamei T, Paegel B M, Scherer J R, Skelley A M, Street R A, Mathies R A, (2003) Anal. Chem., 75-20:5300 - 5305
16. Webster J R, Burns M A, Burke D T, Mastrangelo C H, (2001) Anal. Chem., 73-7:1622 - 1626
17. Nakanishi H, Nishimoto T, Arai A, Abe H, Kanai M, Fujiyama Y, Yoshida T, (2001) Anal. Chem., 22(2):230 - 234
18. Xu C, Li J, Wang Y, Cheng L, Lu Z, Chan M, (2005) IEEE Electron Device Letters, 26(4):240 - 242
19. Misiakos K, Kakabakos S E, Petrou P S, Ruf H H, (2004) Sens. and Act. B, 76(5):1366 - 1373
20. Fixe F, Chu V, Prazeres D M F, Conde J P, (2004) Anal. Chem., 32(9):e70
21. Thrush E , Levi O , Ha W, Wang K, Smith S J, Harris J S, (2006) Journal of Chromatography A, 1013:103 - 11
22. Romualdi C, Trevisan S, Celegato B, Costa G, Lanfranchi G, (2003) Nucl. Acids Res., 31(23):e149-1 — e149-8
23. Tinoco I, Sauer K, Wang J C, Puglisi J D (2001) Chemistry:Principles and Applications in Biological Sciences. Prentice Hall, New York.
24. Liu M, Jie L, (1999) J Phys. Chem. B, 103(51):11393 - 11397
25. Pei R, Cui X, Yang X, Wang E, (2001) Biomacromolecules, 2(2):463 - 468
26. Shi X, Sanedrin R J, Zhou F, (2002) J. Phys. Chem. B, 106(6):1173 - 1180
27. Kennedy G C, Matsuzaki H, Dong S, Liu W M, Huang J, Liu G, Su X, Cao M, Chen W, Zhang J, Liu W , Yang G, Di X, Ryder T, He Z, Surti U, Phillips M S, Boyce-Jacino M T, Fodor S PA, Jones K W, (2003) Nature Biotechnology, 10(10):1233 - 1237
28. Wang D, Urisman A, Liu Y, Springer M, Ksiazek T, Erdman D, Mardis E, Hickenbotham M, Magrini V, Eldred J, Latreille J, Wilson R, Ganem D, DeRisi J, (2003) PLoS Biology, 1(2):257260
29. Abruzzo L, Lee K, Fuller A, Silverman A, Keating M, Medeiros L, Coombes K, (2005) Biotechniques, 38(5):785 - 792

30. Conzone S D, Pantano C G, (2004) Materials Today, 7(33):20 - 26
31. Caputo D, de Cesare G, Irrera F, Tucci M, (1998) J. of Non-Cryst. Solids, 2:1316 – 1320
32. de Cesare G, Caputo D, Nascetti A, Guiducci C, Riccó B, (2006) Appl. Phys. Lett., 88: 083904-1 – 083904-3
33. Rogers Y, Jiang-Baucom P, Huang Z, Bogdanov V, Anderson S, Boyce-Jacino M T, (1999) Anal Biochem, 266(1):23 - 30
34. Lanzoni M, SuiiC J, Olivo P, Riccó B, (1993) IEEE Trans. on Electron Devices, 30(5):951 – 957
35. Katznelson R D, Frohman-Bentchkowsky D, (1980) IEEE Trans. on Elect. Dev., 27(9):1744 – 1752
36. http://www.biotek.com/products/docs/Nucleic_Acid_Quantitation_Tech_Note.pdf

Electronic Detection of DNA Adsorption and Hybridization

Ulrich Bockelmann

Laboratoire de Nanobiophysique
UMR Gulliver CNRS-ESPCI 7083
10 rue Vauquelin, 75005 Paris (France)
ulrich.bockelmann@espci.fr

Abstract. We consider a transistor based approach for DNA detection. Immobilization of DNA at different positions of poly(l-lysine) coated field effect transistor arrays is achieved by deposition with a microspotting device or specific hybridization between complementary oligonucleotide sequences. The current voltage characteristics of the transistors are measured with the sample surface immersed in aqueous solution. The chapter provides a brief overview of our experimental technique and of its applications to the detection of DNA adsorption and hybridization.

1 Introduction

The large majority of nucleic acid analysis techniques presently rely on detection schemes which involve fluorescence, radioactivity or enzyme based labelling [1]. In many cases the labelling steps are expensive, time-consuming and error-prone and the corresponding equipment is expensive. These disadvantages are major motivations for research on alternative detection methods [2–4].

After early work by P. Bergveld [5], several papers addressed the detection of biomolecules binding on field effect transistor (FET) structures [6–8]. DNA is a strong negatively charged polyelectrolyte in aqueous solutions. Capacitive impedance and FET detection of DNA have been achieved over the last years [9–14].

We use a differential technique based on dc measurements with arrays of integrated FET devices [12, 15, 16]. Our approach allows for efficient miniaturisation and parallelisation through on-chip integration of the detection and readout circuits [17]. It involves adsorption of the biomolecules to part of the array (achieved by local deposition of sub-nanoliter volumes of solutions with a microspotting device) and measurement of the transistors (with the surface of the microchip covered by an aqueous solution).

The chapter is organized as follows. The experimental techniques are presented in Sect. 2. Electronic detection of DNA immobilized by direct adsorption on poly(L-lysine) coated FET arrays is considered in Sect. 3. Sect. 4 is devoted to the detection of specific hybridization between DNA oligonucleotides. Salt effects are briefly considered in Sect. 5. Sect. 6 contains the concluding part of the chapter.

Please use the following format when citing this chapter:

Bockelmann, U., 2007, in IFIP International Federation for Information Processing, Volume 249, VLSI-SoC: Research Trends in VLSI and Systems on Chip, eds. De Micheli, G., Mir, S., Reis, R., (Boston: Springer), pp. 55–67

2   Experimental techniques

2.1   Fabrication of the FET arrays

Arrays of silicon p-channel field effect transistors have been used in this work. The samples were fabricated in the clean room facility of the Max Planck Institut für Biochemie, Martinsried, Germany. The process is based on standard silicon microtechnology and is described in [18, 19]. The active areas of the individual transistors carry a 10 nm thick SiO_2 oxide without metal gate. Typically, 96 FETs are linearly arranged with a period of 20 μm, the active surface area of an individual FET amounts to 36 or 100 μm^2. The drain of each transistor is individually connected, while the whole array shares a common source contact. On the silicon chip these connexions are realized by boron p^+ implantation. The chips are mounted on ceramics sockets, wire bonded and finally a plastic well is glued on the surface to protect the wiring during the subsequent manipulations where the surface is incubated with liquids.

2.2   Surface preparation

Prior to DNA immobilization, we perform the following global treatment of the SiO_2 surface. Incubation in sulfochromic acid, rinsing with H_2O, incubation in a NaOH/ethanol solution, H_2O rinsing and drying with air. Afterwards, the FET array is incubated in a poly(L-lysine) dilution, again followed by H_2O rinsing and drying with air. Immobilization of DNA by a poly(L-lysine) layer is widely used in the field of DNA microarrays, although the resulting attachment is less stable than the one obtained by covalent immobilization strategies. DNA immobilization with poly(L-lysine) is relatively simple and allows us to use a FET array several times. In fact, we can clean the SiO_2 surface after use and repeat DNA immobilization and detection.

2.3   Local deposition of DNA

A piezo spotting device equiped with a custom microscopic imaging system is used to deposit DNA solutions on the FET array, as shown in Figure 1. Typically spot diameters of 50-200 μm are used, corresponding to volumina in the 0.1-1 nl range.

2.4   Electronic measurements

For the electronic measurements the surface of the chip is immersed in aqueous electrolyte solution, together with a reference electrode. We measure the drain current I_D of each transistor as a function of a dc voltage U_{SD} applied between source and drain and a dc voltage U_{SE} applied between source and the reference electrode. These 2D characteristics are recorded at room temperature. An analog amplifier circuit is used to bias the FETs and to measure the drain current I_D, as shown in Figure 2. The individual FETs of the array are multiplexed by a

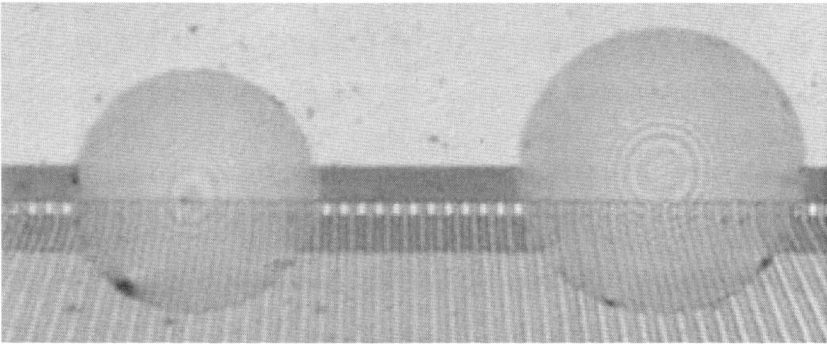

Fig. 1. Microscopic image of a transistor array with two local deposits. The spots have diameters of about 200 μm and each one covers about 10 FETs of the linear array. The active regions of the individual transistors appear as white squares in the image. The gray lines in the bottom part are the individual drain connexions. The homogeneous region in the upper part is the common source contact.

switch unit, the latter being computer controlled by a digital I/O board. Two dc voltages $\tilde{U}_{SE}$, $\tilde{U}_{SD}$ are generated by 16 bit D/A converters of a multifunction I/O board and are lowpass filtered in the input stage of the amplifier. The drain current is converted to a voltage that is measured by a 16 bit A/D converter of the I/O board. The acquisition of the bidimensional characteristics $I_D(\tilde{U}_{SD}, \tilde{U}_S)$ for all FETs is computer controlled.

Dependending on the design of the FET array device the implanted on-chip connexions to the individual drains and the common source may lead to non-negligible serial resistances. In this case, we consider the lateral geometry of the connexions and derive the serial resistances of the drain lead R_D and the source lead R_S for each FET. The transformation

$$U_{SE} = \tilde{U}_{SE} - R_S\, I_D$$
$$U_{SD} = \tilde{U}_{SD} - (R_D + R_S)\, I_D$$

finally gives the intrinsic characteristics $I_D(U_{SD}, U_{SE})$. To study shifts in U_{SE} we convert the $I_D(U_{SD}, U_{SE})$ characteristics, to $U_{SE}(I_D, U_{SD})$ by numerical interpolation.

2.5 Microfluorescence

For comparison with the electronic signals we use microscopic fluorescence measurements. The corresponding optical setup is schematically presented in Figure 3. The beam of a red (632.8 nm) HeNe cw laser or a green (532 nm) solid state cw laser is expanded, introduced in the illumination path of an upright microscope and focused with a microscope objective to a spot of adjustable diameter (0.5-10 μm). Fluorescence is collected by the same objective and passes two

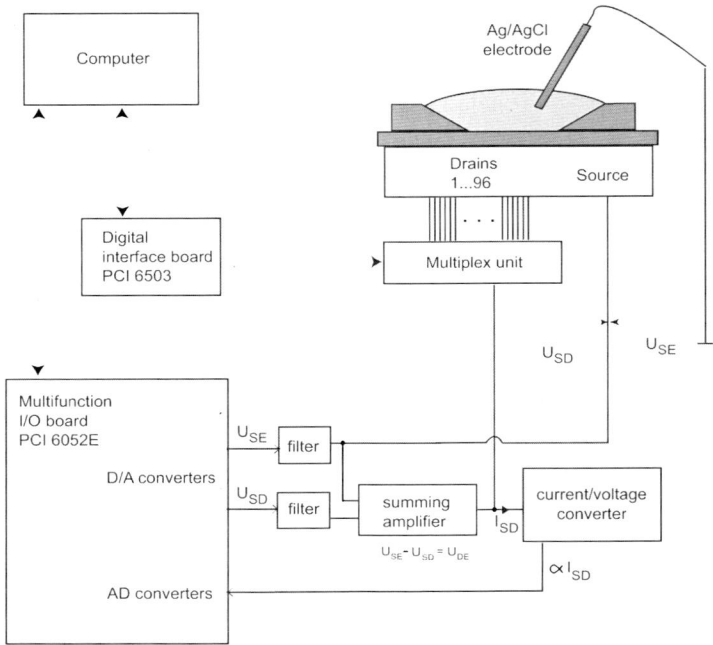

Fig. 2. Recording setup used for the differential detection of biopolymers. The FET array contains one drain connection for each of the 96 transistors and one common source.

crossed piezo-adjustable slits, positioned in an image plane of the sample. The two-slit arrangement defines a rectangular detection region and allows us to optimise this region with respect to the size of the excitation spot. Excitation light is blocked by an emission filter and the fluorescence signal is measured with a cooled photomultiplier tube. A motorized xy translation stage with integrated position sensors operating in a feedback loop is used to scan the sample laterally.

3   Detecting DNA adsorption

A typical electronic measurement of DNA adsorption is presented in Figure 4. One deposit of pure water (left) and four deposits of an oligonucleotide solution (right) are done on the surface of a poly(L-lysine) coated array. The horizontal axis indicates the index of the different transistors. The 96 individual transistors of this array are arranged along a line of about 2 mm in length. Each transistor has an active surface of 100 μm^2. In this experiment, spots of about 200 μm in diameter were deposited. The difference in U_{SE} between a first measurement done prior to the deposition and a second measurement done directly afterwards

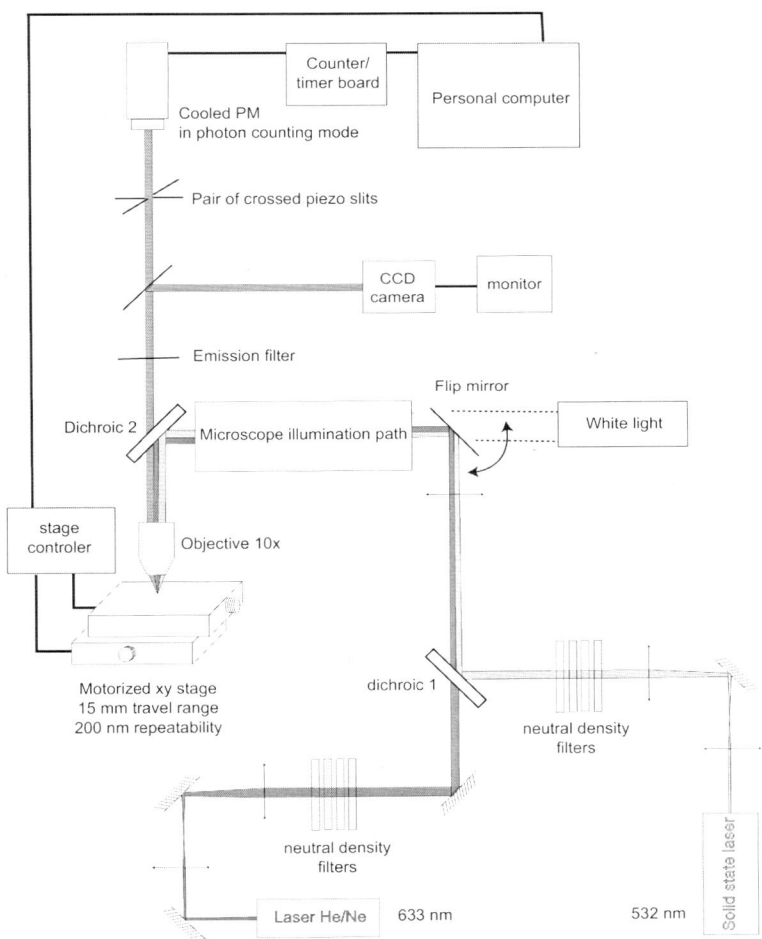

Fig. 3. Dual color fluorescence setup for imaging the transistor array surfaces.

is presented. The adsorption of DNA gives rise to negative shifts ΔU_{SE}, while for the transistors below the reference spot no significant shift relative to the baseline is measured.

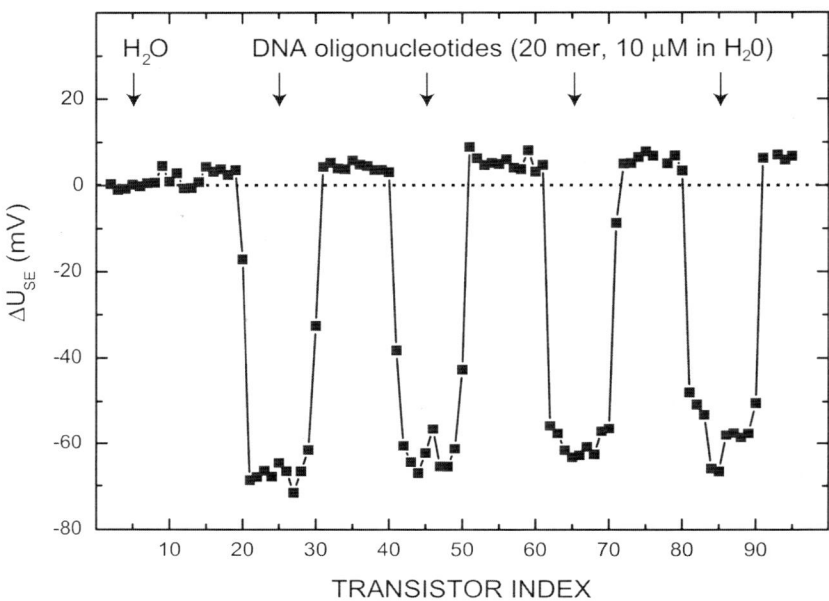

Fig. 4. Electronic detection of DNA adsorbed on a poly(L-lysine) coated array of 96 FETs linearly arranged with a period of 20 μm. Four local deposits of a DNA oligonucleotide solution are performed and a H_2O spot is added for comparison. The positions of these spots are indicated by arrows. Each data point correspond to one transistor of the array (bottom axis). The quantity ΔU_{SE} gives the shift of the current/voltage characteristics at a given (I_D, U_{SD}) working point, observed between two measurements. The first measurement is done prior to the deposition, the second one immediately afterwards, in both cases the sample surface was covered by an aqueous electrolyte solution (0.1 mM KNO_3). The DNA spots induced negative shifts ΔU_{SE} of about 60 mV, while no corresponding signal is observed on the reference spot.

In an earlier publication [12], we presented a comprehensive investigation of the electronic detection of poly(L-lysine) and DNA. Adsorption of DNA on poly(L-lysine) coated SiO_2 induces negative shifts in U_{SE}, while adsorption of poly(L-lysine) on SiO_2 leads to positive shifts. This can be attributed to the fact that we use p-channel FET arrays and that DNA carries negative charge in aqueous solution around pH 7, opposite to the positively charged poly(L-lysine). The electronic signals induced by adsorption of these charged polymers were studied as a function of electrolyte salt and polymer concentrations and an analytical model which accounts for screening of the adsorbed charge by mobile ions was developed. The microfluorescence setup described above was used in combination with Cy3 and/or Cy5 modified oligonucleotides to quantify the local DNA adsorption. Non-modified, Cy3-modified and Cy5-modified oligonucleotides are electronically detected without noticable differences. It is also possible to detect the adsorption of double stranded DNA. This was shown with double stranded molecules synthesized in-vitro by PCR (polymerase chain reaction).

In another study, we showed that the FET based measurement is compatible with enzymatic technology and complex DNA samples [15]. We combined the electronic detection with an allele-specific polymerase chain reaction and thus detected a single-basepair mutation in genomic DNA. The approach was applied to test human DNA for the 35delG mutation, a frequent mutation related to prelingual nonsyndromic deafness.

4   Detecting DNA hybridization

Hybridization involves the specific interaction between complementary base sequences and the transition from two single stranded to one double stranded molecule. In general, the single stranded molecule carries a different effective charge than the double stranded one [20, 21]. Therefore hybridization between a surface bound probe molecule and a target molecule from solution can change the electrostatic potential of the SiO_2/electrolyte interface and it is conceivable to electronically detect this specific interaction with our FET arrays. Specific recognition of target sequences by hybridization to surface bound probe DNA is the basis of microarray based DNA analysis, a rapidly developping technology with broad applications in research and diagnostics.

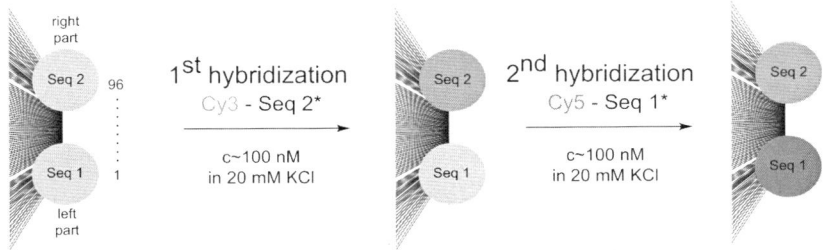

Fig. 5. Detection of hybridization. Oligonucleotide probes with two different base sequences are spotted to the left (transistors 1-31) and right (transistors 54-96) parts of an array of 96 transistors. Each transistor exhibits an active surface of 100 μm^2. The first hybridization is done with Cy3 labeled target oligonucleotides, matching the probes on the right part of the array. Afterwards, the second hybridization is done with Cy5 labeled targets matching the probes on the left part of the array.

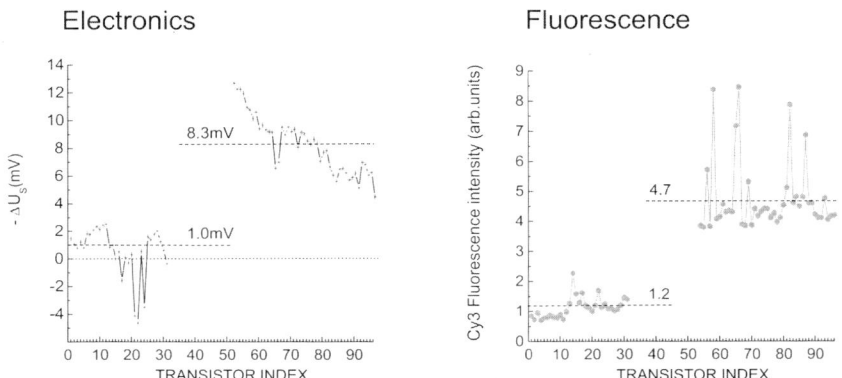

Fig. 6. Detection of the first hybridization. Both electronic (left, $-\Delta U_S$) and fluorescence (right) signals are higher on the right part of the array, where specific hybridization is expected.

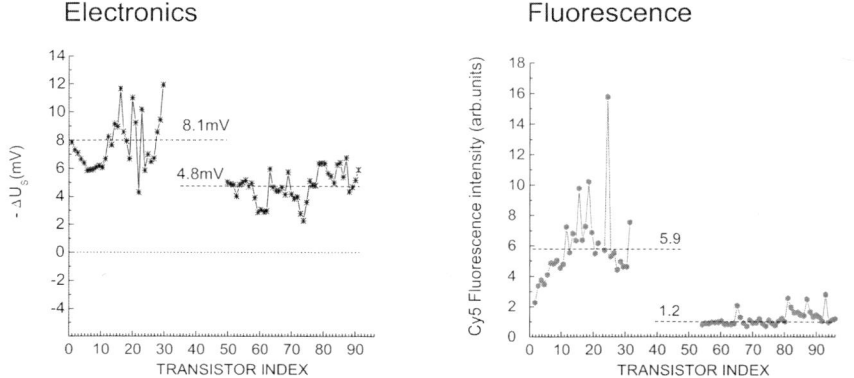

Fig. 7. Detection of the second hybridization. This time electronic and fluorescence signals are higher at the left, as expected.

Figures 5, 6 and 7 illustrate an experiment showing electronic and fluorescence detection of hybridization between DNA oligonucleotides. After initial cleaning we coated the surface of our planar device with poly(L-lysine). Using a micropipette we then deposited two different oligonucleotide solutions (0.3 μl of 1 μM solutions of 20 mer oligonucleotides in 20 mM KCl) to the left and right parts of the FET array (transistors 1-31 and 54-96, respectively). Incubation was done for 15 min at room temperature, before H_2O rinsing and drying with air.

For hybridization, the sample is incubated for 5 min at room temperature in 20 mM KCl buffer containing target oligonucleotides to a concentration of 100 nM. The hybridization is stopped by rinsing with fresh 20 mM KCl buffer (without target DNA and using a micropipette). The left part of figure 6 shows the difference ΔU_S between two electronic measurements at [KCl]=20 mM, done immediately before and after hybridization 1. The base sequence of the target molecules is complementary to the sequence of the probes deposited on the right part of the array; the targets are Cy3 labeled. In hybridization 2 (figure 7), performed afterwards on the same sample, targets are Cy5 labeled and match the probes on the left. Hybridization 1 leads to a stronger negative shift in U_S on the right (8.3 compared to 1 mV in average), while for hybridization 2 we observe a stronger negative shift on the left (8.1 compared to 4.8 mV). This is consistent with specific hybridization in both directions.

The specificity of both hybridizations is subsequently confirmed by fluorescence imaging. For this purpose the sample is rinsed with H_2O and dried. Under excitation at 532 nm (figure 6, right) we observe 4 times more fluorescence on the right part than on the left part of the array. This way we measure the fluorescence arising from the Cy3 labeled targets used in the first hybridization (targets matching the probes on the right part of the array). Under excitation at 633 nm (figure 7) we observe 5 times more fluorescence on the left part than on the right part of the array. This way we measure the fluorescence arising from the Cy5 labeled targets used in hybridization 2 (targets match probes on the left part of the array). Each fluorescence data point corresponds to the average of 3 measurements taken across the active surface of a transistor with a spatial resolution of about 500 nm. The fluorescence measurements are in qualitative agreement with the electronic signals and the signal-to-noise ratios observed with the two different techniques are not very different.

Residual non-specific interactions are expected under our experimental conditions and might explain the signals observed even on the non-matching probes. In particular, here we didn't introduce a blocker step to neutralize the positive charges in the poly(L-lysine) layer that still remain after the deposition of the DNA probes.

The hybridization conditions used in the present experiment (salt, temperature, duration, oligonucleotide length, surface concentrations of probes and hybridized targets) are similar to the conditions used by Fritz et al [10] in ac measurements of oligonculeotide hybridization on capacitive Si/SiO_2 devices and our differences in the average shifts ΔU_S (7.3 mV for hybridization 1 and 3.3 mV for hybridization 2) are close to their result (3 mV).

It is possible to use different salt concentrations for hybridization and detection in order to optimise both the specificity of the sequence recognition (preferentially high salt) and the sensitivity of the field effect detection (preferentially low salt to reduce screening by mobile ions). We have shown experimentally that the differential signals of hybridization can be enhanced significantly (from about 3 mV to 20 mV) by changing the salt concentration between hybridization and detection [16].

5   Salt dependence of the electronic signals

Descriptions of the silica surface in aqueous environment [23] and of its modifications induced by biomolecule binding are complex, combining surface chemistry, statistical mechanics and electrostatics. In these systems the concentration of the mobile ions in the electrolyte is an important parameter, controlling the biomolecule interaction with the surface as well as the sensitivity of the electronic detection.

Towards a quantitative understanding of the FET based electronic detection, we performed combined experimental and theoretical studies of the salt dependencies of the electrostatic potentials at the silica/biomolecule/electrolyte interface. This section only gives a glimpse at our study of the salt dependence of a bare silica surface in contact with a KCl electrolyte. The interested reader is referred to the original references [12, 22]. The topic is related to the modelisation of pH measurement by field effect devices [24, 25] and to recent publications on the mechanisms of field effect detection of surface bound biomolecules [26–28].

To prepare the experimental study of the salt dependence of the silica/electrolyte interface, we prepare the FET array surface according to Sect. 2.2 without poly(L-lysine) coating. In figure 8, a measurement of the salt dependence of U_{SE} at fixed I_D, U_{SD} working point is shown. Starting at 5×10^{-6} M, a series of increasing concentration has been obtained by successively adding concentrated KCl to the electrolyte. The experimental data are compared to two different theoretical descriptions. In the first description (constant charge model, dashed line) we assume an effective interface charge σ_{eff} which does not depend on the salt concentration [KCl] of the electrolyte. In the second case (salt-dependent charge model, solid line) the dependence of σ_{eff} on [KCl] is taken into account considering the chemical equilibrium of the ionisable sites at the silica surface. This introduces two fitting parameters, the surface density of ionisable site N_S and the pH value. We find that, although the constant charge model can describe the global trend of the measured salt dependence, the salt-dependent charge model provides a better description of the experimental data.

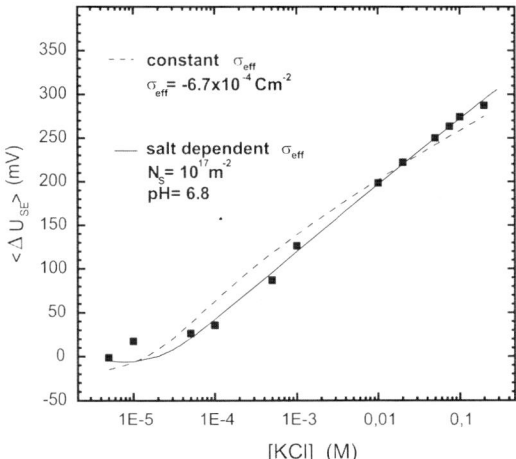

Fig. 8. Average shifts ΔU_{SE} of an array of 29 transistors as a function of the salt concentration of the buffer solution. Zero voltage shift corresponds to the reference measurement performed with a concentration [KCl] of 5×10^{-6} M. The experimental data (squares) are compared to two different theoretical descriptions as explained in the text.

6   Concluding remarks

The electronic detection is sensitive to charged molecules in general. Two different concepts have been used to obtain sequence specific transistor based DNA detection.

In a first approach, point-mutations have been detected in human DNA by allele-specific PCR followed by electronic detection of the double stranded PCR product [15]. The PCR and the electronic detection are separated by a purification that retains PCR product and substrat DNA, but eliminates nucleotides, primers, proteines and salt. As very few substrat DNA is used, the PCR product represents the only species susceptible to induce a significant signal in the electronic detection. Comparative analysis of two different samples on the FET array provides reproducible mutation detection, the overall specificity is simply that of the allele-specific PCR. A combination of FET based DNA detection and sequence specific extension of surface bound oligonucleotides has been reported by Sakata and Miyahara [29].

The second concept consists in obtaining a sequence specific binding on the surface of the FET device by hybridization. Electronic detection of hybridization with probe DNA covalently bound on the surface of FET devices has been reported by several groups [9, 13, 14]. In the work reviewed in Sect. 4, we immobilize DNA probes on poly(L-lysine) and hybridize oligonucleotide targets at low salt, without preceeding blocker step. The low salt hybridization on a charge-compensated surface has initially been introduced to obtain faster hybridization [30]. It has been shown that under these conditions a single base mismatch can be detected in 12 mer oligonucleotides [10], although more standard hybridization protocols use higher concentrations of monovalent salt (50 mM-1M) to improve stringency [31]. For FET array based electronic detection it has been shown experimentally, that the differential signals of hybridization can be enhanced significantly by changing the salt concentration between hybridization and detection [16].

We use devices with multiple transistor structures to detect DNA. Silicon FET arrays can be made with an individual structure size of a few μm^2. Using silicon CMOS technology, Infineon fabricated 2D sensor arrays with 16384 FET structures and integrated readout electronics on a chip surface of 1 mm^2 [17]. As the signals from the individual active FET structures do not decrease with sensor surface and because the dc measurement does not require sophisticated on-chip electronics, this suggests that an electronic dc detection based on a two-dimensional transistor array even integrated with on-chip multiplexing would not limit the number of different DNA probes that could be used in parallel. In the field of DNA microarrays, robotic spotting of probe molecules typically gives spot diameters of about 100 μm, while smaller diameters down to about 5 μm are achieved by direct on-surface oligonucleotide synthesis.

The planar SiO$_2$ surface of the FET arrays exhibits sufficiently uniform wetting properties to allow for local spotting and microscopic fluorescence measurements, common in the field of DNA microarrays. In this geometry, integrations with microfluidics approaches and even sophisticated "lab on chip" devices are conceivable and are also potentially interesting in terms of applications.

Acknowledgments

The contributions of my collaborators C. Gentil, F. Pouthas, D. Côte, G. Philippin and C. Guiducci are gratefully acknowledged. We thank M. Ulbrich, M. Völker, G. Zeck, B. Straub and P. Fromherz for stimulating discussions and help in the preparation of the transistor arrays.

References

1. Csako G (2006) Clin Chim Acta 363:6–31
2. Wang J, Nielsen P E, Jiang M, Cai X, Fernandes J R, Grant D H, Ozsoz M, Beglieter A, Mowat M (1997) Anal. Chem. 69:5200–5202
3. Steemers F J, Ferguson J A, Walt D R (2000) Nat. Biotech. 98:91–94.
4. McKendry R, Zhang J, Arntz Y, Strunz T, Hegner M, Lang H P, Baller M K, Certa U, Meyer E, Güntherodt H J, Gerber C (2002) Proc. Natl. Acad. Sci. USA 99:9783–9788
5. Bergveld P (1972) IEEE Trans. Biomed. Eng. BME-19:342-351
6. Bergveld P (1996) Sens. Actuators A 56:65–73
7. Kharitonov A B, Wassermann J, Katz E, Willner I (2001) J. Phys. Chem. B 105:4205–4213

8. Cui Y, Wei Q, Park H, Lieber C M (2001) Science 293:1289–1292
9. Souteyrand E, Cloarec J P, Martin J R, Wilson C, Lawrence I, Mikkelsen S, Lawrence M F (1997) J. Phys. Chem. B 101:2980–2985
10. Fritz J, Cooper E B, Gaudet S, Sorger P K, Manalis S R (2002) Proc. Natl. Acad. Sci. USA 99:14142–14146
11. Hahm J, Lieber C M (2004) Nano Lett. 4:51–54 (2004).
12. Pouthas F, Gentil C, Côte D, Zeck G, Straub B, Bockelmann U (2004) Phys. Rev. E 70:031906
13. Uslu F, Ingebrand S, Mayer D, Böcker-Meffert S, Odenthal M, Offenhäuser A (2004) Biosens. Bioelectron. 19:1723–1731
14. Sakata T, Miyahara Y (2005) ChemBioChem 6:703–710
15. Pouthas F, Gentil C, Côte D, Bockelmann U (2004) Appl. Phys. Lett. 84:1594–1596
16. Gentil C, Philippin G, Bockelmann U (2007) Phys. Rev. E 75:011926
17. Eversmann B, Jenkner M, Hofmann F, Paulus C, Brederlow R, Holzapfl B, Fromherz P, Merz M, Brenner M, Schreiter M, Gabl R, Plehnert K, Steinhauser M, Eckstein G, Schmitt-Landsiedel D, Thewes R (2003) IEEE J. Solid-St. Circ. 38:2306–2317
18. Kiessling V, Müller B, Fromherz P (2000) Langmuir 16:3517–3521
19. Zeck G, Fromherz P (2001) Proc. Natl. Acad. Sci. USA 98:10457-10462
20. Manning G S (1978) Quart. Rev. Biophys. 11:179–246
21. Pack G R, Wong L, Lamm G (1999) Biopolymers 49:575–590
22. Gentil C, Côte D, Bockelmann U (2006) Phys. Stat. Sol. A 203:3412–3416
23. van Hal R E G, Eijkel J C T, Bergveld P (1996) Adv. Coll. Interf. Sci. 69:31–62
24. Siu W M, Cobbold R S C (1979) IEEE Trans. El. Dev. 26:1805–1815
25. Fung C D, Cheung P W, Ko W H (1986) IEEE Trans. El. Dev. 33:8–18
26. Landheer D, Aers G, McKinnon W R, Deen M J, Ranuarez J C (2005) J. Appl. Phys. 98:044701
27. Neff P A, Wunderlich B K, Lud S Q, Bausch A R (2006) Phys. Stat. Sol. A 203:3417–3427
28. Poghossian A, Cherstvy A, Ingebrandt S, Offenhäuser A, Schoning M J (2005) Sens. Actuators B 111:470–480
29. Sakata T, Miyahara Y (2006) Angew. Chem. Int. Ed. 45:2225–2228
30. Belosludtsev Y, Belosludtsev I, Iverson B, Lemeshko S, Wiese R, Hogan M, Powdrill T (2001) Biochem. Biophys. Res. Commun. 282:1263-1267
31. Sambrook J, Russel D W (2000), Molecular Cloning, A Laboratory Manual. Cold Spring Harbor Laboratory Press, New York

Probabilistic & Statistical Design—
the Wave of the Future

Shekhar Borkar

Intel Corp, 2111 NE 25th Ave, Hillsboro, OR 97124, USA
Shekhar.Y.Borkar@intel.com

Abstract. As technology scales, variability will continue to become worse. Random dopant fluctuations, sub-wavelength lithography, and dynamic variations due to circuit behavior will look like inherent unreliability in the design. Increasing soft errors will increase intermittent error rate by almost two orders of magnitude. As transistors scale even further, degradation due to aging will become worse. We discuss these effects and propose solutions in microarchitecture, design, and testing, for designing with billions of unreliable components to yield predictable and reliable systems, with probabilistic and statistical design methodology.

1 Introduction

VLSI system performance has increased by five orders of magnitude in the last three decades, made possible by continued technology scaling. This treadmill will continue, providing integration capacity of billions of transistors; however, power, energy, variability, and reliability will be the barriers to future scaling.

Die size, chip yields, and design productivity have so far limited transistor integration in a VLSI design. Now the focus has shifted to energy consumption, power dissipation and power delivery [1]. Transistor sub-threshold leakage continues to increase, and leakage avoidance, leakage tolerance, and leakage control techniques for circuits have been devised [2]. As technology scales further we will face new challenges, such as variability [3], single event upsets (soft errors), and device (transistor performance) degradation—these effects manifesting as inherent unreliability of the components, posing design and test challenges. We will discuss these effects and propose solutions in microarchitecture, circuit, and testing, for designing with many unreliable components (transistors) to yield reliable system designs in the future.

This problem is not new; even today we design systems to comprehend reliability issues. For example, error correcting codes are commonly used in memories to detect and correct soft errors. Careful design, and testing for frequency binning, copes with variability in transistor performance. What is new is that as technology scaling continues, the impact of these issues keeps increasing, and we need to devise techniques to deal with them effectively.

Please use the following format when citing this chapter:

Borkar, S., 2007, in IFIP International Federation for Information Processing, Volume 249, VLSI-SoC: Research Trends in VLSI and Systems on Chip, eds. De Micheli, G., Mir, S., Reis, R., (Boston: Springer), pp. 69–79

2   Sources of Variation

Primarily there are two types of variations: static and dynamic. Static variations are caused by variations induced during processing, and the behavior does not change over time. Dynamic variations, on the other hand, are caused by the behavior of the chip while it is functioning. An example of static variations is Vt mismatch between two adjacent transistors caused during fabrication. Supply voltage droop generated during operation of a chip causing circuit to slow down is an example of a dynamic variation.

The variations could be within a die, or between different dies; they could be systematic or random. Systematic variations are primarily induced by limitations in the processing, and random variations are caused by randomness in physical dimensions such as line edge roughness and discreteness of dopant atoms.

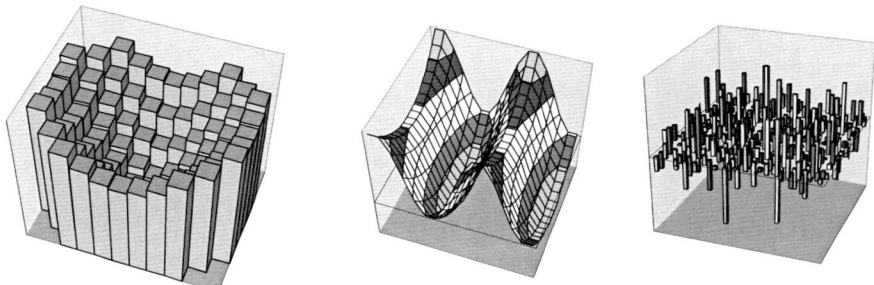

Fig. 1. Resist Thickness, Lens Aberrations, and Random placement of dopant atoms.

Figure 1 shows variation in photo resist thickness resulting in die to die variations, lens aberrations causing systematic variations, and random placement of dopant atoms causing random variations.

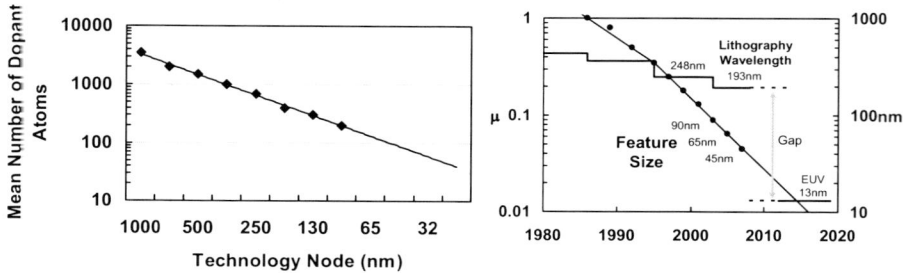

Fig. 2. (a) Mean number of dopant atoms in transistor channel decreases, and (b) sub-wavelength lithography until EUV.

Random Dopant Fluctuations results from discreteness of dopant atoms in the channel of a transistor [4]. Transistor channels are doped with dopant atoms to control their threshold voltage. Figure 2(a) shows dopant atoms in the channel of several generations of transistors. As a transistor scales in size each technology generation, its area is reduced by half, and thus the number of dopant atoms in the channel reduces exponentially over generations. In one micron technology generation there were thousands of dopant atoms, whereas in 32 to 16 nm generation there will be only 10's of dopant atoms left in the channel, and the law of large numbers does not apply. Therefore, two transistors sitting side by side will have different electrical characteristics due to randomness in small number of dopant atoms, resulting in variability.

Another source of variability is due to sub-wavelength lithography shown in figure 2(b). Since 0.25μ technology generation, sub-wavelength lithography is being used for patterning transistors. For example, 248nm wavelength of light was used to pattern 0.25μ (250nm) and 0.18μ (180nm) transistors. The wavelength reduced to 193nm for 130nm technology, and since then it has remained constant until today for even 65nm transistors. There may be some additional breakthroughs to effectively reduce this wavelength (157nm light source or immersion technology) but the difference in the wavelength of light and the patterning width will continue to get wider until EUV (Extreme Ultra-violet, 13nm) technology becomes available. This sub-wavelength lithography is the primary reason for line edge roughness in transistors, and several other effects, resulting in variations.

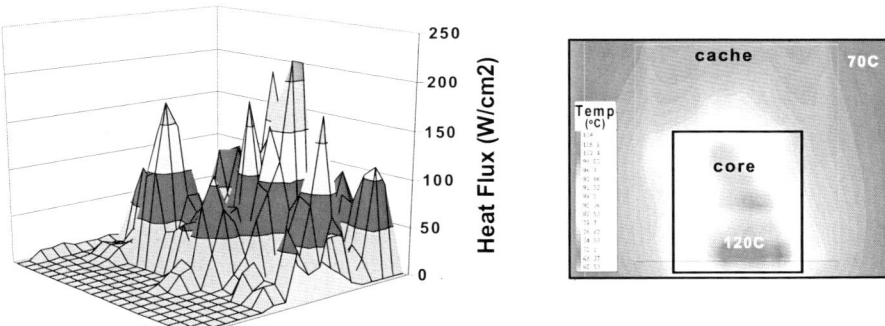

Fig. 3. Heat flux, and temperature variation across a microprocessor die.

Figure 3 shows heat flux (power density) across a microprocessor die, varying depending on the functionality of the circuit block. For example, the cache has less heat flux than an execution unit, and it also depends on the activity and compute load at any given time. Higher heat flux also puts more demand on the power distribution grid resulting in resistive and inductive voltage drops, creating time dependant, dynamic, supply voltage variations. Higher heat flux results in higher temperature, creating hot spots, and thus temperature variations across the die affecting circuit performance. This also results in higher sub-threshold leakage, variations in the leakage across the die, and dynamic variations in power delivery demand across the power distribution grid.

3   Impact of Variations on Products

Variations have been with us for a long time, and their impact on product was not profound in the past, but now it impacts performance, power, and yield. As technology scales, this impact will probably become worse [3].

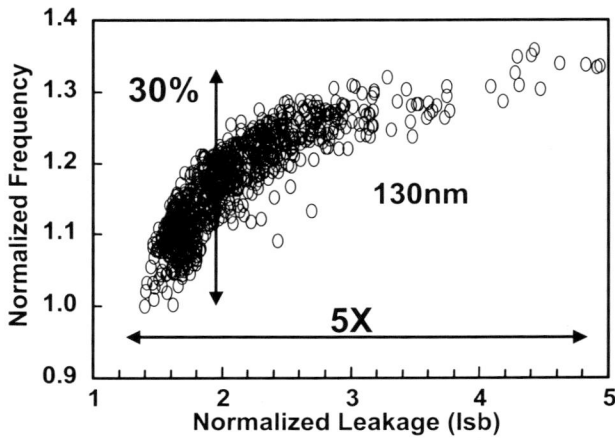

Fig. 4. Impact of variations on microprocessor frequency and leakage power.

Impact of variations on a product is shown in figure 4. It shows frequency and sub-threshold current measurements of about thousand microprocessors fabricated in 130nm technology, with 30% variation in the frequency distribution and about 5 to 10X spread in the sub-threshold leakage current. Since sub-threshold leakage power is a major portion (30-50%) of the total power consumption, 5-10X variation in the leakage power alone contributes to almost 50% variation in the total power. The behavior of the fabricated design in power and performance is different from what was intended, and hence the effect of variations looks like inherent unreliability in the design.

4   Variations in transistors

Variation in transistor length, width, and threshold voltage affect circuit performance, and thus overall performance of the VLSI chip. It also impacts yields and probability of meeting performance targets with yields.

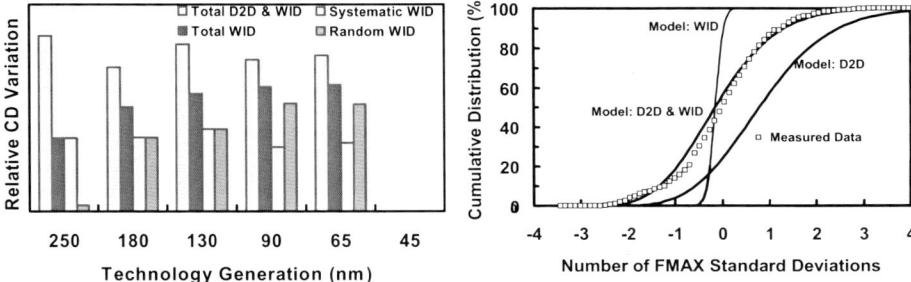

Fig. 5. (a) Gate length variation trend, and (b) within and die-to-die variation.

Figure 5(a) shows systematic and random, within-die and die-to-die gate length variation. The first bar shows the aggregate effect, and notice that it is almost a fixed percentage of the nominal gate length. However, random variations within die increase with gate length scaling. Figure 5(b) shows impact of die to die and within-die variations. Within-die variations impact Fmax mean, that is the mean of maximum frequency of operation, and die-to-die variations impact the variance of Fmax. So, loosely speaking, within-die variations reduce maximum frequency of operation, and die-to-die variations reduce the yield of the design at the maximum frequency.

5   Design Considerations

Design practice too impacts performance in the presence of variations as shown in figure 6.

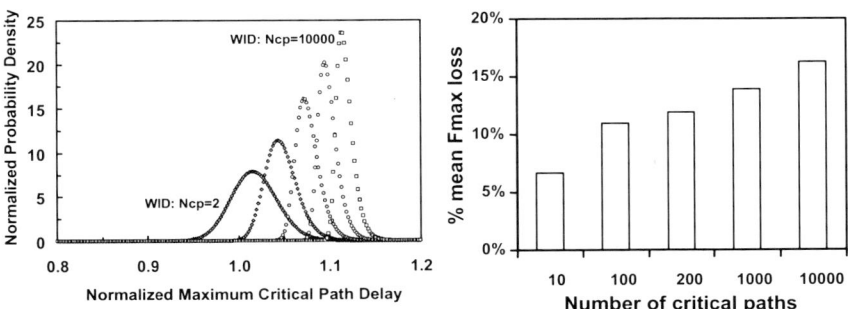

Fig. 6. Large number of time critical paths impact design performance.

Typically, the maximum frequency of operation of a design depends on a handful of time critical paths, and the number of such critical paths in the design plays important role in variation tolerant design. Notice that with a large number of critical paths in a design, the maximum critical path delay increases, thus lowering mean Fmax, but the variance also decreases. Reduction in the mean Fmax with the number of critical paths is logarithmic, that is, the reduction in Fmax slows down as the number of critical paths increase.

Microarchitecture also plays an important role in tolerating variations in a design. For example, microarchitecture with small number of gate stages in a clock period yields higher frequency of operation, compared to one with large number of gates. Both may provide equivalent logic throughput; however, with different variation tolerance. Large numbers of gates in a clock cycle tend to average out the effect of random variations, as shown in figure 7.

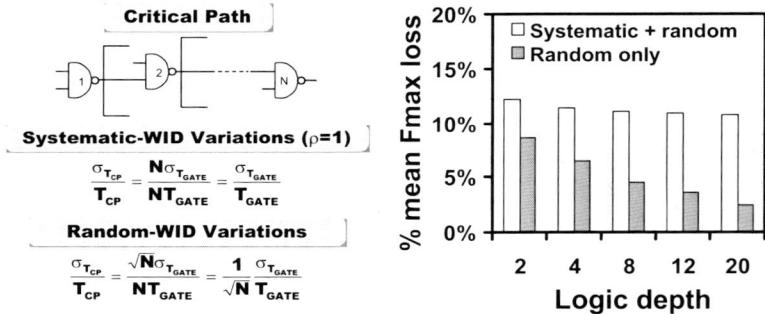

Fig. 7. Impact of random variations reduces with increasing logic depth.

6   Variation Tolerant Design

Numerous process technology, circuit, and architectural solutions have been proposed to deal with variations [3,5,6], which may require radical changes in our design methodology. For example, forward and reverse body bias can be used to tighten sub-threshold leakage and frequency distributions. Chips with higher leakage tend to be faster, hence reverse body bias may be applied to reduce the leakage and reduce the frequency. Similarly, slow chips can benefit from forward body bias to improve their speed at the expense of moderate increase in sub-threshold leakage power. Similarly, adaptive supply voltage too can be used in conjunction with body bias to tighten the distribution [6].

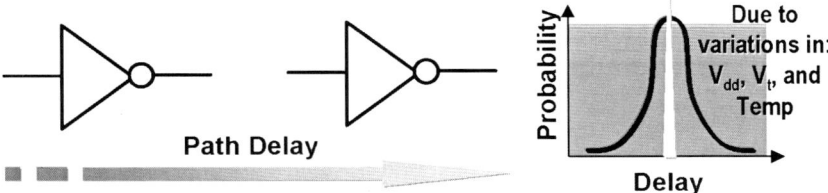

Fig. 8. Modeling and simulation of a critical path.

The chip frequency depends on the speed of the critical paths. Hence a critical path is typically modeled to have deterministic delay as predicted by a circuit simulator, as shown in figure 8; however, due to static and dynamic variations discussed before, the delay of the circuit is probabilistic. When the design is complete and conventional design methods are applied, transistors are typically down sized to save active power. As a result, transistors in the critical paths will also get down sized adequately. While every attempt is made during manufacturing to maintain the deterministic behavior, the increased variability, or not fully comprehended variability in transistor performance, can make these path delays probabilistic. Therefore, we need to deviate from conventional methodology of down-sizing transistors indiscriminately to reduce active power, because down-sizing indiscriminately makes many non-critical paths critical, and reduces probability of meeting the frequency goal.

Similarly, low threshold voltage transistor usage does not have to be minimal to reduce leakage power. With reduced low Vt usage across the design, the transistors near the critical path too may get replaced with high Vt transistors, and due to variations in the threshold voltage these paths could become slower, resulting in wider frequency distribution. Design tools and methodologies need to comprehend variations, and optimize the design not for frequency alone, but for active and leakage powers and their distribution.

When a micro-architecture is designed, the tendency is to improve frequency of operation by creating more critical paths, which reduces the probability of meeting the increased frequency goal. Furthermore, to meet higher frequency goals, micro-architecture tends to employ less number of gate-delays in a clock cycle. Since less number of gates in a clock cycle does a poor job of averaging and canceling the effects of variations, it results in reducing the probability of meeting the frequency goal. This once again, is contrary to conventional thinking and design methodology.

We need to evolve from today's deterministic design to probabilistic and statistical design for the future, comprehending variations, and optimizing for yield, performance, and power (figure 9).

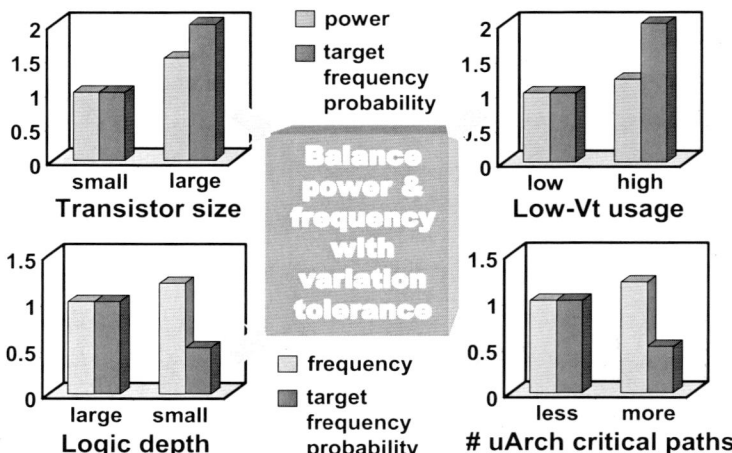

Fig. 9. Variation tolerant design methodology to optimize yield, performance, and power.

7   Longer Term Outlook

As technology continues to scale further, both static and dynamic variations will continue to get worse for the reasons discussed before, resulting in wider distribution of characteristics of the transistors. These variations in the transistors could be severe enough that it would be impossible to correct for them during design—they will have to be compensated somehow at the whole system level.

Single event upsets (soft-errors) are another source of concern. These errors are caused by alpha particles and more importantly cosmic rays (neutrons), hitting silicon chips, creating charge on the nodes to flip a memory cell or a logic latch. These errors are transient and random. These errors are relatively easy to detect and correct in memories by protecting the memory bits with parity, and employing error correcting codes. However, if such a single event upset occurs in a logic flip-flop then it is difficult to detect and correct.

We expect a modest increase in soft-error rate per logic state bit each technology generation [7]. Since the number of logic state bits on a chip double each technology generation (following Moore's Law), the aggregate effect on soft-error rate FIT (failure in time) could be almost 100X by the next few generations.

Aging has had significant impact on transistor performance. Studies have shown that transistor saturation current degrades over years due to oxide wear-out and hot carrier degradation effects. So far, the degradation is small enough such that it can be accounted for as an upfront design margin in the specification of a VLSI component. We expect this degradation to become worse as we continue to scale transistor geometries beyond. It may become so bad that it would be impractical to absorb degradation effects upfront in a system design.

As gate dielectric scales, gate leakage will increase exponentially, and we fear that burn-in power will become prohibitive, making burn-in testing obsolete. Therefore, screening for defects and infant mortalities in VLSI chips will become increasingly difficult if not impossible. One-time factory testing will be insufficient, and what you need is the test hardware embedded in the design, to dynamically detect errors, isolate and confine the faults, reconfigure using spare hardware, and recover on the fly.

8   Paradigm Shift

There are several potential solutions in sight in all discipline to tackle most of the problems discussed before; however, all disciplines of VLSI will have to make concerted efforts to make this successful.

A shift from deterministic design to probabilistic and statistical design would ease impact of transistor variations on circuit performance. Today's design optimizations are performed with only one or two objectives, namely performance and power. This will have to change, with multi-variable design optimizations, comprehending performance, active & leakage power, reliability, yield, and bin-splits. Design tools to implement such optimizations, and statistical and probabilistic methodologies to go along with the tools need development.

In circuit design, replacing regular flip-flops by soft-error tolerant hardened flip-flops will improve soft-error rate tolerance by almost 10X. To catch dynamic errors, innovative techniques such as Razor [8] need serious consideration which will not only detect and correct errors, but will also allow the design to operate at optimum power and performance. This technique is power efficient because it replicates only those flip-flops that are critical and need to be checked for correctness, and is also capable of catching circuit marginalities arising from variations.

At the system architecture level, functional redundancy check may work; however, it may not be power and energy efficient, since it almost doubles the hardware and power consumption for the same performance. Any redundancy and checking hardware must be used judiciously to dynamically catch errors and take corrective action, and should not burden the system with excessive power consumption and complexity. An interesting microarchitecture is proposed in [9], where a traditional processor core is accompanied by a small, yet robust, core as a checker. The checker core is correct by construction, may be over designed to be variation tolerant and is made immune from any further errors—both static and dynamic. Since the checker core is small it consumes very little power, and can dynamically detect and correct any errors made by the large core, thus providing reliable operation of the system.

9   Multi—a Solution to Variability and Reliability

The key to the variability and reliability problems may be to exploit abundance of transistors using Moore's Law to your advantage. Instead of relying upon higher and higher frequency to deliver higher performance, a shift towards parallelism to deliver higher performance is in order, and thus Multi- may be the solution at all levels—from multiplicity of functional blocks in a design to multiple processor cores in a system [10].

Multiple functional blocks, operating at lower voltage and frequency, provide the same logic throughput, but at much reduced power, and can be used for redundancy and error checking. For example two ALUs (Arithmetic and Logic Units) could be used to provide higher throughput when needed, and can be used to check and correct results produced by each other (figure 10). Multiple cores in a system will provide similar performance and redundancy benefit with functional redundancy checking employed at a coarse level of granularity.

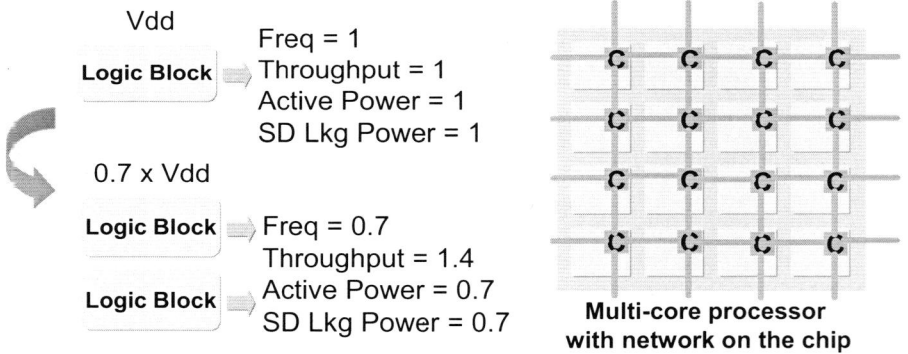

Fig. 10. (a) Multiple functional blocks, and (b) Multi-core processor for resiliency.

Test functionality can either be distributed as a part of the hardware, to dynamically detect errors, correct, and isolate aging & faulty hardware, or a sub-set of cores in the multi-core design can be used to do this task. This microarchitecture strategy, with Multi-cores to assist in redundancy, is called Resilient Microarchitecture, to continually detect errors, isolate faults, confine faults, reconfigure the hardware, and adapt. If such a strategy is made to work, then there is no need for one time factory testing or burn-in, since the system is capable of testing and reconfiguring itself to make itself work reliably throughout its lifetime.

All this is possible, but all disciplines from fabrication to software will have to cooperate and make the system reliable in spite of unreliable components. A lot of research and development needs to be done, however, to make this concept into a reality.

Acknowledgements

The author would like to thank Vivek De, Jim Tschanz, Ali Keshavarzi, Keith Bowman, Tanay Karnik, Peter Hazucha, and Jose Maiz, for their help and insightful discussions.

References

[1] Shekhar Borkar, "Design Challenges of Technology Scaling, IEEE Micro", July-August 1999.

[2] Shekhar Borkar, "Circuit Techniques for Subthreshold Leakage Avoidance, Control, and Tolerance", IEDM 2004.

[3] Shekhar Borkar et al., "Parameter Variations and Impact on Circuits and Microarchitecture", Proceedings of Design Automation Conference, 2003.

[4] Xinghai T et al., Intrinsic MOSFET parameter fluctuations due to random dopant placement, IEEE Transactions on Very Large Scale Integration (VLSI) Systems, Dec. 1997.

[5] Shekhar Borkar, "Probabilistic and Statistical Design—The wave of the future", VLSI-SOC, 2006.

[6] Jim Tschanz et al., "Effectiveness of adaptive supply voltage and body bias for reducing impact of parameter variations in low power and high performance microprocessors", IEEE Journal of Solid States Circuits, Volume 38, Issue 5, May 2003 Page(s):826-829

[7] Peter Hazucha et al., "Neutron Soft Error Rate Measurements in a 90-nm CMOS Process and Scaling Trends in SRAM from 0.25-μ to 90-nm Generation", IEDM 2003.

[8] Ernst D et al., Razor: a low-power pipeline based on circuit-level timing speculation, International Symposium on Microarchitecture, 2003. MICRO-36.

[9] Austin T, DIVA: a reliable substrate for deep submicron microarchitecture design, 32nd Annual International Symposium on Microarchitecture, 1999. MICRO-32.

[10] Shekhar Borkar, "Designing reliable systems from unreliable components: The challenges of transistor variability and degradation", IEEE Micro, Nov-Dec 2005.

A CMOS Mixed-Mode Sample-and-Hold Circuit for Pipelined ADCs

Shan Jiang, Manh Anh Do, and Kiat Seng Yeo

Nanyang Technologies University,
School of Electrical & Electronic Engineering,
Block S1, 50 Nanyang Avenue,
Singapore 639798
shanjiang@pmail.ntu.edu.sg

Abstract. This paper describes the design of a high-speed CMOS sample-and-hold (S/H) circuit for pipelined analog-to-digital converters (ADCs). This S/H circuit consists of a switched-capacitor (SC) amplifier and a comparator to generate the mixed-mode sampled output data, which are represented both in analog and digital forms. The mixed-mode sampling technique reduces the operational amplifier (op amp) output swing. As a result, the requirements on op amp DC gain, slew rate and bandwidth are relaxed; the linearity of the SC amplifier is also improved. The reduction of signal swing in the front-end also brings benefit to the pipelined stages in speed and power consumption. The aperture errors at high frequency are minimized by time constant matching and digital error correction logic in the pipelined ADC. Designed in a 0.18-μm CMOS process, the proposed S/H circuit operates up to 200-MSample/s with a total harmonic distortion (THD) less than -60 dB and a signal-to-noise and distortion ratio (SNDR) larger than 59 dB in the worst-case simulation. The power consumption of the mixed-mode S/H circuit is 3.6-mW with 1.8-V supply voltage.

1 Introduction

The pipelined analog-to-digital converter (ADC) is one of the most popular converter architectures and has been widely used in many high-speed applications in wireless communication and video systems. Most pipelined ADCs include a sample-and-hold (S/H) circuit at the front-end to minimize the high frequency errors and to improve system performance. The performance of the S/H circuit dominates the overall ADC dynamic characteristics and plays a major role in determining the spurious free dynamic range (SFDR), signal-to-noise and distortion ratio (SNDR) of the system.

The stringent performance requirements of the S/H circuit make it the most design-challenging and power-hungry block in a pipelined ADC. Although a dedicated front-end S/H circuit can be removed and the sampling function is performed in the first stage of a pipelined ADC, the input capacitance of the pipelined ADC in this implementation is significantly increased due to the multi-bit first stage configuration [1]. A large input capacitance stresses the driving

Please use the following format when citing this chapter:

Jiang, S., Do, M.A. and Yeo, K.S., 2007, in IFIP International Federation for Information Processing, Volume 249, VLSI-SoC: Research Trends in VLSI and Systems on Chip, eds. De Micheli, G., Mir, S., Reis, R., (Boston: Springer), pp. 81–99

circuit of the ADC, which usually is a variable gain amplifier (VGA). In some solutions, the full scale input range of the ADC is reduced and so is the achievable dynamic range [2]. The time constant matching is another concern in a pipelined ADC without a front-end S/H circuit. Although the aperture errors due to the time constant mismatch can be treated as comparator offset errors and removed by the digital error correction logic, due to the high gain of the multi-bit first stage, it is easy for the aperture errors to saturate the S/H circuit output swing. This sets a limit on the highest working frequency of the pipelined ADC without a front-end S/H circuit. Therefore, a dedicated front-end S/H is necessary in these considerations.

The performance of an S/H circuit is usually determined by an operational amplifier (op amp) in a switched-capacitor (SC) implementation. In an S/H circuit, the op amp must have enough DC gain to guarantee the accuracy requirement; and a fast slew rate and large bandwidth to meet the speed demand. The op amp performance also has an effect on the linearity of the S/H circuit, and consequently the overall ADC dynamic performance. Beyond these, in order to achieve a desirable signal-to-noise ratio (SNR), a large signal swing is required at the output of op amp even in the low voltage environment.

In high-speed low-power applications, it is always desirable to use the single-stage op amp architecture with the minimum transistor size whenever possible. This is due to the high-speed, low-noise, low-power and stability characteristics of the single-stage op amp. In practice, one of the challenges is that the DC gain of the single-stage op amp is limited unless larger transistors are use to increase the output impedance. However, a larger transistor size transfers into a lower speed because of the parasitic capacitance load. Another problem associated with single-stage op amp is the limited output signal swing due to the cascode operation, which leads to a limitation on the SNR and linearity. These challenges become more and more significant as the supply voltage and device size scale down.

This paper proposes a mixed-mode S/H circuit that reduces the requirements of op amp DC gain, slew rate, bandwidth, and output swing. The relaxation on performance requirements enables the use of single-stage cascode op amp in low voltage environment without degrading the system performance. The reduced requirements are accomplished through a built-in comparator in the S/H circuit. The aperture errors are minimized by time constant matching and digital error correction logic.

2   The Op Amp in the S/H Circuit

The schematic of a conventional fully differential switched-capacitor S/H circuit is shown in Fig. 1 [3]. In the sampling mode, switches S_1, S_2 and S_4 are on and S_3, S_5 off, the input signal is sample in C_S and the feedback capacitor C_F is reset. At the end of the sampling mode, S_2 turns off first, leaving the top plate of C_S floating to eliminate the signal-dependent charge-injection caused by S_1. Subsequently, S_1, S_4 turn off and S_3, S_5 turn on. The charges in C_S are transferred to the feedback capacitor C_F. The output is given by

$$V_{out} = V_{in} \frac{C_S}{C_F + \frac{C_F + C_S + C_{in}}{A_0}} \tag{1}$$

where C_{in} is the input gate capacitance of op amp and A_0 is the op amp DC gain. If A_0 is large,

$$V_{out} \approx V_{in} \frac{C_S}{C_F} (1 - \frac{1}{\beta \cdot A_0}) \tag{2}$$

where $\beta = C_F/(C_F + C_S + C_{in})$ is the feedback factor in the holding mode. Equation (2) shows that there is a gain error of $1/(\beta \cdot A_0)$ due to the finite op amp DC gain.

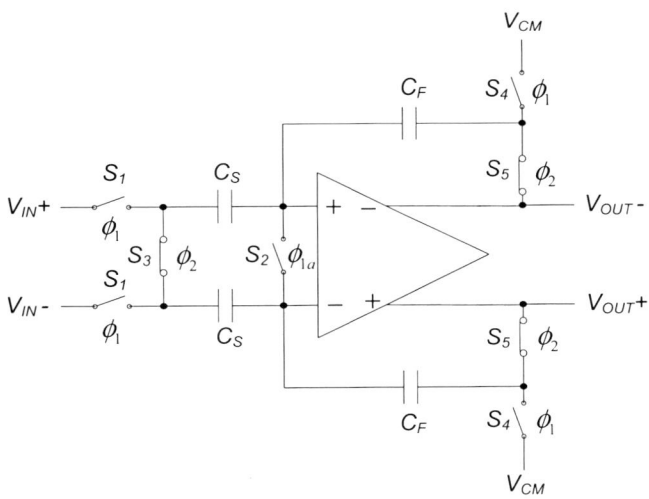

Fig. 1: SC implementation of a sample-and-hold circuit.

Assume this S/H circuit was used in an N-bit ADC, for a full scale step input, the output error due to finite op amp DC gain must be less than half of the least-significant-bit (LSB) in order to avoid introducing any error to the following pipeline stages This determines the minimum DC gain requirement of the op amp, which is

$$A_0 > \frac{1}{\beta} \times 2^{N+1} \tag{3}$$

Although an S/H gain error can be tolerated in some applications, the gain error drift must be minimized, which requires a high op amp DC gain across temperature and process corners.

2.1   Op Amp DC Gain Variation with Input Signals

The op amp DC gain depends on input and output signal swing, which can be illustrated by the differential amplifier shown in Fig. 2. In this amplifier, the current difference $\Delta I = I^+ - I^-$ can be expressed as [4]

$$
\begin{aligned}
\Delta I &= \frac{1}{2}\mu_n C_{ox} \frac{W_1}{L_1} V_{in} \sqrt{\frac{I}{\mu_n C_{ox}\frac{W_1}{L_1}} - V_{in}^2} \\
&= \frac{1}{2}\mu_n C_{ox} \frac{W_1}{L_1} V_{in} \sqrt{4(V_{GS} - V_{TH})_1^2 - V_{in}^2} \\
&= \mu_n C_{ox} \frac{W_1}{L_1}(V_{GS} - V_{TH})_1 V_{in} \times \sqrt{1 - \frac{V_{in}^2}{4(V_{GS} - V_{TH})_1^2}} \\
&= g_{m1} V_{in} \sqrt{1 - \frac{V_{in}^2}{4(V_{GS} - V_{TH})_1^2}}
\end{aligned}
\tag{4}
$$

where L_1, W_1 and g_{m1} are the channel length, width and transconductance of M_1, respectively.

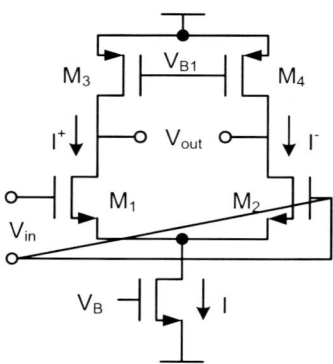

Fig. 2: A differential amplifier.

The output impedance of the amplifier is $r_{o1}\|r_{o3}$. Assume $r_{o1} = r_{o3}$, the gain of the amplifier equals to

$$
\begin{aligned}
A &= \Delta I \frac{r_{o1}}{2}/V_{in} \\
&= g_m \frac{r_{o1}}{2} \sqrt{1 - \frac{V_{in}^2}{4(V_{GS} - V_{TH})_1^2}} \\
&\approx A_0 (1 - \frac{V_{in}^2}{8V_{eff1}^2})
\end{aligned}
\tag{5}
$$

where $A_0 = g_{m1}r_{o1}/2$ is the amplifier DC gain when the output is zero. $V_{eff1} = (V_{GS} - V_{TH})_1$ is the input transistor overdrive voltage.

The discussion so far on the op amp DC gain dependence on input and output signals has been based on the assumption that the output impedance r_o of MOS transistor in the saturation region is constant during signal swing. This is approximately true for long channel device working at high voltage environment. In short channel device, however, r_o varies very much with the drain-to-source voltage V_{DS}. In saturation region, this dependence can be approximated as

$$r_o = \frac{2L}{1 - \frac{\Delta L}{L}} \frac{1}{I} \sqrt{\frac{qN_B}{2\epsilon_{si}}(V_{DS} - V_{dsat})} \qquad (6)$$

The variation of r_o gives rise to nonlinearity in an op amp. The amount of nonlinearity is heavily depends on how much the output signal swing, i.e. how much the V_{DS} changes. In addition, the transistor transconductance g_m also varies with V_{DS}, which further exacerbates the op amp nonlinearity since the voltage gain is determined by $g_m \cdot r_{o1}$. This phenomenon becomes significant in cascode op amp as V_{DS} of the cascode devices change significantly during the operation. The g_m, r_o, and voltage gain of an op amp vary with the output is illustrated in Fig. 3, 4, and 5, respectively.

Because the voltage gain varies during the operation, the gain requirement in (3) should be the gain when a largest output swing is applied. At this condition, the required zero output op amp DC gain is usually much larger than (3).

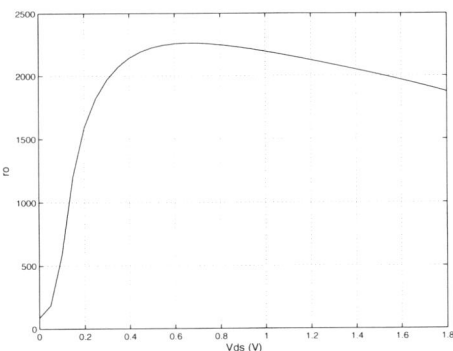

Fig. 3: g_m variation with output.

2.2   Nonlinearity Errors

The nonlinearity of the op amp DC gain is an important cause of nonlinearity errors in the S/H circuit. This harmonic distortion can be analyzed via charge

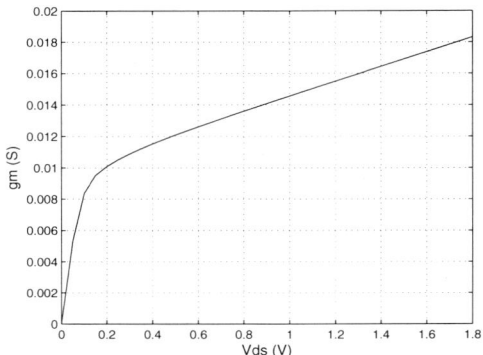

Fig. 4: r_o variation with output.

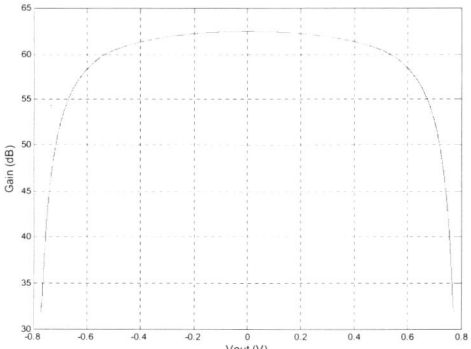

Fig. 5: A_0 variation with output.

conservation in a switched-capacitor circuit. If A_0 in (2) is replaced with A in (5), the transfer function of the S/H circuit in Fig. 1 can be written as

$$
\begin{aligned}
V_{out} &\approx V_{in}\frac{C_S}{C_F}(1 - \frac{1}{\beta \cdot A}) \\
&\approx V_{in}\frac{C_S}{C_F}(1 - \frac{1}{\beta \cdot A_0} \cdot \frac{1}{1 - \frac{V'^2}{8V_{eff1}^2}})
\end{aligned}
\tag{7}
$$

V' is the voltage at the op amp input, which can be approximated as

$$
V' \approx -\frac{V_{out}}{A_0} \approx -\frac{V_{in} \cdot \frac{C_S}{C_F}}{A_0}
\tag{8}
$$

Substitute (8) into (7), Vout can be expressed as

$$V_{out} \approx V_{in} \frac{C_S}{C_F}[1 - \frac{1}{\beta \cdot A_0} \cdot \frac{1}{1 + \frac{V_{in}^2 \cdot (C_S/C_F)^2}{A_0^2}}]$$

$$\approx V_{in} \frac{C_S}{C_F}[1 - \frac{1}{\beta \cdot A_0} \cdot (1 - \frac{V_{in}^2 \cdot (C_S/C_F)^2}{A_0^2})]$$

$$= V_{in} \frac{C_S}{C_F}(1 - \frac{1}{\beta \cdot A_0}) + \frac{(C_S + C_F)^3}{\beta \cdot A_0^3} V_{in}^3 \qquad (9)$$

Equation (9) indicates the dependence of the S/H nonlinearity on the input signal. The above analysis doesn't take into account the nonlinearity due the device transconductance g_m and output impedance r_o variations with the output swing. This also contributes a large amount of nonlinearity but its derivation is tedious.

2.3   Op Amp Slew Rate and Bandwidth

The speed of an S/H circuit is determined by the settling time of the op amp in holding mode. The op amp settling time can be divided into two phases as shown in Fig. 6. The first phase is the nonlinear settling which is contributed by the slewing behavior of the op amp. The second phase is the quasi-linear settling which is contributed by the bandwidth of the op amp. For speed consideration, the slewing phase should be minimized, which requires a high op amp slew rate. The op amp with a high slew rate settles to the final value in the linear settling phase. Therefore even if the op amp is not fully settled at the end of the hold mode, there is only a linear error and an improvement of the dynamic performance can be expected. Since the slew rate of the op amp is determined by the output swing and the sampling frequency, a reduced output swing can improve the S/H accuracy and linearity.

In the linear settling phase, the op amp output settles exponentially towards its final value. If the op amp has one dominated pole and the second pole frequency is much higher, the output of the S/H circuit in Fig. 1 can be expressed as

$$V_{out}(t) = (1 - e^{-\frac{t}{\tau}})V_{in} \frac{C_S}{C_F} \qquad (10)$$

For the op amp incomplete settling introduced error to be less than LSB/2 of an N-bit ADD, the minimum value of op amp time constant τ should be

$$\tau < \frac{1}{2 \cdot F_s \cdot (N + 1) \cdot ln2} \qquad (11)$$

where F_S is sampling frequency. The time constant τ can be further related to the op amp unity-gain bandwidth ω_u as

$$\tau = \frac{1}{\beta \cdot \omega_u} \qquad (12)$$

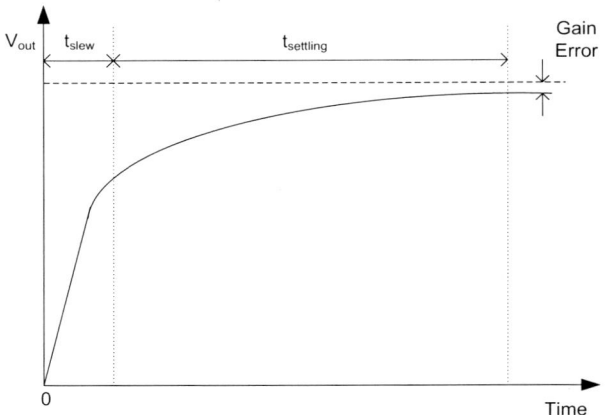

Fig. 6: Settling behavior of the op amp.

As a result, the op amp bandwidth requirement for an incomplete settling error smaller than LSB/2 is

$$\omega_u > \frac{2 \cdot (N+1) \cdot F_s \cdot ln2}{\beta} \tag{13}$$

3   The Mixed-Mode S/H Architecture

From above discussions, the variation of op amp DC gain with respect to the output swing introduces both gain error and nonlinearity in an S/H circuit. These errors become severe as the headroom left for the cascode device operation decreases in a low voltage environment. Although the single-stage op amp is preferred in high-speed applications, the error and nonlinearity often deprive its use ability in low voltage environment.

In order to exploit the single-stage op amp in low voltage environment, we propose a mixed-mode S/H circuit as shown in Fig. 7. One comparator is added to the conventional S/H circuit. The S/H operation is controlled by two non-overlapped clock phases, namely sampling phase ϕ_1 and holding phase ϕ_2. ϕ_{1a} is a copy of ϕ_1 but with an earlier falling edge. During the sampling phase, switches controlled by ϕ_1 and ϕ_{1a} are on, the sampling capacitor C_S is charge to $V_{ir} - V_{os}$ with the aid of the virtual ground formed by the op amp in the unity-gain configuration. V_{os} is the offset voltage of op amp. Meanwhile, the feedback capacitor C_F is reset. The sampling phase ends at the falling edge of ϕ_{1a}. At the same clock edge, the comparator quantizes the input signal and generates the digital output D_{OUT}. Subsequently, ϕ_1 turns off the input switches and the bottom plate of C_S is connected to either $+V_{ref}/2$ or $-V_{ref}/2$, determined by the value of D_{OUT}. The S/H circuit is in the holding mode and the op amp offset is eliminated by the correlated double sampling [5]. As a result, for $C_S = C_F$,

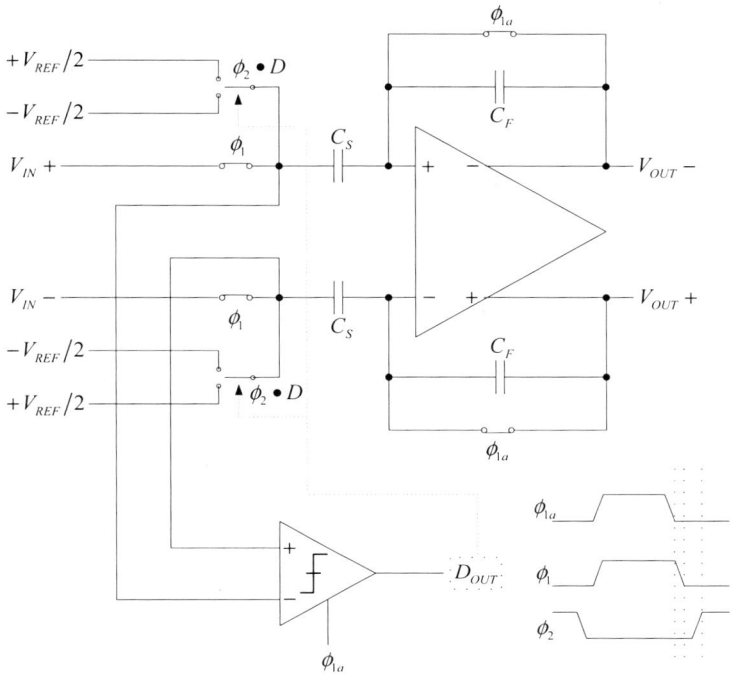

Fig. 7: The proposed mixed-mode sample-and-hold circuit.

the transfer function of the mixed-mode S/H circuit is given as

$$V_{out} \approx (1 - \frac{1}{\beta \cdot A_0})[V_{in} + (-1)^{D_{OUT}} \times V_{ref}/2] \qquad (14)$$

where

$$D_{OUT} = \begin{cases} 1 \text{ for } V_{in} \geq 0 \\ 0 \text{ for } V_{in} < 0 \end{cases} \qquad (15)$$

$$\beta = C_F/(C_F + C_S + C_{in}) \qquad (16)$$

The sampled data is represented both in analog and digital forms. The transfer curve of this mixed-mode S/H circuit is illustrated in Fig. 8. Also shown is the one bit digital output send to the digital error correction logic. The dashed line shows the transfer curve of the conversional S/H circuit in Fig. 1 with C_S equals to C_F. As expected, the output swing of the proposed S/H circuit is the half of the full scale input range. The reduced analog signal swing doesn't degrade the SNR or stress the following pipelined stage since now the information is stored both in analog and digital forms and the full scale range is maintained.

Although the full scale input is unchanged, the effective input signal to the op amp is reduced by $V_{ref}/2$ as can be seen in (14). Therefore, the maximum error introduced by finite op amp DC gain is $F_S/2\beta A_0$ in the proposed S/H circuit. If this S/H circuit is used in an N-bit ADC, for gain error to be less than LSB/2, we need

$$A > \frac{1}{\beta} \times 2^N \tag{17}$$

which is 6 dB lower than the requirement for the conventional S/H circuit as shown in (3). Although this configuration increases the error caused by finite op amp gain when the input is in the vicinity of zero as shown in Fig. 8, this error is still within the range of LSB/2 of the full scale. In addition, since the output swing is reduced, the DC gain is more stable, therefore an improvement of the S/H circuit linearity is expected.

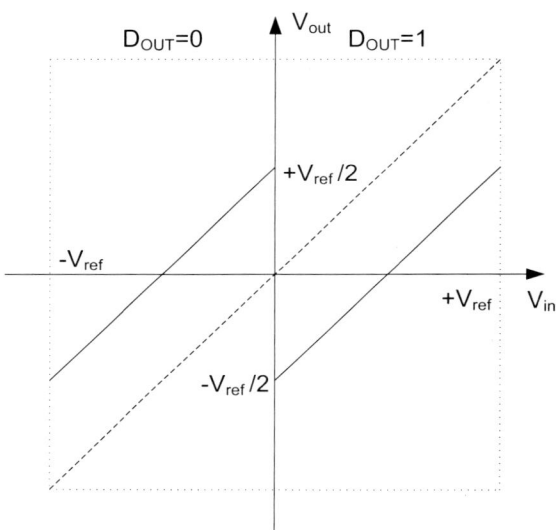

Fig. 8: The transfer curve of the proposed and conventional S/H circuit.

Since the output swing is halved, the required op amp unity-gain bandwidth in the mixed-mode S/H circuit becomes

$$\omega_u > \frac{2 \cdot N \cdot F_s \cdot ln2}{\beta} \tag{18}$$

which is smaller compare to the requirement of conventional S/H circuit expressed in (13).

The analog output of the mixed-mode S/H circuit will be processed by the following pipelined ADC stages. The 1-bit digital output will be combined with ADC outputs to generate the final output. Fig. 9 shows an example of the mixed-mode S/H circuit used in a 3-bit pipelined ADC. The S/H circuit has a 1-bit output F_1. The final ADC digital output $D_2 D_1 D_0$ is the combination of the S/H output F_1 and pipeline stages' output $F_2 F_3 F_4 F_5$.

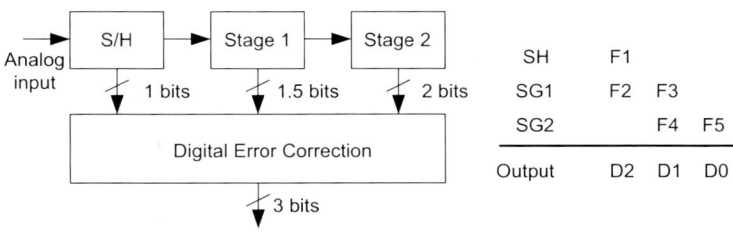

Fig. 9: The proposed S/H circuit used in a 3-bit pipelined ADC.

In a pipelined ADC, the op amp in the first pipeline stage has the most stringent requirements. Another benefit of the mixed-mode S/H is that since the output swing, which is the input of the first pipeline stage, is now reduced, the op amp gain and slew rate requirements in this stage are also relaxed based on the above observation.

Due to the additional comparator, in the sampling mode there are two signal paths in this S/H circuit. One is formed by the sampling switch, the sampling capacitor C_S and the op amp. The other goes through the comparator. Because of the time constant difference between these two paths, there exists a voltage error, i.e. the sampling capacitor C_S and the comparator see different input signals. The voltage difference is known as aperture error and will be increased with the input frequency. For an input signal of $V_{in} = V_{ref} sin(2\pi f_{in} t)$, the maximum slope of this signal can be presented as

$$\frac{dV_{in}}{dt} \big|_{max} = 2\pi f_{in} V_{ref} \tag{19}$$

Assuming the unmatched time constant between the two signal paths is $\Delta\tau$, the aperture error voltage, V_e, can be calculated as

$$V_e = 2\pi f_{in} V_{ref} \Delta\tau \tag{20}$$

Despite the existence of the aperture error, particularly at high frequency input, it is possible to minimize this error by matching the two signal paths in terms of topology and time constant. Fig. 10 shows the two signal paths during the sampling period. Instead of connecting the comparator directly to the input signal, it is connected to the output of the sampling switch. Thus the two paths see the same delay caused by the sampling switch. In addition, the op amp and comparator both use the falling edge of ϕ_{1a} to sample and quantize the input signal.

In actual implementation, it is difficult to match these two signal paths particularly at high frequency due to parasitic components and second-order effects.

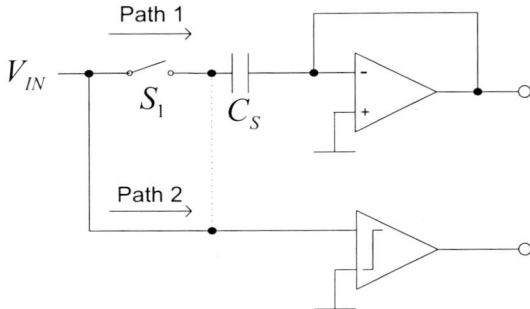

Fig. 10: The two signal paths in the sampling mode.

However, based on the characteristic of pipelined ADCs, the aperture error due to time constant mismatch can be treated as comparator offset error and eliminated by the digital error correction logic.

Although op amp and comparator offset errors can be tolerated in pipelined ADC by using digital error correction, these errors will increase the output swing, occupy a large offset correction range, and degrade the effective of the mixed-mode sampling technique. Therefore, in this design, auto-zero technique is used for the op amp in the sampling mode to eliminate its offset error. This is realized by connecting the op amp in the unity-gain configuration during the sampling mode. The comparator does not employ the auto-zero configuration because of the speed consideration. Otherwise the comparator has to quantize the input signal during the hold mode, which leads to a higher op amp speed requirement, or complicates the timing scheme that uses a shorter sampling phase to increase the time slot for amplifying.

4   Building Blocks

4.1   Operational Amplifier

Because the mixed-mode sampling technique reduces the signal swing and relaxes op amp gain and slew rate requirements, a gain-boosted single-stage telescopic op amp is use in the proposed S/H circuit as shown in Fig. 11. Transistors $M_1 \sim M_9$ forms the main telescopic op amp. Transistors $M_{B1} \sim M_{B7}$ provide the biasing voltages for the op amp. To improve the gain, transistors M_{10}, M_{12} and M_{11}, M_{13} form two common-source amplifiers and introduce negative feedback loops that make the source voltages of the common-gate transistors M_3 and M_4 less sensitive to the output signal. The gain boosting circuit increases the output impedance without adding more cascode devices. The gain boosting is only applied to the NMOS cascode transistors. The output impedance of the PMOS active load are increased by increasing the channel length of M_7 and M_8 since the size of these two devices have less effect on the op amp frequency response. The overall DC gain of the gain-boosted telescopic op amp can be shown as

$$A_0 = g_{m1}\{[g_{m3}r_{o3}r_{o1}g_{m10}(r_{o10}||r_{o12})]||(g_{m5}r_{o5}r_{o7})\} \qquad (21)$$

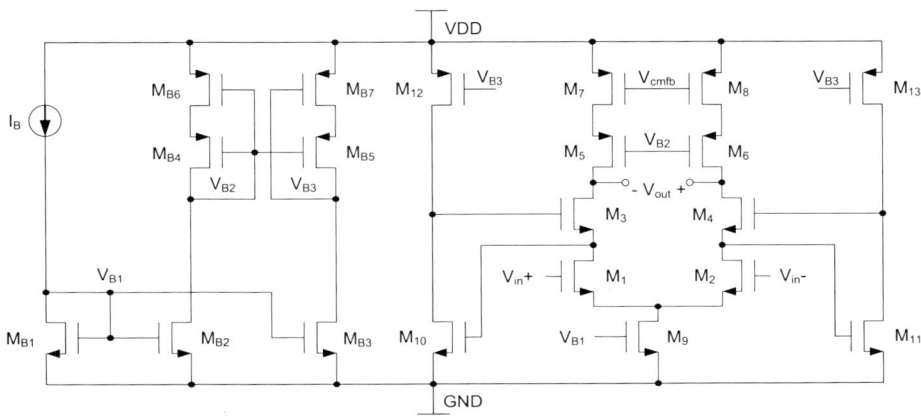

Fig. 11: The telescopic op amp with gain boosting.

It is clearly shown in (21) that the op amp DC gain depends on the transconductance of transistor M_1 M_3 M_5 and output impedance of M_1 M_3 M_5 M_7. In short channel devices, both g_m and r_o increase with V_{DS} before the devices reach drain-introduce barrier lowering (DIBL). In the mixed-mode S/H circuit, since the op amp output swing was reduced, the gate voltages of M_3 and M_5 are made high and low enough respectively to archive a higher DC gain. Fig. 12 shows the dependence of DC gain on output swing simulated with the slow MOS model. The simulation with the slow model gives the lowest available output swing which is the worst-case for accuracy and linearity. The simulated op amp has a DC gain of 64 dB, a phase margin of 70 degree and a unity-gain bandwidth of 1.1-GHz with 1-pF capacitive load as plotted in Fig. 13. The power consumption of the op amp is 2.8-mW at 1.8-V supply voltage.

For stability and settling considerations, the gain boosting circuit doesn't have to be very fast as long as its unity-gain frequency can satisfy

$$\beta\omega_{um} < \omega_{ug} < \omega_{np} \qquad (22)$$

where β is the close-loop feedback factor, ω_{um} is the unity-gain bandwidth of the main amplifier, ω_{ug} is the unity-gain bandwidth of the gain-boosting amplifier and ω_{np} is the second-pole frequency of the main amplifier [6]. Special attention is paid to the position of the pole-zero doublet introduced by the gain-booting amplifier. Although this doublet does not deteriorate the op amp stability, it can affect the op amp close-loop settling time [7].

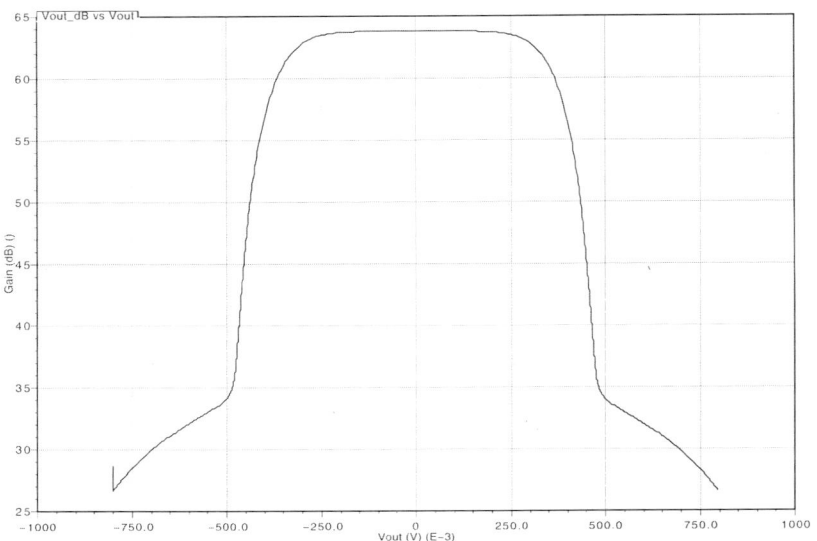

Fig. 12: Worst-case simulation of op amp gain variation.

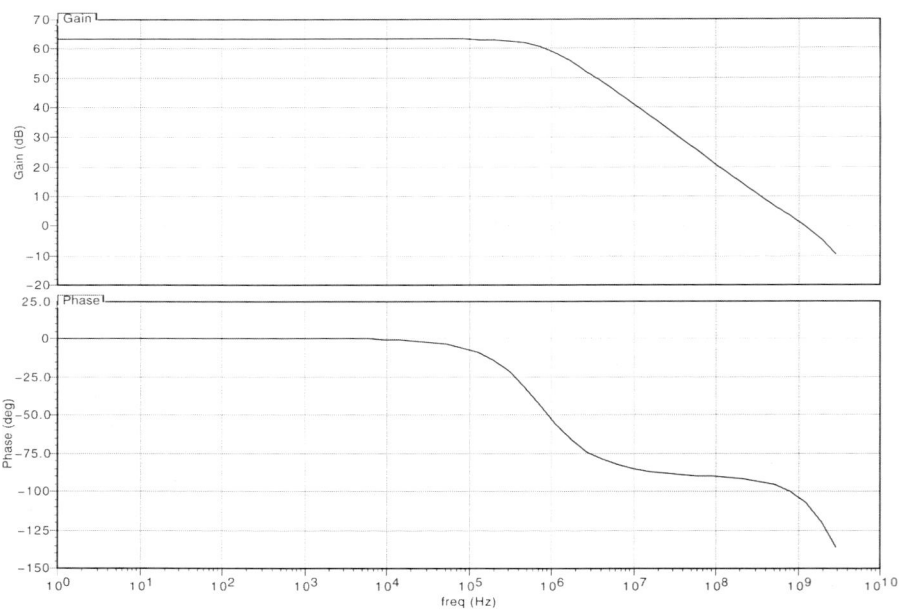

Fig. 13: The op amp gain and phase plot.

4.2   Comparator

The comparator used in the mixed-mode S/H circuit is shown in Fig. 14. The differential pair M_1 and M_2 amplify the input signal and transistors $M_4 \sim M_7$ form a regeneration latch. When ϕ_{1a} is high, $V_{out}+$ and $V_{out}-$ are reset to V_{DD} via M_8 and M_9. When ϕ_{1a} goes low, the differential pair M_1 and M_2 compare the input $V_{in}+$ and $V_{in}-$ and generate voltage difference at the drain of transistors M_4 and M_5. This voltage difference is amplified by the positive feedback of the latch therefore $V_{out}+$ and $V_{out}-$ goes to V_{DD} or ground according to the input voltages.

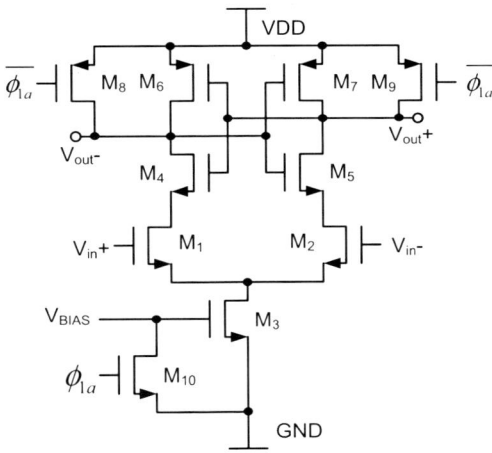

Fig. 14: The comparator used in the S/H circuit.

The offset of this comparator can be expressed as

$$V_{os} = \Delta V_{TH1,2} + \frac{(V_{GS} - V_{TH})_{1,2}}{2}\left(\frac{\Delta S_{1,2}}{S_{1,2}} + \frac{\Delta R}{R}\right) \qquad (23)$$

$V_{TH1,2}$ is the threshold voltage mismatch of transistors M_1 and M_2. $\Delta S_{1,2}$ is the physical dimension mismatch between M_1 and M_2. ΔR is the load resistance mismatch, which is contributed by transistors $M_4 \sim M_7$. The offset voltage in this comparator is dominated by the mismatch between transistors M_1 and M_2. The mismatches caused by other transistors are reduced by the gain of M_1 and M_2. Besides increasing the size of M_1 and M_2, the offset can also be reduce by decrease $(V_{GS} - V_{TH})_{1,2}$, which is controlled by the tail current of the differential pair.

With manually introduced 20% device dimension mismatch, this comparator has an offset voltage less than 7-mV in worst-case simulation. This comparator achieves less than 300-pS regeneration time for a 1-mV differential input signal in the worst-case simulation. The power consumption is 0.4-mW at 1.8-V supply voltage.

4.3   Bootstrapped MOS Switch

Besides the op amp, the input sampling switches also introduce nonlinearity due to its signal-dependent on-resistance. In order to investigate the error introduced by the op amp only. Bootstrapped MOS switches are used in the proposed S/H circuit. Fig. 15 illustrates the bootstrapped switch circuit [8].

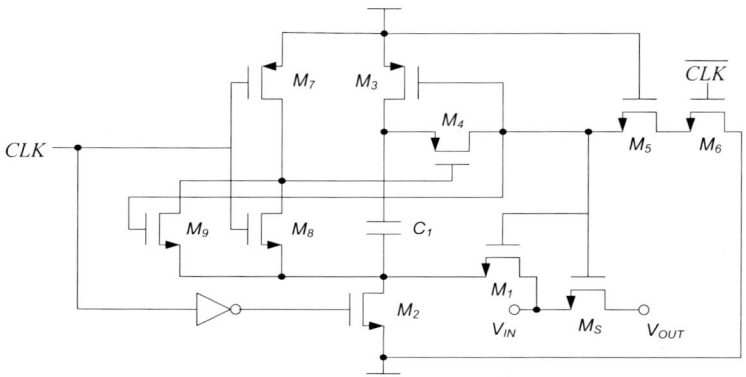

Fig. 15: The bootstrapped switch

The bootstrapped voltage is realized with the capacitor C_1, which is precharged to V_{DD} during the switch-off period. At the switch-on period, C_1 is connected between the gate and source terminals of the switch M_S via the switches M_1 and M_4. As a result, a constant gate-to-source voltage of V_{DD} applies to M_S, making its on-resistance independent of input signals.

5   Simulation Results

The proposed mixed-mode S/H circuit has been designed using a 0.18-μm CMOS process and simulated at worst-cast corner with manually introduced 7-mV comparator offset. The simulation is performed at 200-MHz sampling frequency and 1.8-V supply. Under these conditions, the S/H circuit consumes 3.6-mW.

Fig. 16 illustrates the digital and analog outputs of the mixed-mode S/H circuit at 200MSample/s with a 1-V_{PP} input sine-wave of 190-MHz (sub-sampling). It can be seen that the output swing is less than 540-mV with the aperture error and 7-mV comparator offset presented.

Plotting in Fig. 17 is the simulated SNDR, SFDR and THD as a function of the input signal frequency while the S/H samples at 200-MHz. The mixed-mode

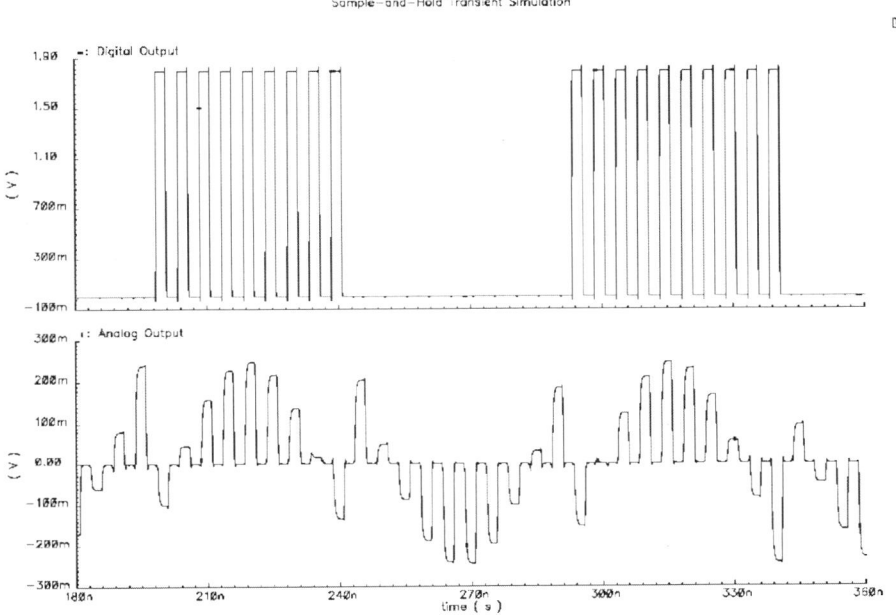

Fig. 16: Output waveform at 200-MSample/s with 190-MHz input.

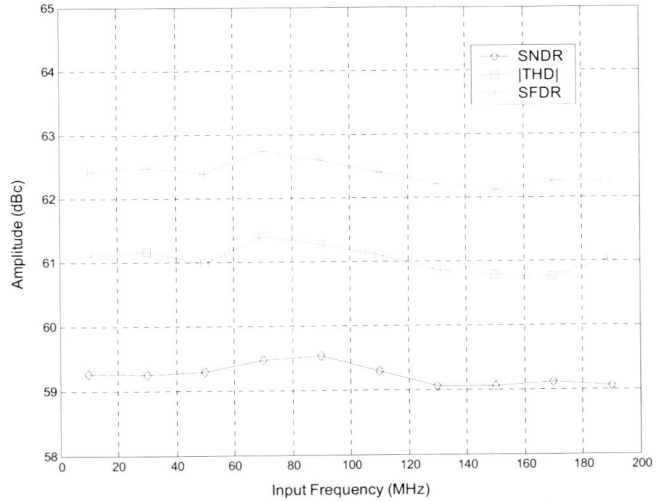

Fig. 17: Dynamic performance of the mixed-mode S/H circuits at 200-MSample/s.

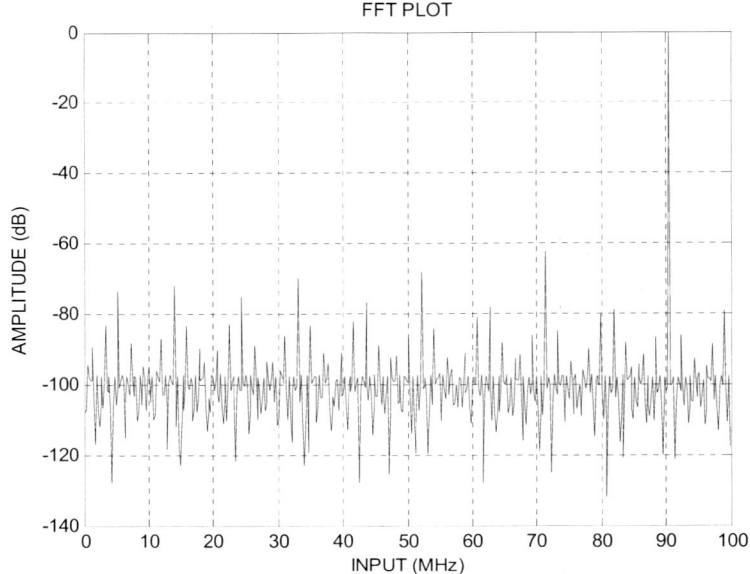

Fig. 18: Output spectrum with 200-MSample/s and 90-MHz input.

S/H circuit exhibits a THD lower than -60 dB and SNDR larger than 59 dB, i.e. better than 9-bit accuracy. The SNDR is larger than theory analysis in (17) because the linear settling behavior of the op amp. Fig. 18 shows the FFT plot of a 90-MHz input with 200-MSample/s, where about -61.2 dB THD and 59.5 dB SNDR are observed.

Table 1 summaries the simulated performance of the mixed-mode S/H circuit and compared it with the conventional S/H circuit in Fig. 1. Both the gain error and nonlinearity of the proposed mixed-mode S/H circuit are less than that of the conventional S/H circuit using the same op amp and switches.

Table 1: S/H circuit performance summary and comparison

Design	Mixed-Mode S/H	Conventional S/H
Technology	0.18-μm CMOS	0.18-μm CMOS
Power Supply	0.18-V	0.18-V
Sampling Frequency	200-MHz	200-MHz
SFDR	62.5 dB @ 90-MHz input	55.2 dB @ 90-MHz input
THD	-61.2 dB @ 90-MHz input	-53.2 dB @ 90-MHz input
SNDR	59.5 dB @ 90-MHz input	49.4 dB @ 90-MHz input
Power Consumption	3.6-mW	3.1-mW

6   Conclusion

In this paper, the design of a high-speed sample-and-hold circuit has been demonstrated by simulation results at 200-MSample/s. The proposed S/H circuits exhibits better linearity and lower power characteristic due to the mixed-mode sampling technique. This technique enables the use of single-stage cascode amplifier in the low voltage environment without degrading the dynamic performances and linearity.

References

1. D.-Y. Chang, "Design Techniques for a Pipelined ADC Without Using a Front-End Sample-and-Hold Amplifier," *IEEE Trans. Circuits and Systems I,* vol. 51, pp. 2123-2132, Nov. 2004.
2. H.-C. Kim, D.-K. Jeong, and W. Kim, "A Partially Switched-Opamp Technique for High-Speed Low-Power Pipelined Analog-to-Digital Converters," *IEEE Trans. Circuits and Systems I,* vol. 53, pp. 795-801, Apr. 2006.
3. I. Mehr and L. Singer, "A 55-mV, 10-bit, 40MSample/s Nyquist-Rate CMOS ADC," *IEEE J. Solid-State Circuits,* vol. 35, pp. 318-325, Mar. 2000.
4. B. Razavi,"Design of Analog CMOS Integrated Circuits," McGraw-Hill Higher Education, 2001.
5. C. C. Enz and G. C. Temes, "Circuit Techniques for Reducting the Effects of Op-Amp Imperfections: Autozeroing, Correlated Double Sampling, and Chopper Stabilization," *Proceedings of the IEEE,* vol. 84, pp. 1584-1614, Nov. 1996.
6. K. Bult and G. J. G. M. Geelen, "A Fast-Settling CMOS Op Amp for SC Circuits with 90-dB DC Gain," *IEEE J. Solid-State Circuits,* vol. 25, pp. 1379-1384, Dec. 1990.
7. B. Y. Kamath, R. G. Meyer, and P. R. Gray, "Relationship Between Frequency Response and Settling Time of Operational Aplifiers," *IEEE J. Solid-State Circuits,* vol. sc-9, pp. 347-352, Dec. 1974.
8. M. Dessouky, A. Kaiser, "Input Switch Configuration for Rail-to-Rail Operation of Switched Opamp Circuits," *Electronics Letter,* vol 35, pp. 8-10, Jan. 1999.

Probabilistic Design: A Survey of *Probabilistic* CMOS Technology and Future Directions for Terascale IC Design

Lakshmi N. B. Chakrapani, Jason George, Bo Marr, and Bilge E. S. Akgul, and Krishna V. Palem

Center for Research on Embedded Systems and Technology
School of Electrical and Computer Engineering
Georgia Institute of Technology
Atlanta, Georgia 30332–0250, USA.

Abstract. Highly scaled CMOS devices in the nanoscale regime would inevitably exhibit statistical or probabilistic behavior. Such behavior is caused by process variations, and other perturbations such as noise. Current circuit design methodologies, which depend on the existence of "deterministic" devices that behave consistently in temporal and spatial contexts do not admit considerations for probabilistic behavior. Admittedly, power or energy consumption as well as the associated heat dissipation are proving to be impediments to the continued scaling (down) of device sizes. To help overcome these challenges, we have characterized CMOS devices with probabilistic behavior (probabilistic CMOS or PCMOS devices) at several levels: from foundational principles to analytical modeling, simulation, fabrication, measurement as well as exploration of innovative approaches towards harnessing them through system-on-a-chip architectures. We have shown that such architectures can implement a wide range of probabilistic and cognitive applications. All of these architectures yield significant energy savings by trading probability with which the device operates correctly—lower the probability of correctness, the greater the energy savings. In addition to these PCMOS based innovations, we will also survey *probabilistic arithmetic*—a novel framework through which traditional computing units such as adders and multipliers can be deliberately designed to be erroneous, while being characterized by a well-defined *probability* of correctness. We demonstrate that in return for erroneous behavior, significant energy and performance gains can be realized through probabilistic arithmetic (units)—over a factor of $4.62X$ in the context of an FIR filter used in a H.264 video decoding—where the gains are quantified through the *energy-performance* product (or EPP). These gains are achieved through a systematic *probabilistic design* methodology enabled by a design space spanning the probability of correctness of the arithmetic units, and their associated energy savings.

1 Introduction and Overview

Device scaling, the primary driver of semiconductor technology advances, faces several hurdles. Manufacturing difficulties in the nanometer regime yield non uniform devices due to parameter variations, and low voltage operation makes them susceptible to

Please use the following format when citing this chapter:

Chakrapani, L.N.B., George, J., Marr, B., Akgul, B.E.S. and Palem, K.V., 2007, in IFIP International Federation for Information Processing, Volume 249, VLSI-SoC: Research Trends in VLSI and Systems on Chip, eds. De Micheli, G., Mir, S., Reis, R., (Boston: Springer), pp. 101–118

perturbations such as noise [18, 25, 32]. In such a scenario, current day circuit design methodologies are inadequate to design circuits, since they depend on devices with deterministic (in terms of their temporal behavior, since they are operated at high voltages) and uniform spatial behavior. To design robust circuits and architectures in the presence of this (inevitable) emerging statistical phenomena at the device level, it has been speculated that a shift in the design paradigm, from the current day deterministic designs to statistical or *probabilistic designs* of the future, would be necessary [2].

We have addressed the issue of probabilistic design at several levels: from foundational models [26, 27] of probabilistic *switches* establishing the relationship between probabilistic computing and energy, to analytical, simulation and actual measurement of CMOS devices whose behavior is rendered probabilistic due to noise (which we term as *probabilistic* CMOS, or PCMOS devices). In addition, we have demonstrated design methodologies and practical system-on-a-chip architectures which yield significant energy savings, through judicious use of PCMOS technology, for applications from the cognitive, digital signal processing and embedded domains [3, 13]. In this paper we present a broad overview of our contributions in the area of *probabilistic* design and PCMOS, by surveying prior publications [3, 5, 6, 13, 26, 27]. The exception is our recent work on *probabilistic* arithmetic, a novel framework through which traditional computing units such as adders and multipliers while erroneous, can be used to implement applications from the digital signal processing domain. Specifically, our approach involves creating a novel style of "error-prone" devices with probabilistic characterizations—we note in passing that from a digital design and computing standpoint, the parameter of interest in a PCMOS device is its probability of correctness p—derived by scaling the voltages to extremely and potentially undesirably low levels [19], referred to as *over-scaling*.

The rest of the paper is organized as follows. In Section 2 we outline the foundational principles of PCMOS technology based on the *probabilistic Switch*. In Section 3 we show approaches through which these abstract foundational models can be realized in the domain of CMOS, in the form of noise susceptible scaled CMOS devices operating at low voltages. The two laws of PCMOS technology using novel asymptotic notions will be the highlights. To help with our exposition, it will be convenient to partition the application domain into three groups (i) applications which benefit from (or harness) probabilistic behavior at the device level naturally, (ii) applications that can tolerate (and trade off) probabilistic behavior at the device level (but do not need such behavior naturally) and (iii) applications which cannot tolerate probabilistic behavior at all. We will briefly sketch our approach towards implementing PCMOS based architectures for application categories (i) and (ii), in Section 4.1 and Section 4.2 respectively. In Section 5, we describe probabilistic arithmetic. In Section 6, we outline other emerging challenges such as design for manufacturability, and present a novel probabilistic approach towards addressing one such problem—the problem of multiple voltage levels on a chip. Finally, in Section 7, we conclude and sketch future directions of inquiry.

2 Foundational Principles

Probabilistic switches, introduced by Palem [27], incorporate probabilistic behavior as well as energy consumption as first class citizens and are the basis for PCMOS devices. A probabilistic switch is a switch, which realizes a *probabilistic one-bit switching function*. As illustrated in Figure 1, the four deterministic one bit switching functions (Figure 1(a)) have a probabilistic counterpart (Figure 1(b)) with an *explicit* probability parameter (probability of correctness) p. Of these, the two constant functions are trivial and the others are non-trivial. We consider an abstract probabilistic switch sw to be the one which realizes one of these four probabilistic switching functions. Such elementary probabilistic switches may be composed to realize primitive boolean functions, such as AND, OR, NOT functions.

Input	output
0	0
1	1

Identity Function

Input	output
0	1
1	0

Complement Function

Input	output
0	0
1	0

Constant Function

Input	output
0	1
1	1

Constant Function

(a)

Input	output	
0	0 (p)	1 (1-p)
1	1 (p)	0 (1-p)

Identity Function

Input	output	
0	1 (p)	0 (1-p)
1	0 (p)	1 (1-p)

Complement Function

Input	output	
0	0 (p)	1 (1-p)
1	0 (p)	1 (1-p)

Constant Function

Input	output	
0	1 (p)	0 (1-p)
1	1 (p)	0 (1-p)

Constant Function

(b)

Fig. 1. (a) Deterministic one bit switching functions (b) Their probabilistic counterparts with probability parameter (probability of correctness) p

The relationship between probabilistic behavior—the probability with which the switching steps are correct—and the associated energy consumed was shown to be an entirely novel basis for energy savings [26]. Specifically, principles of statistical thermodynamics were applied to such switches to quantify their energy consumption, and hence the energy consumption (or energy complexity) of a network of such switches. While a switch that realizes the deterministic non-trivial switching function consumes at least $\kappa t \ln 2$ Joules of energy [24], a probabilistic switch can realize a probabilistic non-trivial switching function with $\kappa t \ln(2p)$ Joules of energy in an idealized setting. For a complete definition of a probabilistic switch, the operation of a network of probabilistic switches and a discussion of the energy complexity of such networks, the reader is referred to Palem [27].

3 The CMOS Domain: Probabilistic CMOS

Probabilistic switches serve as a foundational model supporting the physical realizations of highly scaled probabilistic devices as well as emerging devices. In the domain of CMOS, probabilistic switches model noise-susceptible CMOS (or PCMOS) devices operating at very low voltages [6]. To show that PCMOS based realizations correspond to abstract probabilistic switches, we have identified two key characteristics of PCMOS: (i) probabilistic behavior while switching and (ii) energy savings through probabilistic switching. These characteristics were established through analytical modeling and HSpice based simulations [6, 19] as well as actual measurements of fabricated PCMOS based devices.

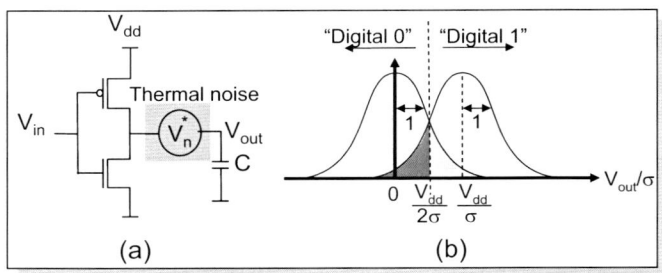

Fig. 2. (a) PCMOS switch (b) Representation of digital values 0 and 1 and the probability of error for a PCMOS switch

For a PCMOS inverter as shown in Figure 2 (a), the output voltage (V_{out}) is probabilistic, in this example, due to (thermal) noise coupled to its output. The associated noise magnitude is statistically characterized by a mean value of 0 and a variance of σ^2. The normalized output voltage $\frac{V_{out}}{\sigma}$ can be represented by a random variable whose value is characterized a Gaussian distribution as shown in Figure 2 (b), where the variance of the distribution is 1. The mean value of the distribution is 0 if the (correct) output is meant to be a digital 0, and $\frac{V_{dd}}{\sigma}$ if the (correct) output is meant to be a digital 1. In this representation, the two shaded regions of Figure 2 (b) (which are equal in area) correspond to the probability of error associated with this PCMOS inverter during each of its switching steps. From this formulation, we determine the probability of correctness denoted as p, by computing the area in the shaded regions and express p as

$$p = 1 - \frac{1}{2} erfc \left(\frac{V_{dd}}{2\sqrt{2}\sigma} \right) \tag{1}$$

where $erfc(x)$ is the complementary error function

$$erfc(x) = \frac{2}{\sqrt{\pi}} \int_{x}^{\infty} e^{-t^2} dt \tag{2}$$

Using the bounds for $erfc$ derived by Ermolova and Haggman [9], we have

$$p < 1 - 0.28 e^{-1.275 \frac{V_{dd}^2}{8\sigma^2}} \tag{3}$$

Using this expression to bound V_{dd} and hence the switching energy $\frac{1}{2}CV_{dd}^2$ from below, we have, for a given value of p, the energy consumed represented by

$$E(p, C, \sigma) > C\sigma^2 \left(\frac{4}{1.275}\right) \ln \left(\frac{0.28}{1-p}\right) \tag{4}$$

Clearly, the energy consumed E is a function of the capacitance C, determined by the technology generation, σ the "root-mean-square" (RMS) value of the noise, and the probability of correctness p. For a fixed value of $C = \hat{C}$ and $p = \hat{p}$, $\tilde{E}_{\hat{C},\hat{p}}(\sigma) = \hat{C}\sigma^2 \left(\frac{4}{1.275}\right) \ln \left(\frac{0.28}{1-\hat{p}}\right)$. Similarly for fixed values of $C = \hat{C}$ and $\sigma = \hat{\sigma}$, $\hat{E}_{\hat{C},\hat{\sigma}}$ a function of p alone: $\hat{E}_{\hat{C},\hat{\sigma}}(p) = \hat{C}\hat{\sigma}^2 \left(\frac{4}{1.275}\right) \ln \left(\frac{0.28}{1-p}\right)$.

We will succinctly characterize these behavioral and energy characteristics of PC-MOS switches using asymptotic notions from computer science [7, 14, 30] in the form of two laws. The notion of *asymptotic complexity* is widely used to study the efficiency of algorithms, where "efficiency" is characterized by the growth of its running time (or space), as a function of the size of its inputs [7, 14, 30]. The O notation provides an asymptotic *upper-bound*, where, for a function $f(x)$ where x is an element of the set of natural numbers

$$f(x) = O\left(h(x)\right)$$

given any function $h(x)$, there exist positive constants c, x_0 such that $\forall x \geq x_0, 0 \leq f(x) \leq c.h(x)$.

Similarly, the symbol Ω is used to represent an asymptotic *lower-bound* on the rate of growth of a function. For a function $f(x)$ as before,

$$f(x) = \Omega\left(h(x)\right)$$

whenever there exist positive constants c, x_0 such that $\forall x \geq x_0, 0 \leq c.h(x) \leq f(x)$. In the classical context, the O and the Ω notation is defined for functions over the domain of natural numbers. For our present purpose, we now extend this notion to the domain of real numbers. For any $y \in (\alpha, \beta)$ where $\alpha, \beta \in \{\Re^+ \cup 0\}$

$$\hat{h}(y) = \Omega_r\left(g(y)\right)$$

whenever there exists a $\gamma \in (\alpha, \beta)$ such that $\forall y \geq \gamma, 0 \leq g(y) \leq \hat{h}(y)$. Intuitively, the conventional asymptotic notation captures the behavior of a function $h(x)$ "for very large" x. Our modified notion Ω_r captures the behavior of a function $\hat{h}(y)$, defined in an interval (α, β). In this case, $\hat{h}(y) = \Omega_r(g(y))$ if there exists some point γ in the interval (α, β) beyond which $0 \leq g(y) \leq \hat{h}(y)$. Thus our current notion can be interpreted to mean "the function $\hat{h}(y)$ eventually *dominates* $g(y)$ in the interval (α, β)". In this paper, we will use this asymptotic approach to determine the rate of growth of energy described in Equation 4, as follows.

Returning to the lower-bound from (4) using the novel asymptotic (Ω_r) notation. Again, fixing $C = \hat{C}$ and $\sigma = \hat{\sigma}$, let us consider the expression $\hat{C}\hat{\sigma}^2\left(\frac{4}{1.275}\right)\ln\left(\frac{0.28}{1-p}\right)$ from Equation 4, and compare it with the *exponential* (in p) *function*, $E^e_{\hat{C},\hat{\sigma}}(p) = \hat{C}\hat{\sigma}^2 e^p$. We note that, when $p = 0.5$,

$$\hat{C}\hat{\sigma}^2\left(\frac{4}{1.275}\right)\ln\left(\frac{0.28}{1-p}\right) < E^e_{\hat{C},\hat{\sigma}}(p)$$

Furthermore, both functions are monotone increasing in p and they have equal values at $p \approx 0.87$. Hence,

$$\hat{C}\hat{\sigma}^2\left(\frac{4}{1.275}\right)\ln\left(\frac{0.28}{1-p}\right) > E^e_{\hat{C},\hat{\sigma}}(p)$$

whenever $p > 0.87$. Then, from the definition of Ω_r, an asymptotic lower-bound for $\hat{E}_{\hat{C},\hat{\sigma}}(p)$ in the interval $(0.5, 1)$ is

$$\hat{E}_{\hat{C},\hat{\sigma}}(p) = \Omega_r(E^e_{\hat{C},\hat{\sigma}}(p)) \tag{5}$$

Let $E^q_{\hat{C},\hat{p}}(\sigma) = \hat{C}\left(\frac{4}{1.275}\right)\ln\left(\frac{0.28}{1-\hat{p}}\right)\sigma^2$. Referring to (4) and considering $\tilde{E}_{\hat{C},\hat{p}}(\sigma)$ for a fixed value of $C = \hat{C}$ and $p = \hat{p}$, using the Ω_r notation,

$$\tilde{E}_{\hat{C},\hat{p}}(\sigma) = \Omega_r\left(E^q_{\hat{C},\hat{p}}(\sigma)\right) \tag{6}$$

Observation 1: For $p \in (0, 1)$, whereas the function $E^e_{\hat{C},\hat{\sigma}}(p)$ grows at least exponentially in p, for a fixed $C = \hat{C}$ and $\sigma = \hat{\sigma}$, the function $E^q_{\hat{C},\hat{p}}(\sigma)$, grows at least quadratically in σ, for fixed values $C = \hat{C}$ and $p = \hat{p}$

Then, from (5) and (6), we have

Law 1: Energy-probability Law: For any fixed technology generation determined by the capacitance $C = \hat{C}$ and constant noise magnitude $\sigma = \hat{\sigma}$, the switching energy $\hat{E}_{\hat{C},\hat{\sigma}}$ consumed by a probabilistic switch grows with p. Furthermore, the order of growth of $\hat{E}_{\hat{C},\hat{\sigma}}$ in p is asymptotically bounded below by an exponential in p since $\hat{E}_{\hat{C},\hat{\sigma}}(p) = \Omega_r\left(E^e_{\hat{C},\hat{\sigma}}(p)\right)$.

Law 2: Energy-noise Law: For any fixed probability $p = \hat{p}$ and a fixed technology generation (which determines the capacitance $C = \hat{C}$), $\tilde{E}_{\hat{C},\hat{p}}$ grows quadratically with σ since $\tilde{E}_{\hat{C},\hat{p}}(\sigma) = \Omega_r\left(E^q_{\hat{C},\hat{p}}(\sigma)\right)$.

Earlier variations of these laws [5, 6, 19] were implicitly based on the asymptotic notions described here explicitly. Together these laws constitute the characterization of probability and its relationship with energy savings in CMOS devices level. We will now show how this characterization helps build architectures composed of such devices and how energy savings as well as the associated performance gains can be extended up to the application level.

4   Implementing Applications Using PCMOS Technology

So far, we have summarized abstract models of probabilistic switches and their implementation and characterization in the domain of CMOS. To harness PCMOS technology to implement applications, we now reiterate that we consider three application categories: (i) applications which benefit from (or embody) probabilistic behavior intrinsically, (ii) applications that can tolerate probabilistic and (iii) applications which cannot tolerate statistical behavior.

4.1   Applications Which Harness Probabilistic Behavior

We will first consider applications from the cognitive and embedded domains which embody probabilistic behaviors. Probabilistic algorithms are those in which computational steps, upon repeated execution *with the same inputs*, could have distinct outcomes characterized by a probability distribution. A well known example of such an algorithm is the celebrated probabilistic test for primality [29, 34].

Input	Output with corresponding probability parameters		
000	00 (0.98)	01 (0.01)	10 (0.01)
001	00 (0.01)	01 (0.98)	10 (0.01)
010	00 (0.01)	01 (0.01)	10 (0.98)
011	00 (0.98)	01 (0.01)	10 (0.01)
100	00 (0.98)	01 (0.01)	10 (0.01)
101	00 (0.69)	01 (0.30)	10 (0.01)

Fig. 3. The probabilistic truth table for a node in a Bayesian network with 37 nodes, where the desired probability parameter p is represented parenthetically

In particular, the applications we have considered are based on *Bayesian inference* [21], *Probabilistic Cellular Automata* [11], *Random Neural Networks* [12] and *Hyper Encryption* [8]. For brevity, these algorithms will be referred to as BN, PCA, RNN and HE respectively. Common to these applications (and to almost all probabilistic algorithms) is the notion of a *core probabilistic step* with its associated probability parameter. An abstract model of such a step is a *probabilistic truth* table. In Figure 3, we illustrate the probabilistic truth table for a step in BN. Intuitively, realizing such probabilistic truth tables using probabilistic switches built from PCMOS is inherently more efficient in terms of the energy consumed when compared to those built from CMOS technology. This is because of the *inherent* probabilistic behavior of the PCMOS switches.

We have constructed *probabilistic system on a chip* (PSOC) architectures for these applications, and as illustrated in Figure 4, probabilistic system on a chip architectures are envisioned to consist of two parts: A *host* processor which consists of a conventional low energy embedded processor like the StrongARM SA-1100 [16], coupled to a co-processor which utilizes PCMOS technology and executes the core probabilistic steps.

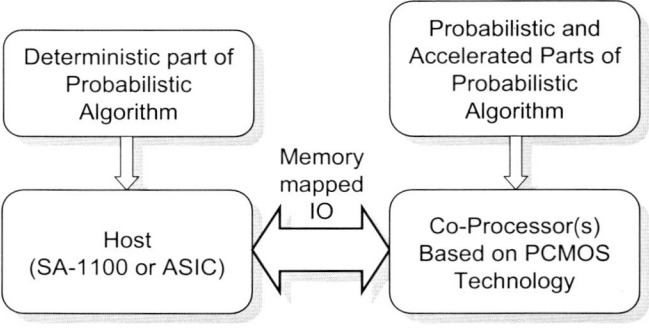

Fig. 4. A canonical PSOC architecture

The *energy-performance product* or EPP is the chief metric of interest for evaluating the efficiency of PSOC based architectures [3]; it is the product of the energy consumed, and time spent in completing an application, as it executes on the architecture. Then, for any given application, *energy-performance product gain* $\Gamma_{\mathcal{I}}$ of its PSOC realization over a conventional (baseline) architecture is the ratio of the EPP of the baseline denoted by the symbol β, to the EPP of a particular architectural implementation $\mathcal{I}$. We note in passing that in the context of the baseline implementation, the source of randomness is a pseudo-random number generator. $\Gamma_{\mathcal{I}}$ is thus:

$$\Gamma_{\mathcal{I}} = \frac{Energy_{\beta} \times Time_{\beta}}{Energy_{\mathcal{I}} \times Time_{\mathcal{I}}} \tag{7}$$

When compared to a baseline implementation using software executing on a StrongARM SA-1100, the gain of a PCMOS based PSOC is summarized in Table 1

Algorithm	$\Gamma_{\mathcal{I}}$	
	Min	Max
BN	3	7.43
RNN	226.5	300
PCA	61	82
HE	1.12	1.12

Table 1. Minimum and Maximum EPP gains of PCMOS over the baseline implementation where the implementation $\mathcal{I}$ has a StrongARM SA-1100 host and a PCMOS based co-processor

In addition, when the baseline is a custom ASIC realization (host) coupled to a functionally identical CMOS based co-processor, in the context of the HE and PCA applications, the gain $\Gamma_\mathcal{I}$ improves dramatically to 9.38 and 561 respectively. Thus, for applications which can harness probabilistic behavior, PSOC architectures based on PC-MOS technology yield several orders of magnitude improvements over conventional (deterministic) CMOS based implementations. For a detailed explanation of the architectures, experimental methodology and a description of the applications, the reader is referred to Chakrapani et. al. [3].

4.2 Applications Which Tolerate Probabilistic Behavior

Moving away from applications that embody probabilistic behaviors naturally, we will now consider the domain of applications that *tolerate* probabilistic behavior and its associated error. Specifically, we considered applications wherein energy and performance can be traded for application-level quality of the solution. Applications in the domain of digital signal processing are good candidates, where application-level quality of solution is naturally expressed in the form of *signal-to-noise ratio* or SNR. To demonstrate the value of PCMOS technology in one instance, we have implemented filter primitives using PCMOS technology [13], used to realize the H.264 decoding algorithm [23].

As illustrated in Figure 5(b), the probability parameter p_δ of correctness can be lowered uniformly for each bit in the adder; which is one of the building blocks of the FIR filter used in the H.264 application. While this approach saves energy, the corresponding output picture quality is significantly degraded when compared to conventional CMOS based and error-free operation. However, as illustrated in Figure 5(c), if the probability parameter is varied *non-uniformly* following the biased method described earlier [28], significantly lower energy consumption can be achieved with minimal degradation of the quality of the image [28]. Hence, not only can PCMOS technology be leveraged for implementing energy efficient filters, but can also be utilized to naturally trade-off energy consumed for application level quality of solution, through novel probabilistic biased voltage scaling schemes [13, 28].

5 Probabilistic Arithmetic

Following our development of characterizing error in the context of probabilistic behaviors induced by noise and considering an adder as a canonical example, we will associate a parameter δ, which indicates the magnitude by which the output of a computing element, an adder for example, can deviate from the correct answer before it is deemed to be erroneous; thus, an output value that is within a magnitude of δ from the correct value is declared to be correct. The rationale for this approach is that in several embedded domains in general and the DSP domain in particular—a topic to be discussed in some detail in the sequel—error magnitudes below a "tolerable"' threshold, quantified through δ, will be shown to have no impact on the perceived quality of an image. The *probability* of correctness p_δ of the *probabilistic adder* is defined, following the frequentist notion, to be the ratio of the number of correct values of the output compared to the total number of values.

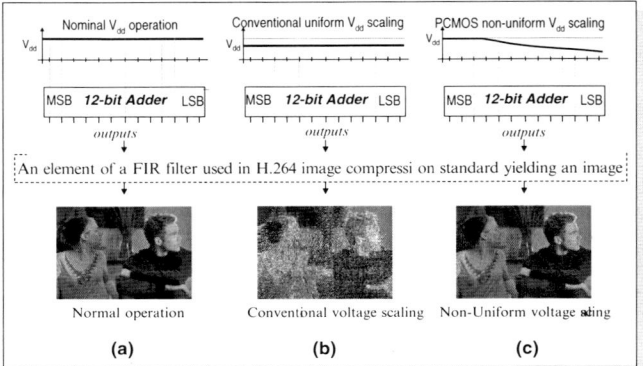

Fig. 5. Comparing images reconstructed using H.264 decoder (a) Conventional error-free operation (b) Probability parameter p_δ lowered uniformly for all bits (c) Probability parameter p_δ varied non-uniformly based on bit significance

Note that our approach to realizing energy savings and performance gains is entirely novel, and can be distinguished from similar aggressive and well-known approaches to voltage scaling: Our approach is aimed at designing arithmetic elements that are *deliberately designed to function in an erroneous manner*, albeit in a regime where such erroneous behavior can be characterized through probabilistic models and methods.

5.1 Probabilistic Arithmetic Through Voltage Overscaling

It is well known that energy savings can be achieved through scaling down supply voltages in a circuit. However, in the past, this approach resulted in increasing propagation times, consequently lowering the circuit's performance. Instead of avoiding bit errors through conventional voltage scaling, we advocate a probabilistic approach through voltage overscaling. We consider an approach here wherein the speed of the system clock is not lowered, even as the switching speed of the data path is lowered by voltage scaling.

Thus, source of error is caused by the gap between the values of the clock period, γ, and the effective switching speed σ. To understand this point better, consider keeping the system clock period, γ, fixed at $6ns$ in the context of a "probabilistic adder"; throughout this paper, we will consider the ripple-carry algorithm [20] for digital addition. Now, consider that as a result of voltage overscaling, for a particular input, the adder switches at a slower clock value of $\sigma = 7ns$, potentially yielding an incorrect result when its output is consumed or read by the system every $6ns$. Thus the output is not completely calculated at $6ns$ (system clock speed) intervals since the adder takes $7ns$ to completely switch in this example. However, by lowering the operating voltage of the adder and thus increasing σ, the energy consumption of the adder is lowered as well. In this case, the relationship of interest is between the rate at which the output value is incorrect and the associated savings in energy. As one would expect, error rates will be increased while yielding greater energy savings and this relationship will be characterized in Section 5.2.

5.2   Energy Savings Through Overscaled PCMOS

Typically, the nominal clock rate for a computing element is set by allowing for the worst case, critical path delay. However, the critical path is not active for most operational data sets, since the active path in the circuit is determined by the input data. In order to maintain correct operation, all potential propagation paths must be considered and the system clock rate must accommodate this worst case. This results in a clock period to delay (*clock-to-delay*) gap necessary to account for worst case. Voltage overscaling, however, attempts to take advantage of this gap by trading deterministic operation in exchange for energy savings.

Empirically Characterizing the Energy-Probability Relationship Through Benchmarks   To demonstrate the potential energy savings through overscaled PCMOS, we consider an 18-bit ripple-carry adder, a 9-bit two's-complement tri-section, array multiplier, and a 6-tap 9-bit FIR filter composed of adders and multipliers. In each of these cases, we will execute three benchmark data sets: (i) uniformly distributed random data, (ii) H.264 data from a low quality video source, and (iii) H.264 data from a high quality video source. As seen in Table 2, all three cases show reductions in energy consumption. However, H.264 data sets yield greater energy reductions when compared to uniformly distributed data. This is due to the fact that H.264 video data tends to have little variance and relatively infrequent output switching and as a result, only small portions of the circuit are active on occasions when there is output switching. As a result, the computation infrequently causes delays greater than the system clock period.

Conversely, uniformly distributed data exercises all portions of the circuit because of an associated larger variance. Accordingly, there is a smaller clock-to-delay gap and as a result, the energy savings are lower for a given probability parameter p.

Table 2. Voltage Overscaled PCMOS Energy Savings for Benchmark Data Sets

Computing Element	Benchmark	p_δ	$E(p_\delta)$	Δp_δ	$\Delta E(p_\delta)$	p_δ Sacrifice	Energy Savings
Adder	Uniform data	0.9999	$0.88pJ$	0.0001	$2.59pJ$	0.01%	75%
$E_{nom} = 3.47pJ$	Low Quality H.264	0.9993	$0.62pJ$	0.0007	$2.85pJ$	0.07%	82%
δ threshold $= 127$	High Quality H.264	0.9998	$0.62pJ$	0.0002	$2.85pJ$	0.02%	82%
Multiplier	Uniform data	0.9998	$8.30pJ$	0.0002	$11.73pJ$	0.02%	59%
$E_{nom} = 20.03pJ$	Low Quality H.264	0.9549	$2.11pJ$	0.0451	$17.92pJ$	4.51%	89%
δ threshold $= 127$	High Quality H.264	0.9862	$2.11pJ$	0.0138	$17.92pJ$	1.38%	89%
FIR	Uniform data	0.9999	$102.89pJ$	0.0001	$34.67pJ$	0.01%	25%
$E_{nom} = 137.56pJ$	Low Quality H.264	0.9998	$37.37pJ$	0.0002	$100.19pJ$	0.02%	73%
δ threshold $= 255$	High Quality H.264	0.9999	$57.46pJ$	0.0001	$80.1pJ$	0.01%	58%

As in the case of PCMOS devices, the energy-probability relationship will be used to characterize our design space. As an illustrative example, we will consider an 18 bit ripple carry adder and its overscaled variants. The design space is characterized by three dimensions. The probability parameter p_δ, the energy and the relationship between γ and σ. For example in Figure 6, consider a specific value of energy. For this fixed energy budget, the probability of correctness is determined by the clock period of the circuit. As a result of these three properties, there exists a 3-dimensional design space where probability of correct output can be traded for energy savings and performance gains. A plot of one possible design space for a PCMOS adder is shown in Figure 6.

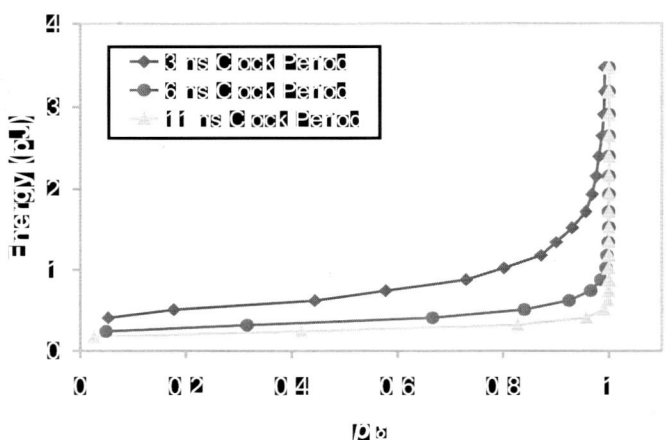

Fig. 6. Energy/performance/probability tradeoff for an 18-bit, ripple-carry adder: at nominal clock rate (11 ns period), at $1.8X$ faster clock rate (6 ns period), and at $3.7X$ faster clock rate (3 ns period)

By extension, energy can be saved and performance improved by increasing the error rate p_δ. This novel approach to achieving significant energy savings are possible since a "small" decrease in the probability of correctness can yield a disproportionate gain in energy savings (Table 3) as well as in the associated EPP. This energy-probability tradeoff is also characterized in Section 5.2 through the energy-probability or E-p relationship of elemental gates used to realize probabilistic arithmetic. Through this relationship, we provide a coherent characterization of the design space associated with probabilistic arithmetic. Specifically, the design space is determined by the parameters γ and σ yielding a probability parameter p_δ, with an associated energy consumption $E(p_\delta)$.

Using this notion of probabilistic arithmetic primitives as building blocks, we implement two widely used DSP algorithms: the fast Fourier transform (FFT) and the finite impulse response (FIR) filter. As a result of the probabilistic behavior of the arithmetic primitives, the associated DSP algorithm computations are also probabilistic. In this paper, we show the EPP gains in the context of the FIR filter in Section 5.2, and extend it to

Table 3. Probability of correctness and energy savings for a PCMOS adder

Benchmark	p_δ Degradation	Energy Savings
Low Quality Video	0.07%	82%

demonstrate gains at the application level in the context of a movie decoded using this filter based on the H.264 standard. Briefly, from the perspective of human perception, the degradation in quality is negligible whereas the gains quantified through the EPP metric were a factor of $3.70X$ as presented in Section 5.2).

There are several subtle issues that have played a role in this formulation, notably the ability to declare a phenomenon—the behavior of adder in our case—to be *probabilistic* based on aposteriori statistical validation. A detailed analysis is beyond the scope of this discussion, and the interested reader is referred to Jaynes's excellent treatment of this topic [17].

Case Study of an FIR   To analyze the value and the concomitant savings derived from voltage overscaled PCMOS, we have evaluated H.264 video decoding algorithm. Motion compensation is a key component of the H.264 decoding algorithm. Within this motion compensation phase a six-tap FIR is used to determine luminosity for the H.264 image blocks using 1, -5, 20, 20, -5, and 1 as the coefficients at taps [0..5] respectively. Video data from a low quality source (military video of ordnance explosion) and a high quality source (video from the 20th Century Fox movie XMen 2) were used for experimentation.

Experimental Framework First, the FIR was decomposed into its constituent adder and multiplier building blocks. These building blocks were then decomposed into full adders classified by type and output loading. Each full adder class was then simulated in HSpice for all input state transitions that result in an output state transition. This was repeated for both the sum and carry out bits of the full adder classes, and the resulting output transition delays were then summarized into a transition-delay lookup table. All input state transitions that did not result in an output state transition were considered to have no delay. HSpice simulation was then repeated with 1000 uniformly distributed random input combinations for each full adder class to determine average switching energy.

Building on this HSpice model and using a C-based simulation, benchmark data was used and using the current and previous states for both input and output at each full adder, the delay is estimated for each model using the look-up table previously developed using the HSpice simulation framework. Individual full adder delays were further propagated to building block outputs, which were then propagated to FIR outputs and compared to a specified clock period γ. Any FIR output delays violating timing constraints were considered to be erroneous and the appropriate bit was deemed incorrect and forced to be erroneous. The results of the outputs of the FIR filter in the fully functional context is then compared to those derived from overscaling to determine p_δ and

SNR. Energy consumption was determined by adding the energy of each individual full adder comprising the FIR and the results were compared to conventional operation (at a supply voltage $V_{dd}=2.5V$). The overall delay in the FIR filter was determined by maximum propagation delay calculated as the sum of worst case delays for each full adder in the critical path.

Finally, H.264 decoding was performed using a program written by Martin Fiedler. The original code was modified to inject bit-errors determined by the C simulation described above. The resulting decoded frames were then compared to originals to determine SNR. Energy consumption was calculated as the FIR energy consumption for the specific voltage overscaling scheme employed.

FIR **Results** As shown in Figure 7, voltage overscaled PCMOS operation yielded a 47% reduction in energy consumption with a $2X$ factor increase in performance, resulting in an EPP ratio of $3.70X$ for high quality video. We also consider a low quality military video, where the primary requirement is object recognition, and larger gains in energy savings and performance are possible. Thus, voltage overscaled PCMOS operation yeilds a 57% reduction in energy consumption and $2.19X$ factor increase in performance gain with an EPP ratio of $4.62X$ in this case where the quality of the output video is not as significant as the high quality case.

Fig. 7. Application level impact of our approach on high quality H.264 video comparing voltage scaled PCMOS [bottom](with an EPP ratio of $3.70X$) to the original H.264 frames [top]

6   Related work and Some Implementation Challenges

The use of voltage scaling in an effort to reduce energy consumption has been explored vigorously in previous work [4, 22, 36, 37]. In each of these papers, increased propagation delay was considered the primary drawback to voltage overscaling. To maintain circuit performance and *correctness* while simultaneously realizing energy savings through voltage scaling, several researchers employ the use of multiple supply voltages by operating elements along the critical path at nominal voltage and reducing supply voltages along non-critical paths [4, 22, 36, 37]. Supply voltage scheduling and its interplay with path sensitization along with task scheduling has been studied as well [4, 22, 36].

Offering a contrasting approach, in [15, 33, 35], propagation delay errors are removed through error correction in a collection of techniques named "algorithmic noise-tolerance (ANT)". In [15], difference-based and prediction-based error correction approaches are investigated and in [35], adaptive error cancellation (AEC) is employed using a technique similar to echo cancellation. In [33], the authors propose reduced precision redundancy (RPR) to eliminate propagation delay errors with no degradation to the SNR of the computed output. Our work can be distinguished from all of these methods through the fact that our designs permit the outputs of the arithmetic units to be incorrect, albeit with a well-understood probability.

The actual implementation and fabrication of architectures that leverage PCMOS based devices poses further challenges. Chief among them is "tuning" the PCMOS devices, or in other words, controlling the probability parameter p of correctness. Additionally, the number of distinct probability parameters is a concern, since this number directly relates to the number of voltage levels [6]. We make two observations aimed at addressing these problems: (i) Having distinct probability parameters is a requirement of the application and the application *sensitivity* to probability parameters is an important aspect. That is, if an application uses probability parameters p_1, p_2, p_3, for example, it might be the case that the application level quality is not affected when only two distinct values, say p_1, p_2 are used. This, however can only be determined experimentally and is a topic being investigated. (ii) Given probability parameters p_1 and p_2, other probability parameters might be derived through logical operations. For example, if the probability of obtaining a 1 from a given PCMOS device is p and the probability of obtaining a 1 from a second PCMOS device is q, a logical AND of the output of the two PCMOS devices produces a 1 with a probability $p.q$. Using this technique, in the context of an application (the case of Bayesian inference is used here), the number of distinct probability parameters may be drastically reduced. Since the probability parameter p is controlled through varying the voltage, this, in turn reduces the number of distinct voltage levels required and is another topic being investigated.

7   Remarks on Quality of Randomness and Future Directions

In any implementation of applications which embodies probability, the *quality* of the implementation is an important aspect apart from the energy and running time. In conventional implementations of probabilistic algorithms—which utilize hardware or software based implementations of *pseudo* random number generators to supply (pseudo) random bits,—it is a well known fact that random bits of "low quality" affect application behavior, from the correctness of Monte Carlo simulations [10] to the strength of encryption schemes. To ensure that application behavior is not affected by low quality random bits, the quality of random bits produced by a particular strategy should be evaluated rigorously. Our approach to determine the quality of random bits, is to use statistical tests to determine the quality of randomness. To study the statistical properties of PCMOS devices in a preliminary way, we have utilized the randomness tests from the NIST Suite [31] to assess the quality of random bits generated by PCMOS devices. Preliminary results indicate that PCMOS affords a higher quality of randomness; a future direction of study is to quantify the impact of this quality on the application level quality of solution.

Acknowledgments

This work is supported in part by DARPA under seedling contract #F30602-02-2-0124, by the DARPA ACIP program under contract #FA8650-04-C-7126 through a subcontract from USC-ISI and by an award from Intel Corporation. This document is an expansion of the survey originally presented at the IFIP international conference on very large scale integration [1] and includes novel results.

References

1. B. E. S. Akgul, L. N. Chakrapani, P. Korkmaz, and K. V. Palem. Probabilistic CMOS technology: A survey and future directions. In *Proceedings of The IFIP International Conference on Very Large Scale Integration*, 2006.
2. S. Borkar, T. Karnik, S. Narendra, J. Tschanz, A. Keshavarzi, and V. De. Parameter variations and impact on circuits and microarchitecture. In *Proceedings of the 40th Design Automation Conference*, pages 338–342, 2003.
3. L. N. Chakrapani, B. E. S. Akgul, S. Cheemalavagu, P. Korkmaz, K. V. Palem, and B. Seshasayee. Ultra efficient embedded SOC architectures based on probabilistic cmos technology. In *Proceedings of The 9th Design Automation and Test in Europe (DATE)*, pages 1110–1115, Mar. 2006.
4. J. Chang and M. Pedram. Energy minimization using multiple supply voltages. In *Proc. of IEEE Transactions on VLSI Systems*, volume 5, pages 436 – 443, Dec. 1997.
5. S. Cheemalavagu, P. Korkmaz, and K. V. Palem. Ultra low-energy computing via probabilistic algorithms and devices: CMOS device primitives and the energy-probability relationship. In *Proceedings of The 2004 International Conference on Solid State Devices and Materials*, pages 402–403, Tokyo, Japan, Sept. 2004.
6. S. Cheemalavagu, P. Korkmaz, K. V. Palem, B. E. S. Akgul, and L. N. Chakrapani. A probabilistic CMOS switch and its realization by exploiting noise. In *Proceedings of The IFIP International Conference on Very Large Scale Integration*, 2005.
7. T. H. Cormen, C. E. Leiserson, R. L. Rivest, and C. Stein. *Introduction to Algorithms, Second Edition*. MIT Press and McGraw-Hill, 2001.
8. Y. Z. Ding and M. O. Rabin. Hyper-Encryption and everlasting security. In *Proceedings of the 19th Annual Symposium on Theoretical Aspects of Computer Science; Lecture Notes In Computer Science*, volume 2285, pages 1–26, 2002.
9. N. Ermolova and S. Haggman. Simplified bounds for the complementary error function; application to the performance evaluation of signal-processing systems. In *Proceedings of the 12th European Signal Processing Conference*, pages 1087–1090, Sept. 2004.
10. A. M. Ferrenberg, D. P. Landau, and Y. J. Wong. Monte carlo simulations: Hidden errors from "good" random number generators. *Phys. Rev. Let*, 69:3382–3384, 1992.
11. H. Fuks. Non-deterministic density classification with diffusive probabilistic cellular automata. *Physical Review E, Statistical, Nonlinear, and Soft Matter Physics*, 66, 2002.
12. E. Gelenbe. Random neural networks with negative and positive signals and product form solution. *Neural Computation*, 1(4):502–511, 1989.

13. J. George, B. Marr, B. E. S. Akgul, and K. Palem. Probabilistic arithmetic and energy efficient embedded signal processing. In *International Conference on Compilers, Architecture, and Synthesis for Embedded Systems CASES*, 2006.

14. J. Hartmanis and R. E. Stearns. On the computational complexity of algorithms. *Transactions of the American Mathematical Society*, 117, 1965.

15. R. Hedge and N. R. Shanbhag. Soft digital signal processing. *IEEE Transactions on VLSI*, 9(6):813–823, Dec. 2001.

16. Intel Corporation. SA-1100 microprocessor technical reference manual, Sept. 1998.

17. E. Jaynes. *Probability Theory: The Logic of Science.* Cambridge University Press, Cambridge, UK, 2003.

18. L. B. Kish. End of Moore's law: thermal (noise) death of integration in micro and nano electronics. *Physics Letters A*, 305:144–149, 2002.

19. P. Korkmaz, B. E. S. Akgul, L. N. Chakrapani, and K. V. Palem. Advocating noise as an agent for ultra low-energy computing: Probabilistic CMOS devices and their characteristics. *Japanese Journal of Applied Physics (JJAP)*, 45(4B):3307–3316, Apr. 2006.

20. M. Lu. *Arithmetic and Logic in Computer Systems.* John Wiley & Sons, Inc., Hoboken, NJ, 2004.

21. D. MacKay. Bayesian interpolation. *Neural Computation*, 4(3), 1992.

22. A. Manzak and C. Chaktrabarti. Variable voltage task scheduling algorithms for minimizing energy/power. In *Proc. of IEEE Transactions on Very Large Scale Integration (VLSI) Systems*, volume 11, pages 270 – 276, Apr. 2003.

23. D. Marpe, T. Wiegand, and G. J. Sullivan. The H.264/MPEG4-AVC standard and its fidelity range extensions. *IEEE Communications Magazine*, Sept. 2005.

24. J. D. Meindl and J. A. Davis. The fundamental limit on binary switching energy for terascale integration (TSI). *IEEE; Journal of Solid State Circuits*, 35:1515–1516, Oct. 2000.

25. K. Natori and N. Sano. Scaling limit of digital circuits due to thermal noise. *Journal of Applied Physics*, 83:5019–5024, 1998.

26. K. V. Palem. Proof as experiment: Probabilistic algorithms from a thermodynamic perspective. In *Proceedings of The International Symposium on Verification (Theory and Practice)*, Taormina, Sicily, June 2003.

27. K. V. Palem. Energy aware computing through probabilistic switching: A study of limits. *IEEE Transactions on Computers*, 54(9):1123–1137, 2005.

28. K. V. Palem, B. E. S. Akgul, and J. George. Variable scaling for computing elements. *Invention Disclosure*, Feb. 2006.

29. M. O. Rabin. Probabilistic algorithms. In J. F. Traub, editor, *Algorithms and Complexity, New Directions and Recent Trends*, pages 29–39. 1976.

30. M. O. Rabin. Complexity of computations. *Communications of the ACM*, 20(9):625–633, 1977.

31. Random Number Generation and Testing. http://csrc.nist.gov/rng/.

32. N. Sano. Increasing importance of electronic thermal noise in sub-0.1mm Si-MOSFETs. *The IEICE Transactions on Electronics*, E83-C:1203–1211, 2000.

33. B. Shim, S. R. Sridhara, and N. R. Shanbhag. Reliable low-power digital signal processing via reduced precision redundancy. In *Proc. of IEEE Transactions on Very Large Scale Integration (VLSI) Systems*, volume 12, pages 497–510, May 2004.

34. R. Solovay and V. Strassen. A fast monte-carlo test for primality. *SIAM Journal on Computing*, 1977.

35. L. Wang and N. R. Shanbhag. Low-power filtering via adaptive error-cancellation. *IEEE Transactions on Signal Processing*, 51:575 – 583, Feb. 2003.

36. Y. Yeh and S. Kuo. An optimization-based low-power voltage scaling technique using multiple supply voltages. In *Proc. of IEEE Internaitonal Symposium on ISCAS 2001*, volume 5, pages 535 – 538, May 2001.

37. Y. Yeh, S. Kuo, and J. Jou. Converter-free multiple-voltage scaling techniques for low-power cmos digital design. In *Proc. of IEEE Transactions on Computer-Aided Design of Integrated Circuits and Systems*, volume 20, pages 172–176, Jan. 2001.

Reliability Issues in Deep Deep Submicron Technologies: Time-Dependent Variability and its Impact on Embedded System Design

Antonis Papanikolaou[1], Hua Wang[1,2], Miguel Miranda[1],
Francky Catthoor[1,2] and Wim Dehaene[2]

[1] IMEC vzw, Kapeldreef 75, 3001 Leuven, Belgium
{papaniko,wanghua,miranda,catthoor}@imec.be
[2] Katholieke Universiteit Leuven, ESAT Dept., Kasteelpark Arenberg 10, 3001
Leuven, Belgium
wim.dehaene@esat.kuleuven.be

Abstract. Technology scaling has traditionally offered advantages to
embedded systems in terms of reduced energy consumption and die cost
as well as increased performance, without requiring significant additional
design effort. Scaling past the 45 nm technology node, however, brings a
number of problems whose impact on system level design has not been
evaluated yet. Random intra-die process variability, reliability degradation mechanisms and their combined impact on the system level parametric quality metrics are prominent issues that will need to be tackled
in the next few years. Dealing with these new challenges will require a
paradigm shift in the system level design phase.

1 Introduction

Embedded system design is especially demanding and challenging in terms of
requirements that need to be satisfied, e.g. real-time processing, cost effectiveness, low energy consumption and reliable operation. These requirements have
to be properly balanced until a financially viable global solution is found. Novel
mobile multimedia and communication applications pose extremely severe requirements on the amount of storage, processing and functionality capabilities
of the system. Near future embedded systems will have to combine interactive
gaming with advanced 3D and video codecs together with leading edge wireless
connectivity standards, like software defined radio front-ends and protocol stacks
for cognitive radio. This will increase the platform requirements by at least a
factor of 10. Meanwhile, battery capacity is only increasing by about 7% per year
and users demand longer times between battery recharges. Optimizing any one
of these requirements by compromising on another is a rather straightforward
design task. However, in embedded system design the solution must obey the
constraints in all four requirement axes.

Products containing some sort of embedded system implementation targeting
safety critical applications (i.e. advanced braking systems and traction control

Please use the following format when citing this chapter:

Papanikolaou, A., Wang, H., Miranda, M., Catthoor, F. and Dehaene, W., 2007, in IFIP International
Federation for Information Processing, Volume 249, VLSI-SoC: Research Trends in VLSI and Systems
on Chip, eds. De Micheli, G., Mir, S., Reis, R., (Boston: Springer), pp. 119–141

of modern cars, biomedical devices, etc.) impose aggressive constraints on the design of embedded systems, especially in terms of meeting reliability and fail-safe operation targets during the guaranteed product lifetime. This translates onto very low field return targets during that time, since failures can lead to dire financial consequences or catastrophic results. On the other hand, systems that belong to the low end consumer electronics market are also subject to tight lifetime and reliability targets. They are usually deployed in very large volume, thus even a small percentage of failures can lead to a large amount of field returns that cost both financially and in consumer loyalty and in company image. For all these reasons fail-safe reliable operation throughout a guaranteed product lifetime becomes a strategically important property for the design of embedded systems.

Technology scaling has traditionally enabled improvements in three of the design quality metrics: increased processing performance, lower energy consumption per task and lower die cost. Reliability targets were also guaranteed at the technology level by using well controlled processes and well characterized materials. Unfortunately this "happy scaling" scenario where technology and design could be kept decoupled is coming to an end [1]. New technologies become far less mature than earlier ones, e.g. the nanometer range feature sizes require the introduction of new materials and process steps that are not properly characterized by the time they start being used in commercial products, leading to potentially less reliable products. On the other hand, progressive degradation instead of abrupt failure of electrical characteristics of transistors and wires becomes reality as an intrinsic consequence of the smaller feature sizes and interfaces as well as increasing electric fields and operating temperatures (see [2] and its references). Effects considered as second-order in the past, become a clear threat now for the correct operation of the circuits and systems since they start affecting their parametric features (e.g., timing but also energy dissipation) while the functionality remains unaltered. Moreover, as we show in this work, the combined impact of manufacturing uncertainty (e.g. process variability) and reliability degradation results in time-dependent variability. The electrical characteristics of the transistors and the wires will vary statistically in a spatial and a temporal manner, directly translating into design uncertainty during fabrication and even during operation in the field, especially as a function of the application's functionality influence in the system as such. Unfortunately, current reliability models based on traditional worst case stress analysis are not sufficient to capture these more dynamic system level interactions, resulting in over-pessimistic implementations [2]. Research in fully integrated analysis models (from technology to full system) is urgently needed.

On the solution side, a number of conventional techniques already exist for dealing with uncertainty. However, most of them rely on the introduction of worst-case design slacks at the process technology, circuit and the system level in order to absorb the unpredictability of the transistor and interconnect performance and to provide implementations with predictable parametric features. But trade-offs are always involved in these decisions, which result in excessive energy consumption and/or cost leading to infeasible design choices. From the designers perspective reliability degradation mechanisms manifest themselves as time-dependent uncertainties in the parametric performance metrics of the devices. In the future sub 45 nm regime, these uncertainties would be way too high to be handled with existing worst-case design techniques without incurring significant penalties in terms of area/delay/energy. As a result, reliability becomes a great threat to the design of reliable complex digital systems-on-chip (SoC) implementations. We believe this will require the development of novel reliability models at all three levels, namely device, circuit and system level. They should be capable of capturing the impact of the application functionality on the system as well as new design paradigms for embedded system design in order to build reliable systems out of technology which will be largely unpredictable in nature. This problem cannot only be solved at the technology and circuit level anymore. A shift toward Technology-Aware Design solutions will be required to keep designing successful systems in future aggressively scaled technologies.

2   Reliability Degradation Mechanisms for Scaled Technology Nodes

Reliability has always been a concern in the technology development community. In the past decades however, technology scaling involved shrinking the feature sizes of transistors and wires as well as the supply voltage with minimal intervention on the materials used. The available reliability margins were quite large and guaranteeing a life time of ten years for each of the transistors in the design was a feasible target, even under worst-case assumptions on the operating conditions. Furthermore, the first transistor to break in the die has been assumed to render the entire die non-functional which is another worst-case assumption that reliability engineers have always made in order to guarantee life-time under all circumstances. Still these conditions were based on reasonable assumptions. But scaling toward Deep Deep Sub-Micron (DDSM) technology nodes is not business as usual. Along with feature miniaturization, process technologists have also introduced a number of novel materials and process steps in the leading edge manufacturing processes. Examples include the high-k materials used for the transistor gate insulation from the channel, the low-k materials for the implementation of the dielectrics in the metal stack, the re-introduction of copper for the implementation of interconnect wires a couple of technology nodes ago etc. Characterizing these materials and their interactions for reliability degradation mechanisms is an extremely complex task. Typically they are used in commercial processes before full understanding of the physical degradation mechanisms is available. At the same time, the supply voltage scaling has been saturating in order to keep enough headroom between the transistor threshold voltage and the supply voltage hence increasing the electric fields and stress conditions for these devices. Furthermore, effects that in the past have been considered second order are now becoming a clear threat for the parametric and functional operation of the circuits and systems in near future technologies. Examples include soft-breakdowns (SBD) in gate oxide of transistors (especially dramatic in high-k oxides) [2], Negative Bias Temperature Instability (NBTI) issues in the threshold voltage of the PMOS transistors [3], Electro-Migration (EM) problems in copper interconnects [4], breakdown of dielectrics in porous low-k materials [5], etc.

The net result is that it becomes increasingly difficult to guarantee the life time of transistors and wires for new technology nodes, as will be discussed in the remainder of this section. Apart from the reliability mechanisms, transistors and wires are also subject to manufacturing imperfections which lead to static manufacturing time variability. This is also aggravated by novel transistor architectures. The development of FinFETs is a good example. Variability due to random dopant fluctuations can be severely reduced by alleviating or reducing the need of dopant atoms in the channel. But implementing FinFETs in a stable and reliable process requires the controlled and precise manufacturing of very complex three-dimensional structures (fins), which leads to a significant increase in the variability contribution due to line edge roughness in all three dimensions.

NBTI effects [6, 3] in PMOS transistors and (Soft) gate oxide Break-Downs (SBD) in NMOS transistors [7] are becoming two of the most important sources of progressive degradation of electrical properties of devices in DDSM technologies. Thinner equivalent gate oxides, due to dimension scaling, and a deficient supply voltage scaling are leading to higher electrical fields in the oxide interfaces, hence in larger tunneling currents that degrade the electrical properties of the oxide, resulting in electric traps in the interfaces. These traps translate in both NBTI and SBD effects. NBTI appears as a progressive drift of the threshold voltage of the PMOS transistors over time, which can partially be recovered once the negative voltage stress between the gate and the drain/source becomes zero or positive. SBD appears when enough traps align in the gate dielectric. A conducting path is created resulting in "micro" tunneling currents through the gate. After some time the path created will "burn out" leading to an electrical short or Hard Break-Down (HBD) resulting in a catastrophic failure of the transistor. The transition from the initial conducting path to the HBD is not abrupt, the gate leakage current will start to progressively increase long before the HBD actually occurs (Fig. 1). Moreover, changes of the stress conditions due to the application usage of the platform, like activity, and the way this is translated into operating conditions of the devices and wires will also have a major impact on the actual dynamics of the degradation phenomena.

Similar effects are predicted for wires from the 45nm technology node on. Both electro-migration in the metal wires and reliability problems in the dielectrics between them are becoming serious concerns for guaranteeing correct and reliable operation during a specified product lifetime. The ever decreasing widths of the local wires combined with the slowing scaling of the supply voltage lead to an increase in current densities along technology nodes, which is accelerating electro-migration problems not only in aluminum but also in the more robust copper interconnects [4]. The problem is not alleviated by assuming a decreasing fan-out condition which would provide a temporary partial solution

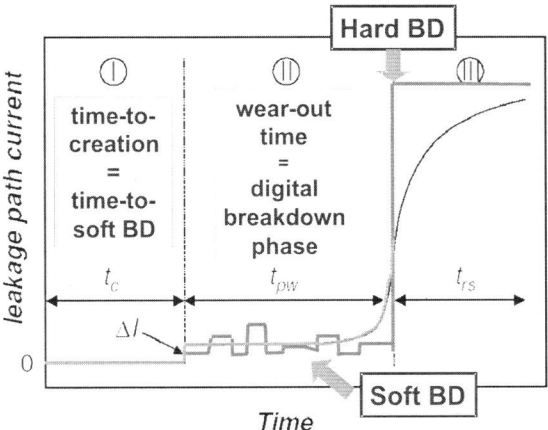

Fig. 1. Wear-out and breakdown model for normal (SiON) and high-k (HfO2) gate oxides [8]

to current densities control. For relatively long local and intermediate interconnects even though the current densities can increase due to large fan-outs, electro-migration is not a considerable problem. System-on-Chip level communication typically has more relaxed constraints on energy consumption per task and performance. Local interconnects, on the other hand, which are used to implement processing elements or local communication between processors and local memories/caches have all the fore-mentioned stringent constraints. Guaranteeing real-time performance and improving density in order to minimize area (die cost) leads to the minimization of the lateral dimensions of the wires [9]. These conditions significantly speed up the electro-migration mechanism in this context.

Similar to SBD effects in transistors, electro-migration is also translated to a progressive degradation of the associated resistance of the wire. The thinner the wire is, the earlier the degradation will start [4] (see Fig. 2). This is aggravated by asymmetries in the printed interconnect features, such as connections between wires and vias. Interfaces between different materials across the conducting path are especially susceptible to electro-migration problems. Also irregularities in the critical dimensions of the interconnects, due to Line Edge Roughness [10] as a consequence of sub-wavelength lithography, will make the whole metal structure far more vulnerable to electro-migration problems. This can lead to uncontrollable (location- and impact-wise), random hot-spots.

A similar case can also be drawn for breakdowns in the dielectrics in the interconnect stack, where the wire pitch is reducing in each new technology node. This leads to reduced thicknesses of the dielectrics between metal wires, while the supply voltage does not reduce at the same pace. As a result the main figure of merit for reliability, Mean Time To Failure (MTTF), drastically reduces [5]

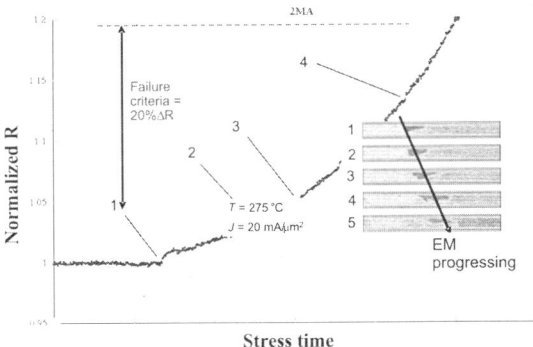

Fig. 2. EM signature in narrow lines (<120 nm line width) [4]

(see Fig. 3). The reason is the combination of the increasing electric fields in active wires due to the insufficient voltage scaling and the introduction of low-k dielectric materials for improving the RC delay of wires based on less electrically robust porous materials. Even when this failure phenomenon manifests itself as catastrophic without an explicit progressive degradation phase, the number of dielectric breaks over time and the time to first break becomes less predictable than earlier. Imperfections of the low-k dielectric material, like granularity of the material grains and/or air gaps, are dramatically increasing the uncertainty on the actual useful life-time of the product.

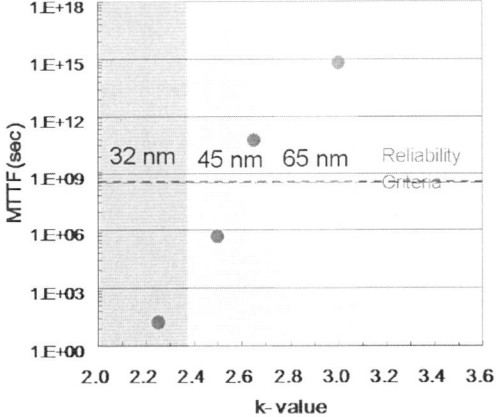

Fig. 3. Reliability targets and projected MTTF in advanced Cu-low-K materials [5]

3   The Impact of Reliability Degradation Mechanisms on the Circuit Level Performance Metrics

For a proper evaluation of the impact that the fore-mentioned reliability problems have in circuit and system design, it is not sufficient to have models representing the mechanism and effect of a particular reliability effect in a single device or interconnect. Not even considering possible interactions with other reliability phenomena is sufficient, e.g. studying the combined impact of NBTI and SBD effects in the behavior of an SRAM cell [11]. The real problems need to be evaluated in the context of the particular circuit where the device/interconnect subject to degradation is situated. The fact that a progressive degradation effect may manifest mildly when looking at each single transistor/wire separately does not provide any information about its impact on the circuit level performance metrics. For instance, oxide breaks manifest themselves as a slight increase in the total gate leakage [12] that may not have strong impact on the transistor current-voltage characteristics [8], since the drain current does not change significantly at the moment the soft oxide breakdown occurs. However, when looking to the interaction that the gate current increase may have with the circuit operation, although small, it can affect the parametric figures of the circuit by affecting the current of another device whose drain is connected to that gate. A typical example where small changes in the gate current of a single transistor can cause major problems at the circuit level are SRAM sense amplifiers or other circuits that work under a common mode rejection mode. Affecting the bias conditions of one of the transistors even slightly may have detrimental effects for the functionality of the circuit. Different types of circuits are much more robust toward breakdowns, for example ring oscillators can tolerate hard breakdowns on several of their transistors before they stop oscillating at the specified frequency [13]. This means that in order to evaluate the impact that reliability degradation mechanisms have on the circuit level performance metrics we need analysis and modeling tools that can take into account the context where the affected transistor/wire is operating in.

In the general case, the gate leakage current of FETs can either impede or favor the charging/discharging process of the output node of a gate leading to longer/shorter delays. In terms of equivalent SBD resistance, previous research has predicted that it is in the order of several hundred kilo-Ohm and above for sub-45nm technologies [14]. Furthermore, the extra leakage contributes directly to the increase of total energy consumption. A lower than nominal voltage swing can be observed at the gate of the output node, due to the soft oxide breakdown induced gate leakage. Such a voltage swing then slows down the downstream logic driven by the defective gate [15]. Delay degradation induced by such a defect has already been observed in simple logic NOR/NAND gates and small data-paths (full adder)[15, 16].

Apart from the standard logic gates, it has recently been shown that SBDs in the NMOS transistors of SRAM components can also bring shifts in their performance. The energy and delay of both sense amplifiers and individual SRAM cells are dramatically affected by having a single SBD in one of their transistors.

A variation of 36% in energy and 22% in delay is reported for the sense amp and a similar variation is reported in the SRAM cell parametric (energy/delay) operation [14]. The amount of drift is mainly due to the impact of soft oxide breakdown on the internal feedback loops of these sub-circuits. Similar to combinational logic, the infected feedback loop can also reduce/increase the delay of the actual component. Such drifts come from the second-order interactions of the gate leakage increase enabled via the circuit topology and a more significant variation in the circuit parametric figures is also expected when the soft gate oxide breakdown starts affecting the first order characteristics of the device. The actual behavior of the associated sub-circuit under SBD effect is more difficult to model than those of logic gates because of these feedback loops.

Moreover, these complex interactions exhibit a multiplicative effect when considered in combination with random intra-die process variability. The time dependent nature of the degradation effects and the uncertainty in the initial parametric figures due to variability lead to time dependent variability that is very difficult to predict and control by countermeasures that are only based on design time analysis and solutions. Given that the breakdown resistance value and location are random in nature [7], it is reasonable to expect a more dramatic impact of this combined effect on the energy and delay of the SRAM in the DDSM era. Figure 4 illustrates the increase in the uncertainty ranges of the sense amplifier performance metrics when a single soft break-down is considered in one of its transistors. The delay and energy consumption ranges increase by more than a factor of two. This additional uncertainty largely prohibits the designers to predict the run-time circuit behavior at design time. Thus it is impossible to steer circuit level optimizations, e.g. timing slacks, device sizing decisions, etc., that make the circuit robust enough to both effects combined.

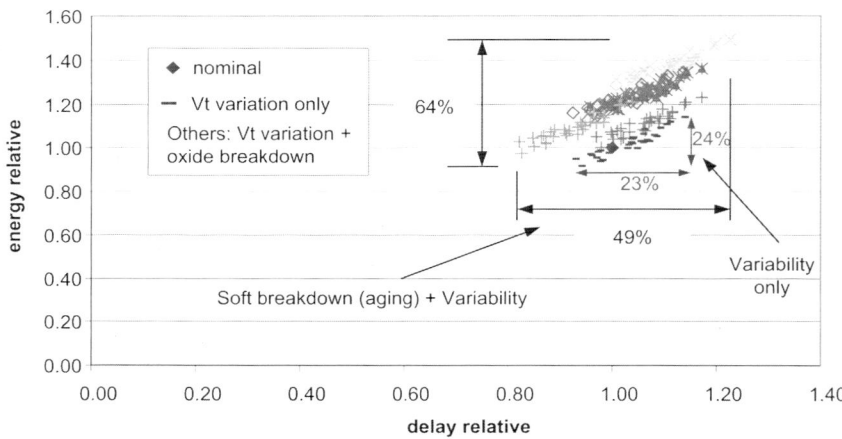

Fig. 4. Impact of variability and gate oxide breakdown in the energy consumption and delay of an SRAM sense amplifier when only one transistor suffers a soft breakdown

In the case of a complete SRAM cell matrix the conclusion is quite different. Only the access delay is greatly affected by SBDs, while the associated access energy is only marginally influenced. The matrix consists of a very large number of cells, where a few of them are accessed in parallel in every memory read or write operation. The matrix static energy consumption is an accumulation of the static energy of all the cells, so a change in the leakage current of one of the transistors in the matrix is unlikely to impact the total static energy consumption significantly. Dynamic energy consumption also exhibits the same trend. The impact on break downs on the dynamic energy consumption of the matrix is also rather small. In the case of delay, however, things are quite different. The break downs have a significant impact on the relative driving strengths of the transistors in the cross-coupled inverter pair, which leads to a significant impact on the delay of reading or writing the activated cells. Transistor level simulations have been carried out to evaluate the impact of soft break downs on the main performance metrics of the SRAM matrix. The difference in dynamic and static energy consumption of the matrix incurred by injecting soft break downs in four individual transistors is around 1%, so it is indeed negligible. The impact of these break downs on delay however are much larger. The standard deviation on the read delay is about 20% of the nominal, while it increases to 60% in the case of the write delay. In addition, the number of soft breakdowns present in the matrix also affects the variation range and distribution of the matrix delay. Such effects can be clearly observed in Fig. 5 which shows the cumulative density functions of the cell matrix delay in the case of one, two or three individual transistors suffering SBDs. The results are obtained via transistor-level simulations of the matrix assuming negligible process variability. The slopes of the cumulative functions indicates the degree of uncertainty, the "slower" the slope the larger the uncertainty and vice versa. Initially no break downs have occurred and the delay of the matrix is completely deterministic. For an increasing number of SBDs it is interesting to note that the delay variation range increases and leads to a more evenly distributed delay over the range. But the mean value of the delay also shifts for a different number of break downs. Moreover, in this case of an SRAM matrix, additional SBDs always increase the mean and the second moment of the delay distribution. The conclusion for this example is that both delay and spread deteriorate for each new SBD suffered. The mechanism behind this is simply due to the increasing interactions between SRAM sub-circuits that have suffered a SBD. For instance, the interaction between a defective SRAM cell and sense amp in the same column during the read operation not only increases the delay variation range, but also leads to a larger uncertainty in delay. Adding the impact of random process variability on delay on top of the fore-mentioned figures gives a perspective on the scale of the real problem. The circuit topology and context are extremely important in determining which circuit metrics will be influenced by degradation mechanisms and which will be unaffected.

Finally, the effect of the application running on the hardware and consequently the bit-level activity that defines the operating voltages of the devices and interconnect is essential to fully characterize the actual impact that the

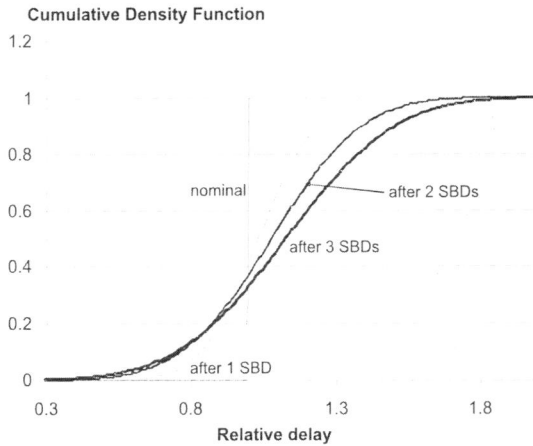

Fig. 5. Cumulative distribution of SRAM matrix delay variation under SBD

reliability effects will have in the time-dependent parametric variations of the system. Trying to characterize this impact at design time becomes extremely difficult, if not impossible in sub-45 nm technologies using existing commercial tools and design flows. Todays worst case analysis and system design paradigms are breaking down in the presence of the increasing dynamism which is present in the modern application in both the multimedia and wireless domains. The way reliability problems appear within the circuit is a rather random process and it depends on the actual operating conditions: time, temperature and stress voltages [7]. This is especially true for large circuit and systems featuring many transistors which can undergo significantly different stress conditions when executing dynamic applications. The actual location of the progressive defect and severity degree is hard to estimate at design time in this case. Moreover due to the varying nature of the stress induced by the application the defect generation rate also becomes very difficult to capture unless this is done at operation time (run-time). These facts simply indicate that innovation in circuit and system level design and analysis has to take place to counteract the impact that progressive parametric degradation mechanisms will have in the actual useful life-time of the system.

For the past decades variations have always existed on critical parameters during the design and operation of electronic systems. The most common such parameters are temperature, activity and other operating conditions. The circuits must always operate within the specified performance constraints for a given range of temperature and humidity conditions. In recent years, variations have also been observed in the electrical parameters, like capacitance, drive current etc., of the transistors and wires due to tolerances during the processing of the wafers. The conventional solution for dealing with these variations is to incorporate worst-case margins so that the circuit will always meet the target constraints under all possible specified conditions. The minimum and maximum value of each varying parameter is characterized and the combinations of these values for all the parameters form the corners of the parameter space which defines the working conditions of the design. Designers typically tune their designs to meet the performance constraints for all the corner-points, this technique is called corner-point analysis.

This technique is still widely used in the industry, but it suffers from a number of disadvantages. The corner points are usually very pessimistic; it is extremely unlikely that all the parameters will have their maximum or minimum values simultaneously. Thus, the design margins required to make the circuits operational under all corner conditions are excessive. Furthermore, the number of parameters affected by time-dependent variability becomes very large. This means that circuit designers will have to deal with parameter spaces of many dimensions and extremely large numbers of corner points. Finally, corner-point analysis techniques cannot handle the impact of intra-die time-dependent variability, which is spatially uncorrelated in nature [17], because the electrical parameters of each transistor would become an additional axis in the parameter space and the complexity would become unmanageable. So similar to the evolution at the system level, also here the worst case design paradigm is breaking down.

The most prominent alternative for corner-point analysis, which is already finding its way into the design flows of the major companies of the consumer electronics segment, is Static Statistical Timing Analysis (SSTA). Instead of just working with the value ranges of each electrical parameter, SSTA works with the statistical distribution of each of the parameters. Standard cell libraries are calibrated in order to correctly reflect the impact of variability on the transistor threshold voltage, beta and other electrical parameters on the delay of the standard cells. Then the delay of the complete circuit is estimated by statistically adding the delays of the critical path standard cells. This opens an entirely new perspective to circuit designers. Instead of blindly trying to achieve functional and parametric compliance in all corner points, they can evaluate the sensitivity of the design margins on the timing yield of the circuit. Thus, designers can trade-off the magnitude of the required design margins against the parametric yield of the circuit in a qualitative manner. Accepting some parametric yield loss can significantly limit the required margins, which is beneficial for energy consumption and area.

Mani et al. [18] have quantified the impact of corner-point analysis and statistical analysis on the power consumption, performance and yield of small logic circuits comprising a few hundred gates for the 130 nm technology node. They have assumed a limited impact of variability on the performance characteristics of the gate, a 25% delay variation in terms of $3\sigma/\mu$, which was reasonable for the 130 nm technology node. In their paper they demonstrate that in order to achieve a yield of 3σ (99.73%) using statistical timing analysis, squeezing the last 5.5% out of the circuit delay to meet the performance constraint incurs a power overhead of about 65% even for a small circuit. The overheads that corner-point analysis incurs, on the other hand, are about 30% larger on average. This illustrates one of the walls that circuit designers have to face due to the increased variability. The larger spreads of the delays due to variability lead to a need to excessively over-design the circuit, so that the nominal or average delay becomes much faster than the target. This headroom between the average and the target delay is there to absorb the spread due to variability. But faster circuits consume more energy, so an implicit energy consumption vs. timing yield trade-off exists for a given performance specification. Furthermore, fundamental limits exist for the maximum speed of circuits. Increasing the transistor sizes, for instance, fails when self-loading exceeds the output load. Further increases in transistor sizes lead to degrade energy consumption and delay.

In the meantime, variability on the electrical characteristics of devices and wires and hence of the circuits is growing in magnitude as technology scales. Moreover it is becoming randomly time-dependent as illustrated in the previous section and verified by the results in Fig. 5 due to the more progressive degradation of the key electrical parameters of devices and wires. As a result the uncertainty region collecting the actual electrical properties of the devices/wire will move randomly in space as time progress. This leads to a new global region of uncertainty resulting from the collection of the local variability "clouds" (see Fig. 6) which becomes far bigger than the corresponding one right after manufacturing. In the conceptual view in Fig. 5, t0 represents manufacturing time and t1,t2 represent moments in time during the product normal operation in the field. The 1 sigma, 2 sigma and 3 sigma contours correspond to iso-yield boundaries. It is clear from the above discussion, that the various degradation mechanisms will force the initial uncertainty cloud to shift in different directions as well as increase in magnitude. The region of uncertainty that is relevant for the designer is not just the initial (t0) cloud, but rather the aggregate area of all the clouds, because the design may be situated in any of these points during its life time. If the total cloud becomes too large, the possibility exists that it will be impossible to design a circuit for a given combination of performance and power budget constraints.

4   Impact of Time-Dependent Variability and Progressive Degradation in System Design

This increase in uncertainty has a very significant impact on system design as well. It effectively means that the system architect should build a system out of components that have unpredictable performance and quality metrics (that cannot even be fully bounded at design time anymore) as well as limited reliability guarantees. Conventional system design optimization techniques include trading-off energy consumption for performance at design-time, where most options are available at the component level. For instance, if a system has to meet a given clock frequency target, memories from a high-speed memory library might be used instead of slower low-power memories to guarantee sufficient timing slack. Typically components that are significantly faster than the given requirement are used in order to guarantee the parametric system target is met with

Energy consumption

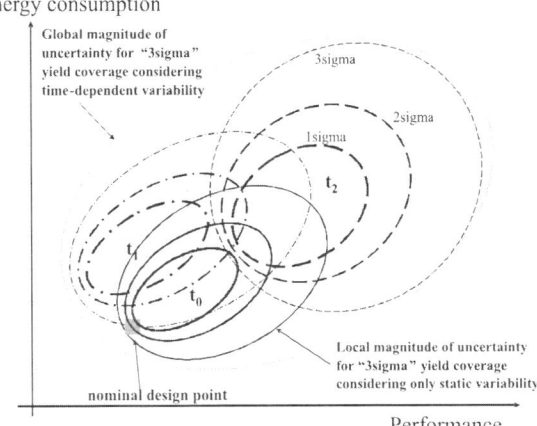

Fig. 6. Evolution of the uncertainty region of the system-level energy consumption and performance

reasonable yield. This is a worst-case margin that is usually added by system designers on top of the worst-case circuit tuning already performed by circuit designers. However, the large performance and energy consumption uncertainty at the component level combined with the requirement for very high yield forces designers at all levels to take increasingly larger safety margins. Stacking all these margins leads to systems that are nominally much faster than required and hence, much more energy hungry and potentially costly as well. It becomes clear that using margins is an acceptable solution only if we can give up on one of the major embedded system requirements (real-time performance, low energy consumption, low cost, high yield). Design margins trade-off energy consumption for performance, redundancy trades off cost for yield, parallelism trades off cost for performance and so on. No solution exists, however, that can optimize all these cost metrics simultaneously.

Furthermore, it is not yet known whether the corner points for each of the varying parameters will be fully characterizable, because they will depend on the detailed operating conditions on each device, like activity and stress conditions on the transistors and wires. Furthermore, these operating conditions heavily depend on the applications that are running on the system and the way they use the system resources. This means that the corner points and the distributions of each parameter, which guide the corner point and the SSTA analysis and optimization techniques respectively, will not be available anymore at design time. The only reasonable way out in the current design flows is to add second order design margins, namely on the place of the corner points to tackle the uncertainty due to time-dependent variability. Putting the fore-mentioned results of the SSTA technique in perspective of this unpredictability of the magnitude of the growing time-dependent variability, we conclude that design-time tuning of the circuit will be impossible for the target constraints of real-time performance, low energy and low cost.

5   Inadequacy of State of the Art Solutions

Even though both variability and reliability mechanisms affect the quality metrics of the same transistors and wires, the communities working on processing and reliability aspects at the technology level are different and usually disjoint. Plenty of literature exists in the process technology community about the sources and impact of variability in devices. At the technology level though little can be done to reduce the magnitude of random intra-die process variability. Random dopant fluctuations, for instance, are an unavoidable side-effect of the shrinking dimensions due to the limited amount of dopant atoms in the channel region of the transistor. Thus this type of variability has to be dealt with by the design community. The reliability community, on the other hand, generally focuses on the impact of the physical breakdown and degradation mechanisms on individual transistors and interconnects in typically small circuits and test structures which are not fully representative of the design reality. The main assumption there is the classical way of reliability lifetime prediction, which is based on extensive accelerated testing and extrapolations toward real operating conditions, design sizes and time scales. But the reliability community typically fails to also take into account the impact of random variability, since few test structures are used and statistics on manufacturing imperfections cannot be extracted with sufficient confidence.

Circuit and system designers have always been confronted with process variability and reliability degradation issues especially in the analog domain. A variety of alternative solutions has been developed in the previous years to deal with them. Good examples of such solutions include the one-time post fabrication tuning and binning technique, adaptive body bias, statistical static timing analysis, asynchronous design styles, architectural error detection and correction techniques and redundancy mechanisms, among others.

Binning has been the most popular technique used in general purpose microprocessors to deal with fabrication process induced inter-die variations. Instead of clocking every chip (of the same design) at the same frequency, the capable frequency of a chip is decided after fabrication with the help of at-speed testing. In parallel, chip-level supply voltage (V_{dd}) and body-bias voltage (V_{bb}) can be adjusted so as to increase the percentage of chips that can meet the design target frequency [19]. As frequency, V_{dd}, and V_{bb} are coarse chip-level controls, this method is not suitable to deal with stochastic intra-die variability, which requires some of control parameter that operates at a much finer granularity level.

Prevailing worst-case design methodologies use best-case and worst-case process corners to predict the impact of intra-die variability and enable potential optimizations. But they also fail in handling the complete problem in a generic manner. Static timing analysis (STA) which computes the critical path delays and hence clock period uses a single worst-case gate delay, which is the result of the most pessimistic corner for delay. As corners move farther and farther apart due to the increasing random intra-die variability component, STA based design incurs significant overheads (in terms of area/delay/energy depending upon the specific design objectives) which could jeopardize the scaling bene-fits. Statistical STA (SSTA) exploits the fact that device parameters and hence gate delays are stochastically distributed. As a result the path delay is much smaller than the sum of worst-case delays due to the averaging effect [20–22] of adding statistical distributions. SSTA calculates path delay distributions and hence the clock period distribution, which allows trade-offs between parametric timing yield and performance. Use of SSTA also improves the efficacy of circuit optimizations, such as circuit sizing under intra-die uncertainty [18]. But it suf-fers from a major drawback: it can only handle sequential or combinational logic circuits comprising standard cells, which is usually only a small part of current embedded system designs.

Razor [23] is a micro-architectural error technique based on dynamic detec-tion and correction of circuit timing errors. The key idea of Razor is to tune the supply voltage by monitoring the error rate during circuit operation, thereby eliminating the need for voltage margins. A Razor flip-flop is introduced that double-samples pipeline stage values, once with a fast clock and again with a time-borrowing delayed clock. A meta-stability-tolerant comparator then vali-dates the latch values sampled with the fast clock. In the event of a timing error, a modified pipeline misspeculation recovery mechanism restores the cor-rect program state. This solution can guarantee correct I/O functional behavior of the processor pipeline. But it works on the principle of error detection and correction, so the timing at the application level cannot be guaranteed because the number of faulty cycles cannot be a priori known. So this is not directly portable to real-time embedded systems.

Asynchronous design styles produce circuit implementations that are inher-ently very robust toward local performance uncertainties [24]. Functionality in terms of correct input/output behavior of the circuit can be easily guaranteed, since no synchronization boundaries exist to create timing violations. Their ma-jor drawback is that their actual performance is completely unpredictable, thus mapping real-time applications on asynchronous circuits is very difficult.

Redundancy has been a popular technique to tackle reliability concerns in the past. Historically designers have been treating reliability degradation mech-anisms as a pure functional concern and hence built reliability support by ex-ploiting one (or some combination) of three forms of redundancy: information, hardware or time [25]. Use of information redundancy, such as parity or error correction codes (ECC), allows detection and/or correction of certain classes of bit errors. Systems achieve hardware redundancy by carrying out the same computation on multiple, independent hardware units at the same time and cor-roborating the redundant results to expose errors. Systems with triple (or higher) redundancy can obtain a correct answer through a majority voting scheme. Time redundancy techniques are based on redundant computation in time, they repeat the same operation multiple times on the same hardware. They mostly target to counteract soft errors, but they cannot handle catastrophic failures in a circuit. All forms of redundancy, however, come with a large associated overhead. Time redundancy incurs a significant delay penalty, which is not acceptable in the domain of real-time performance embedded systems. Hardware redundancy, on the other hand, incurs significant area overheads and does not provide adequate solutions. Time-dependent variability influences both the performance characteristics of processing elements and memories as well as those of communication networks. Thus communication becomes the weak link of the system. Existing redundancy solutions rely on perfect communication between the various degrading blocks in order to find an optimal assignment of tasks to system resources. Moreover, the new degradation mechanisms incur parametric drifts in all the utilized system components, thus they will all degrade uniformly. This makes it impossible to detect which redundant component has a "defect". Finally, existing testing fault models are not appropriate for dealing with the parametric degradations, because they have been developed for catastrophic defects that impact one or a few of the redundant layers [26]. In the case of parametric time-dependent variability all the layers will be affected, thus conventional redundancy solutions cannot be applied. In conclusion, redundancy techniques are only suited to partly deal with functional reliability issues, not with parametric ones.

All the fore-mentioned techniques, however, were developed to tackle the manifestation of variation and degradation mechanisms of past technology nodes. Post-fabrication tuning and binning techniques, for instance, are very successful at recovering dies that suffer from systematic variations, like die-to-die and wafer-to-wafer variations etc. Coarse-grain redundancy mechanisms based on majority voting can easily overcome malfunctions in limited parts of the design, due to failures related to sudden break downs of parts of the design. But the nature of the currently prominent process variability and reliability degradation effects has changed significantly by scaling feature dimensions into the DDSM regime. Systematic process variations are being overshadowed by random spatially-uncorrelated intra-die variability. Binning and adaptive body bias techniques cannot tackle the impact of variability on the quality metrics of the design, because they operate at a very coarse-grain level thus failing to deal with the spatial dynamics of variability. Reliability degradation mechanisms, on the other hand, are shifting from effects causing abrupt failures which are catastrophic for the circuit operation to gradual and graceful degradations of the circuit performance and energy consumption during normal operation. Redundancy mechanisms fail to provide adequate solutions for these new effects, since all the redundant components of the design will also degrade along with the original ones if they are used in parallel, thus providing negligible improvements in the product life-time.

It becomes clear that even though partial solutions for intra-die process variability and reliability issues are being worked out, solutions that can deal with the combined impact of time-dependent variability have not gained attention yet by the research community. On the other hand, both effects manifest themselves as parametric drifts in the timing and the energy consumption of the devices. Their combined impact can also be described as time-dependent variability. For any solution to be adequate, especially for real-time embedded systems, it will have to deal with the run-time temporal shifts in the performance metrics of the devices and circuits.

6 A Paradigm Shift in System Design Solutions

One of the main reasons why the existing solutions are breaking down in the case of time-dependent variability is that they try to tackle both the functional and the parametric issues at the circuit level with clear performance constraints on meeting the target clock period. This means that all the system components are designed so as to be functional and satisfy the frequency performance constraints with minimal performance variations to achieve maximum parametric yield. This forces the designers to design for the worst-case, since all the components should meet the common clock period constraint. In reality, the performance of each system component will follow a statistical distribution if margins are not embedded in its design, see [27] for a case study on on-chip memories. Some components will be faster than the mean performance and some will be slower, due to the nature of the statistics of their performance. This variation is not exploited in state-of-the-art techniques dealing with variability issues at the system level. Instead all the components are designed to have a predictable performance, even though this incurs a significant energy overhead. Meeting the constraints of low energy, low cost and real-time performance for maximum yield will become impossible with the conventional techniques, if the magnitude of uncertainty due to time-dependent variability increases. A paradigm shift will be required both in the design of the circuits and at system level design to overcome these limitations.

Current commercially available design and modeling flows are just starting to incorporate SSTA techniques to incrementally reduce the required design margins. For the transition to technology nodes where time-dependent variability becomes prominent, these flows will have to be extended significantly. Specifically, new statistical techniques will have to be developed to cover two main holes of the existing techniques. The first hole is the lack of dynamic energy calculation in the existing SSTA techniques, currently they can only estimate timing and static energy consumption. Total energy consumption is an extremely important metric for the design of battery-powered embedded systems, even more important than timing in some cases. The second required extension is a move to a higher abstraction level [28]. SSTA today deals with combinational or sequential logic blocks. Systems, however, are heterogeneous in their composition, memories and other IP blocks take up a very significant part of the die. Statistical techniques should move one level higher and they should be able to provide complete modeling of the entire die and an estimation of the timing, dynamic and static energy consumption as well as parametric yield for the complete system. An initial attempt to cover this gap has been outlined in [29]. A Variability and Reliability Aware Modeling (VRAM) framework exhibiting all the fore-mentioned attributes is required, which can be used in parallel to the existing design flow. A potential instance of such a framework can be seen in Fig. 7. It will aid designers in characterizing the impact of random variability and degradation mechanisms on the specific design and evaluate whether the impact on the design performance and quality metrics is severe. Such a framework would enable the quantitative evaluation of the magnitude of the potential problem and supply all the relevant information for designers to decide whether the problem is significant and which solutions are appropriate.

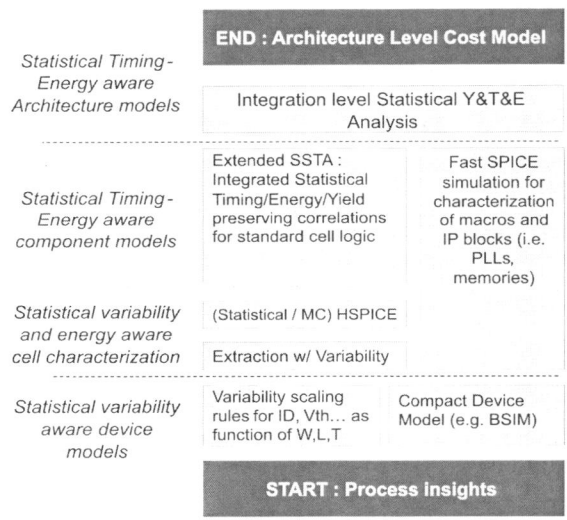

Fig. 7. An instance of a complete modeling flow for propagating variability and reliability information from the technology level to the complete system level.

If the problem is deemed significant enough to require a solution, one of the necessary steps is to separate the functional issues from the parametric issues, like performance and energy consumption. Circuit designers should deal with making circuits that are robust enough to remain functionally correct independently of the degree of time-dependent variability impact, because it may be impossible anyhow to fully characterize that at design time. The previous section has already outlined a number of existing methods for tackling functional issues. Solutions for functional degradations due to reliability based on redundancy and other techniques that enhance robustness are already available. Another example of a circuit level technique to design robust SRAMs cells under variability can be found in [30]. Asynchronous logic is another way of implementing functionally robust circuits against time-dependent variability. The parametric constraints can be ignored at this phase in favor of finding a functional solution for larger uncertainty ranges. This approach relieves the circuit designers of the pressure to meet performance requirements; the target is to design functional circuits under potentially extreme time-dependent variability with minimal overhead in energy consumption and delay. The only additional requirement from the circuit designers is that they should equip their designs with circuit level configuration/tuning parameters, which can trade-off performance for energy consumption at the circuit level, see [31] for an example.

Meeting the performance and energy budget constraints is the responsibility of the system itself. Only when the exact impact of time-dependent variability on the performance of the individual components and the short-term performance constraints are known, can an optimal solution be found. This implies that the actual performance of all the components will have to be measured after fabrication and at regular intervals via in-situ monitors in order to implement the required system observability. In a second step, if the actual performance of some components is lower that the required local timing constraint, the system should be able to influence it via the supplied tuning parameter. A very popular system level tuning parameter in current electronic systems is V_{dd} scaling. By lowering the supply voltage a system or component can decrease its energy dissipation while also reducing its performance and vice versa. But V_{dd} scaling is losing its efficiency due to the reduction of the voltage headroom, thus the required tuning parameters should be designed in the circuits to be more effective. An additional advantage of circuit level parameters is their local scope which is necessary in order to compensate for random variability, as opposed to parameters of global scope like V_{dd} scaling. Such tuning parameters, which we call knobs, provide the necessary controllability over the performance of the individual components and the system overall. The existence of knobs and monitors (K&M) in all, or a few critical, system components along with a simple algorithm for the knob control enables the system to find at run-time the optimal configuration setting for each of the components in order to minimize any given cost function, like timing violations or excessive energy consumption. This eliminates the need for allocating large design time margins so as to make sure that components always meet the most aggressive timing constraints, which is common practice today. Figure 8 illustrates an example architecture which utilizes configurable memories, monitors and an instance of a hardware controller for the tuning of the memories.

If this simple control algorithm does not provide enough range in the timing axis for mitigating the impact of time-dependent variability on performance, a more elaborate solution is needed which involves a more complex control algorithm. Namely the timing constraints can be moved from the level of a clock cycle to the level of application deadlines. Given that the components are designed without unnecessary design margins, their average performance will be faster than the one of components with margins but much more unpredictable. Some components will be faster than the clock frequency and some will be slower. Even though some components will violate the nominal frequency target, the average performance of all the components could still be faster than the target. Thus,

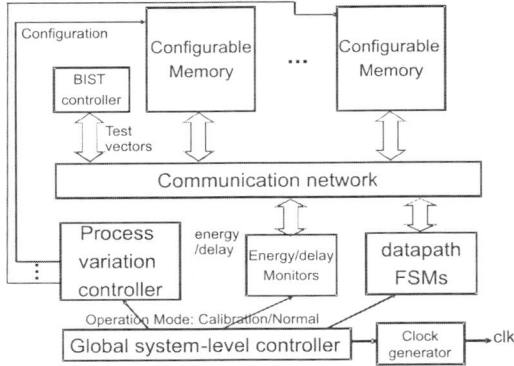

Fig. 8. Instance of a system architecture employing configurable components (memories), monitors for in-situ measurements and a controller for tuning the components.

over a number of cycles the application deadline can still be met, even though some clock level "deadlines" will be violated. This solution does not require system designers to resort to asynchronous logic. The conventional synchronization boundaries can be preserved as long as the clock frequency can be slowly adapted to the speed of the slowest component that is used at each moment in time. This can be achieved via dynamic frequency scaling or fine grain frequency islands, similar to the Globally Asynchronous Locally Synchronous principle. In combination with the use of the knobs that can fine-tune the component performance, a solution that globally meets the application deadline constraints can be achieved.

Energy consumption minimization is equally important to meeting the real-time performance constraints for embedded systems. It is mainly influenced by two factors, time-dependent variability and design margins. Variability introduces side-effects like unnecessary switching overhead and additional standby energy and its impact can only be partially mitigated at the technology level, so the system will have to live with these overhead situations. The second source of additional energy consumption is the design margins themselves. Designers control the magnitude of the margins; separating the functional from the parametric issues will allow the use of smaller margins which will result into more energy efficient system implementations.

A solution method based on the above principles has been outlined in [32] and an implementation in [33]. It is based on the assumption that the performance unpredictability is not completely tackled at the component level. When this unpredictability can be tackled at the circuit level with acceptably small overheads it makes sense to provide circuit solutions. But in many cases the resulting overheads are unacceptably high, especially in energy. In that case the system has to be exposed to the performance unpredictability to enable a reduction of the circuit energy and delay overhead due to the margins, by providing system level solutions for variability. The individual component performance and energy consumption is measured after fabrication by on-chip monitors and relevant component level configuration options are adapted by the system in order to meet the real-time performance requirements of the application with minimal energy and area overhead. In [32] the solution is only activated after processing to increase the initial processing yield. But once it is in place the same approach can be used infrequently, e.g. once every few seconds, to check whether the energy or timing is lower for the alternative path. This is still a reactive approach though and it will not solve all degradation problems in a fully optimal way. But the big advantage is that is it not that difficult to implement in existing design flows. Future work should look at more optimal global paradigm shifts. A further extension of this technique tackling the impact of time-dependent variability in the context of dynamic application has been reported in [34]. It uses the concept of application scenarios to handle the unpredictability coming from the intrinsic dynamism of the application or the user interaction.

In summary, time-dependent variability will require a paradigm shift in the design of electronic systems in order to benefit from the area scaling opportunities offered by technology scaling without excessive energy and performance overheads. A shift toward Technology-Aware Design solutions, which take into account the process imperfections early in the design cycle, will be required to design and fabricate embedded systems that will meet the constraints in all four major cost criteria: energy consumption, real-time performance, area/cost and yield/guaranteed lifetime.

7   Conclusions

Scaling to sub 45 nm technology nodes changes the nature of reliability effects from abrupt functional problems to progressive degradation of the performance characteristics of devices and system components. Process technology can no longer alleviate their impact on the performance and energy consumption at the design level. Moreover, existing design flows cannot evaluate this impact due to the lack of modeling tools, let alone provide adequate solutions. Tackling time-dependent variability will necessitate a paradigm shift for embedded system design in order to meet the power, timing and cost constraints with acceptable yield and life-time guarantees.

Acknowledgments

The authors would like to acknowledge the insightful suggestions and contributions of Guido Groeseneken, Ben Kaczer, Christophe Bruynseraede, Zsolt Tokei, Robin Degraeve, Michele Stucchi, Philippe Roussel, Satya Munaga, Pol Marchal and Bart Dierickx.

References

1. Maex, K., Stucchi, M., Bamal, M., Grossar, E., Dehaene, W., Papanikolaou, A., Miranda, M., Catthoor, F.: Technology aware design and design aware technology. Proc. of Intl. Conf. on Integrated Circuit Design and Technology, 77-81 (2005)
2. Groeseneken, G., Degraeve, R., Kaczer, B., Roussel, R.: Recent trends in reliability assessment of advanced CMOS technologies. Intl. Conf. on Microelectronic Test Structures, 81–88 (2005)
3. Reddy, V., Krishnan, A., Marshall, A., Rodriguez, J., Natarajan, S., Rost, T., Kirshnan, S.: Impact of Negative Bias Temperature Instability on Digital Circuit Reliability. Intl. Reliability Physics Symp., 248–254 (2002)
4. Bruynseraede, C., Tokei, Z., Iacopi, F., Beyer, G., Michelon, J., Maex, K.: The impact of scaling on interconnect reliability. Intl. Reliability Physics Symp., 7–17 (2005)
5. Tokei, Z., Li, Y., Beyer, G.: Reliability challenges for copper low-k dielectrics and copper diffusion barriers. J. of Microelectronics Reliability, 1436–1442 (2005)
6. Schroder, D., Babcock, J.: Negative bias temperature instability: road to cross in deep submicron silicon semiconductor manufacturing. J. Applied Physics, **94**, 1–18 (2003)
7. Stathis, J.: Physical and predictive models of ultrathin oxide reliability in CMOS devices and circuits. IEEE Trans. on Device and Materials Reliability, **1**, 43–59 (2001)
8. Kaczer, B., Degraeve, R., O'Connor, R., Roussel, P., Groeseneken, G.:Implications of progressive wear-out for lifetime extrapolation of ultra-thin SiON films. Intl. Electron Devices Meeting, 713–716 (2004)
9. International Technology Roadmap for Semiconductors, Interconnect chapter, www.itrs.net (2005)
10. Croon, J., Storms, G., Winkelmeier, S., Pollentier, I., Ercken, M., Decoutere, S., Sansen, W., Maes, H.: Line Edge Roughness: Characterisation, Modelling and Impact on Device Behaviour. Intl. Electron Device Meeting, 307-310 (2002)
11. Ramadurai, V., Rohrer, N., Gonzalez, C.: SRAM operational voltage shifts in the presence of gate oxide defects in 90 nm SOI. Intl. Reliability Physics Symp., 270–273 (2006)
12. Kaczer, B., Degraeve, R., Crupi, F., De Keersgieter, A., Groeseneken, G.:"Understanding nMOSFET characteristics after soft breakdown and their dependence on the breakdown location. European Solid-State Device Research, 139–142 (2002)
13. Kaczer, B., Degraeve, R., Rasras, M., De Keersgieter, A., Van de Mieroop, K., Groeseneken, G.: Analysis and modeling of a digital CMOS circuit operation and reliability after gate oxide breakdown: a case study. Microelectronics Reliability, **42**, 555-564 (2002)
14. Wang, H., Miranda, M., Catthoor, F., Dehaene, W.: On the combined impact of soft and medium gate oxide breakdown and process variability on the parametric figures of SRAM components. Intl. Wsh. on Memory Technology, Design and Testing, 71–76 (2006)
15. Carter, J., Ozev, S., Sorin, D.: Circuit-Level Modeling for Concurrent Testing of Operational Defects due to Gate Oxide Breakdown. Design Automation and Test in Europe, 300–305 (2005)
16. Avellan, A., Krautscneider, W.: Impact of soft and hard breakdown on analog and digital circuits. IEEE Trans. on Device and Materials Reliability, **4**, 676–680 (2004)

17. Najm, F.: On the need for statistical timing analysis. Design Automation Conf., 764–765 (2005)
18. Mani, M., Orshansky, M.: A new statistical optimization algorithm for gate sizing. Intl. Conf. on Computer Design, 272–277 (2004)
19. Tschanz, J., Kao, J., Narendra, S., Nair, R., Antoniadis, D., Chandrakasan, A., De, V.: Adaptive body bias for reducing impacts of die-to-die and within-die parameter variations on microprocessor frequency and leakage. IEEE Journal of Solid-State Circuits, **37**, 1396–1402 (2002)
20. Viswewariah, C.: Death, taxes and failing chips. Design Automation Conference, 343–347 (2003)
21. Keutzer, K., Orshansky, M.: From blind certainty to informed uncertainty. Wsh. on Timing Issues in the Specification and Synthesis of Digital Systems (TAU), 37–41 (2002)
22. Kang, K., Paul, B., Roy, K.: Statistical timing analysis using levelized covariance propagation. Design Automation and Test in Europe, 764–769 (2005)
23. Austin, T., Blaauw, D., Mudge, T., Flautner, K.: Making typical silicon matter with Razor. IEEE Computer, **37**, 57–65 (2004)
24. Sparso, J., Furber, S.: Principles of Asynchronous Circuit Design: A Systems Perspective. Kluwer Academic Publishers (2001)
25. Iyer, R., Nakka, N., Kalbarczyk, Z., Mitra, S.: Recent advances and new avenues in hardware-level reliability support. IEEE Micro, **25**, 18–29 (2005)
26. Brahme, D., Abraham, J.: Functional testing of microprocessors, IEEE Trans. on Computers, **C-33**, 475–485 (1984)
27. Wang, H., Miranda, M., Dehaene, W., Catthoor, F., Maex, K.: Impact of deep submicron (DSM) process variation effects in SRAM design. Design Automation and Test in Europe, 914–919 (2005)
28. Blaauw, D., Chopra, K.: CAD tools for variation tolerance. Design Automation Conference, 766 (2005)
29. Papanikolaou, A., Grabner, T., Miranda, M., Roussel, P., Catthoor, F.: Yield Prediction for Architecture Exploration in Nanometer Technology Nodes: A Model and Case Study for Memory Organizations. Intl. Symp. on HW/SW Co-design and System Synthesis, 253–258 (2006)
30. Grossar, E., Stucchi, M., Maex, K., Dehaene, W.: Statistically aware SRAM memory array design. Intl. Symposium on Quality Electronic Design, 25-30 (2006)
31. Wang, H., Miranda, M., Papanikolaou, A., Catthoor, F., Dehaene, W.: Variable tapered Pareto buffer design and implementation allowing run-time configuration for low power embedded SRAMs. IEEE Trans. on VLSI Systems, **13**, 1127–1135 (2005)
32. Papanikolaou, A., Lobmaier, F.., Wang, H., Miranda, M., Catthoor, F.: A system-level methodology for fully compensating process variability impact of memory organizations in periodic applications. Intl. Symp. on HW/SW Co-design and System Synthesis, 117–122 (2005)
33. Papanikolaou, A., Starzer, F., Lobmaier, F., Miranda, M., Catthoor, F., Huemer, M.: A system architecture case study for efficient calibration of memory organizations under process variability. Wsh. on Application-specific Processors, 42–49 (2005)
34. Sanz, C., Papanikolaou, A., Miranda, M., Prieto, M., Catthoor, F.: System-level process variability compensation on memory organizations of dynamic applications: a case study. Intl. Symp. On Quality Electronic Design, 376–382 (2006)

Soft Error Resilient System Design through Error Correction

Subhasish Mitra[*], Ming Zhang[+], Norbert Seifert[+], TM Mak[+], Kee Sup Kim[+]

[*]Stanford University
[+]Intel Corporation

Abstract. This paper presents an overview of the Built-In Soft Error Resilience (BISER) technique for correcting soft errors in latches, flip-flops and combinational logic. The BISER technique enables more than an order of magnitude reduction in chip-level soft error rate with minimal area impact, 7-11% chip-level power impact, and 1-5% performance impact (depending on whether combinational logic error correction is implemented or not). In comparison, several classical error-detection techniques introduce 40-100% power, performance and area overheads, and require significant efforts in designing and validating corresponding recovery mechanisms. Design trade-offs associated with the BISER technique and other existing soft error protection techniques are also analyzed.

1 Who Cares about Soft Errors?

Soft errors are radiation-induced transient errors caused by neutrons generated from cosmic rays and alpha particles from packaging material. Traditionally, soft errors were only a major concern for space applications. That scenario has changed. Terrestrial radiation has been a growing concern, and many designs today implement extensive error detection and correction by way of Error Correcting Codes (ECC) mainly for on-chip SRAMs. However, memory protection alone is not enough for designs in sub-65nm technologies. Most future designs targeting enterprise computing and communication applications require soft error protection of latches and flip-flops, in addition to on-chip SRAMs. While combinational logic protection may not be an immediate necessity, it may eventually be required as more and more transistors are integrated in future technologies. There are multiple ways to minimize system-level soft error rate, applied at various levels of design hierarchy and manufacturing process.

The soft error rate of a design is generally quantified in terms of Failure-in-time, or FIT, where 1 FIT corresponds to one error per billion device hours. According to recent data discussed at the 2006 SELSE workshop (2006 IEEE System Effects of Logic Soft Errors Workshop, www.selse.org), a typical value for latch soft error rate may be assumed to be 10^{-3} FIT. Note that, there is a lot of variance in latch soft error rates depending on

Please use the following format when citing this chapter:

Mitra, S., Zhang, M., Seifert, N., Mak, T.M. and Kim, K.S., 2007, in IFIP International Federation for Information Processing, Volume 249, VLSI-SoC: Research Trends in VLSI and Systems on Chip, eds. De Micheli, G., Mir, S., Reis, R., (Boston: Springer), pp. 143–156.

specific latch designs. Assuming that a design contains 1 million flip-flops (and each flip-flop consists of two latches), the contribution of all flip-flops to the overall soft error rate of the design can be conservatively estimated as 1,000 FITs. In this estimate, a 50% latch timing vulnerability factor (TVF) [Ngyuen 03, Seifert 04] is assumed based on the fact that a latch is vulnerable to soft errors when it holds a logic value (i.e., when its clock input is 0).

Soft error rates of 1,000 FITs may not sound too high. However, it is not uncommon for enterprise systems to contain between 500 – 20,000 processors. For the 500 processor system, the system-level soft error rate contribution of the flip-flops will be 500,000 FITs (if our previously discussed design is a processor). This means, roughly once every 3 months some flip-flop in the system will be erroneous. For a system with 20,000 processors, the system-level soft error rate contributions of flip-flops will be 20 Million FITs – i.e., roughly once every 2 days there will be an error in some flip-flop of the system.

Fortunately, some soft errors do not have any impact on system operation. For example, an error in a flip-flop whose output is AND-ed with another signal with logic value 0 has no effect on the system. As another example, an error in an operand of a speculatively executed instruction which is finally not committed (and becomes a dead instruction) does not impact system operation. However, a significant percentage of errors in flip-flops can result in data corruption without being detected by the system or the user. As a result, system data integrity is compromised. This situation is referred to as *Silent Data Corruption (SDC)*, and is of great concern. Depending on the design and the application, between 10-40% of soft errors can result in SDC [Mukherjee 03, Nguyen 03, Wang 04]. Imagine the significance of SDC caused by a 1 to 0 bit flip in the most significant bit of the register storing the balance of a bank account.

Suppose that we optimistically assume that only 10% of soft errors cause system-level SDC. Continuing our previous analysis, for a 500-processor system, flip-flops will contribute to SDC roughly once in 30 months. For a 20,000-processor system, the latch contribution to system-level SDC is roughly once every 20 days. These numbers are unacceptable for enterprise system installations such as banks and stock markets. That is why future designs will require adequate protection to prevent such unacceptable situations.

SDC protection in terms of error detection alone is not enough. Suppose that we have a perfect way to detect all the soft errors that can potentially cause SDC. Once an error is detected, the system must recover from the detected error. If there is no user transparent way to recover it, it results into the so called Detected but Uncorrected Errors (DUE). Depending on how recovery is implemented, a part or the entire system may be down. (It is possible to implement efficient recovery in a transparent way without having to bring the entire system down [Spainhower 99].) Downtimes are very expensive in the order of $10K to $10M per hour [Hennessy 02]. Hence, it is not enough to simply employ error detection to prevent silent data corruption – it is absolutely necessary to ensure that system downtime is also minimized.

2   Soft Error Scaling Trend

The importance of soft error protection techniques is best understood by analyzing radiation-induced soft error rate trends for SRAM and logic over technology generations. Figure 1 shows the scaling trend of the soft error rate per SRAM memory cell for Intel designs. Alpha-particle and neutron induced soft error rates both show a clear decrease over the last two generations. This trend is consistent with what TI has also observed [Baumann 05]. Since SRAMs are typically protected by ECC for several reasons (soft errors, infant mortality, etc.), this trend does not have a major impact on most system designs targeting applications requiring high data integrity and availability.

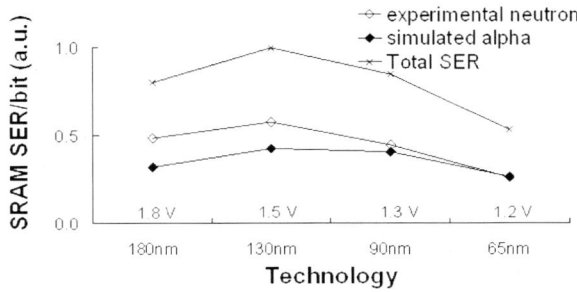

Figure 1. Technology trend of per bit SRAM soft error rates from Intel [Seifert 06].

In contrast to SRAM soft error rates, Baumann of TI [Baumann 05] has observed a steep increase in per latch soft error rate with technology scaling. Intel, on the other hand, has observed a relatively flat trend of per latch soft error rates for the last three generations (Fig. 2). In Fig. 2, the soft error rates of 20-30 most frequently used latches from Intel technology libraries elements are summarized (plotted are the mean and standard deviation). The soft error rate of an actual product depends on the use of specific kinds of latches from the technology library. Soft error rates of various latches in the same technology library can vary by more than an order of magnitude [Seifert 06].

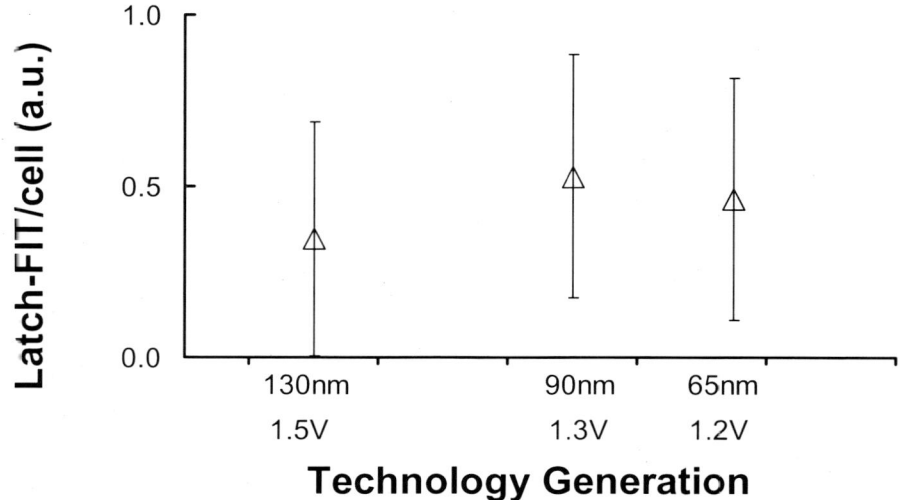

Figure 2. Latch per device relative error rates. Error bars indicate standard deviation within population of 20-30 selected library elements [Seifert 06].

Even if the soft error rate of a single latch or a single SRAM cell stays constant or increases over technology generations, chip-level soft error rates will increase significantly with technology scaling because of increased integration per constant area. We emphasize another soft error rate scaling trend that may be very important for future technology generations. Neutrons do not directly ionize Si but generate electron hole pairs via secondary ions created in neutron – Si spallation reactions. If those secondary ions generate sufficient charge over a region larger than a device, more than one single device maybe affected, creating a so-called multi-bit upsets (MBU). We call this phenomenon charge sharing and it can affect different devices or more than one node in one device. Figure 3 underlines that charge sharing among different SRAM cells is exponentially increasing with process scaling. This trend is expected to grow and we may not be able to ignore the effects of charge sharing in future designs, in particular for radiation hardened designs that are known to be immune to single node upsets only.

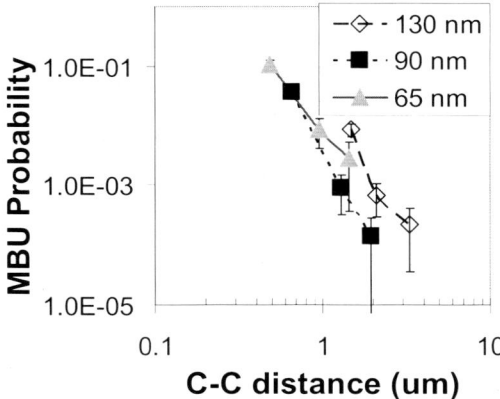

Figure 3. SRAM multiple bit upset probabilities plotted as a function of cell pitch [Seifert 06]. MBU probability is defined as the ratio of the number of MBUs to single bit upsets (SBUs).

3   How to Protect Systems from Soft Errors?

By now the need for logic soft error protection should be clear. The question is how future systems should be protected from soft errors. We will focus on *logic soft error protection*: soft errors in latches, flip-flops and combinational logic. Soft errors in SRAMs are protected using parity or Error Correcting Codes with interleaving (there are some open issues involving efficient error protection of small SRAM arrays and register-files).

Several techniques for logic soft error protection are available in the literature. Each of these techniques has its own advantages and disadvantages. The purpose of this section is to put together a set of metrics that can help distinguish these techniques and understand their pros and cons from an overall system design perspective. We hope that these metrics will help designers understand trade-offs associated with the adoption of one protection technique over another. The metrics are:

- **SDC reduction:** A technique that reduces silent data corruption by a small amount (e.g., by 50%) may be useful in helping a specific design meet its soft error rate goals, but is not scalable with increased integration in future technologies.

- **DUE reduction:** SDC reduction techniques can significantly increase DUEs. Consider a situation where every flip-flop in a design is checked for errors and recovery actions are initiated based upon types of errors detected. All errors that can cause SDC are detected for most practical purposes (double errors may not be detected). However, this approach can significantly increase DUEs. Any error in any flip-flop manifests as a DUE even though only a portion of these errors will actually cause SDC.

- **Cost:** It is extremely important to understand power, performance and area penalties associated protection techniques.

- **Recovery mechanism design and validation effort:** Designing proper error recovery mechanisms and validating them are non-trivial tasks. Costs associated with the design and validation of recovery mechanisms can limit the advantages associated with soft error protection techniques.

- **Configurability:** Soft error protection in future technologies will be significantly impacted by the industry trend to reuse the same design for multiple applications with a wide range of power, performance and reliability requirements. For example, the use of a specific protection technique may incur acceptable power overhead for an application that requires soft error protection; however, the incurred power overhead may be excessive for another application that intends to reuse the same core, but doesn't require soft error protection. One option is to build in two operation modes – an *error resilient mode* in which the protection mechanisms are turned on, and an *economy mode* when the protection mechanisms are turned off reducing the power overhead.

- **Applicability:** Several soft error protection techniques are optimized for specific applications such as processors, signal processing applications, etc. While such techniques are very useful, they have limited applicability for many designs.

- **Flip-flop and combinational logic protection:** It is desirable for protection techniques to address soft errors in both flip-flops and combinational logic using the same soft error protection technique. Otherwise, separate protection techniques for flip-flops and combinational logic introduce additional penalties and design complexity.

4   Built-In Soft Error Resilience (BISER) for Logic Soft Error Correction

We first illustrate the BISER technique for latch-based designs. We will also discuss the use of BISER for flip-flop based designs. Soft errors in latches are corrected using a C-element as shown in Fig. 4 [Mitra 05a, Mitra 05b]. During normal operation, when the clock signal Clock = 1, the latch input is strongly driven by the combinational logic and the latch is not susceptible to soft errors. This is illustrated in the Timing Vulnerability Factor discussion [Nguyen 03, Seifert 04]. When Clock = 0, C-OUT already has the correct value –any soft error in either latch will result in a situation where the logic value on A will not agree with B. As a result, the error will not propagate to C-OUT and the correct logic value will be held at C-OUT by the keeper. The cost associated with the redundant latch is minimized by the reusing on-chip resources such as scan or scanout for multiple functions at various stages of manufacturing and field use [Mitra 05a, Mitra 05b].

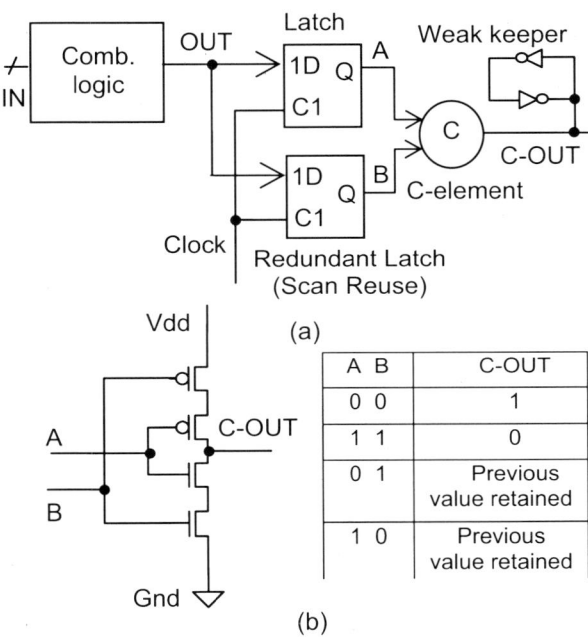

Figure 4. Latch error correction using C-element: (a): Overall technique; (b) C-element.

Extensive simulations in a sub-90nm process technology using a state-of-the-art simulation tool validated by radiation experiments [Nguyen 03] show that the design in Fig. 4 can achieve more than 20-fold reduction in the soft error rate compared to that of an unprotected latch. Note that, a soft error in the keeper does not have a major effect because the C-element output will be strongly driven by the latch contents assuming single error.

Fault injection simulations have been conducted on an Alpha-like microprocessor to evaluate the system-level effectiveness of the BISER technique for latch error correction. The results show that the BISER technique improves system-level soft error rate by 10 times over an unprotected design with negligible area or performance penalty and 7-11% power penalty [Zhang 06].

Soft errors in combinational logic can be corrected using two techniques – Error Correction using Duplication, and Error Correction using Time-Shifted Outputs. Figure 5 shows the soft error correction technique using duplication. Instead of comparing the contents of the latches storing duplicated outputs, we insert a C-element. This technique results in significant reduction (> 60-fold) in combinational logic soft error rate [Mitra 06]. Moreover, this technique also corrects soft errors in latches when Clock = 0. However, there can be significant cost – power and area costs of combinational logic duplication. The Error Correction using Time Shifted Outputs technique, described next, doesn't require combinational logic duplication, but imposes additional performance penalty.

The Time Shifted Outputs technique for error correction is shown in Fig. 6. This technique takes advantage of the fact that soft errors in combinational logic manifest as glitches. Instead of duplicating combinational logic, we sample the combinational logic output (OUT3), and a delayed version of OUT3 called OUT4. In Fig. 6, OUT3 is delayed by τ time units to obtain OUT4. The clock must be slowed down by τ units compared to Fig. 5. The latch outputs are connected to a C-element. The major advantage of the Error Correction using Time Shifted Outputs technique is that the power and area penalties incurred by the duplication scheme are minimized. Note that, τ is a design parameter that can be tuned based on the reliability requirement. Moreover, this technique also corrects soft errors in latches when Clock = 0. Simulation results in [Mitra 06] show that this technique can reduce combinational logic soft error rate by more than an order of magnitude when $\tau = 21$ps. Note that the incremental power penalty of protecting combinational logic using the Time-shifted outputs technique over latch error correction is very little – less than approximately 7% of the power penalty for latch error correction.

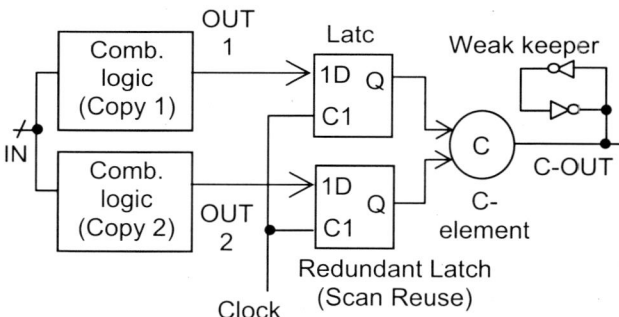

Figure 5. Combinational Logic Soft Error Correction using Duplication.

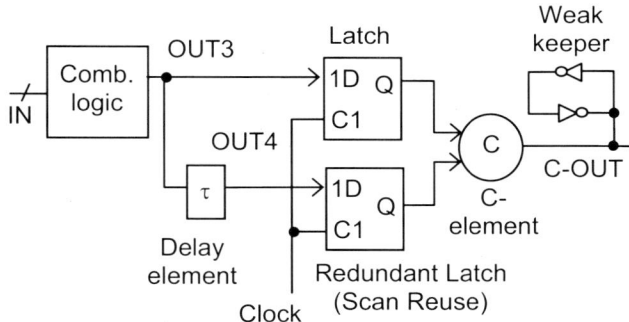

Figure 6. Combinational Logic Soft Error Correction using Time Shifted Outputs.

While the BISER technique has been illustrated for latch-based designs, it is also applicable for flip-flop based designs. Figure 7 shows flip-flop designs for the BISER techniques discussed earlier. Depending on whether duplication or time-shifted-outputs technique is used for combinational logic soft error correction, IN2 in Fig. 4b will be connected to the duplicated logic output (Fig. 5) or the delay element output (Fig. 6), respectively.

5   Comparison of Soft Error Protection Techniques

Table 1 presents a comparative analysis of the trade-offs associated with major soft error protection techniques, in terms of power, performance and area overheads, and the amount of soft error protection that can be obtained. The focus is on latches and flip-flops since they require immediate attention. The protection techniques include: (1) BISER technique; (2) selective node engineering technique, which increases the capacitances of selective nodes of a circuit [Karnik 02]; (3) transistor sizing technique [Zhou 06]; (4) circuit hardening [Calin 96]; and, (5) classical hardware and time redundancy fault-tolerance techniques [Bartlett 04, Mukherjee 02, Oh 02a, 02b, 02c, Saxena 00].

The circuit-level comparison between BISER and circuit hardening techniques is conducted by a unified timing and power characterization methodology [Zhang 06]. While optimizing the various flip-flop designs, the objective is to match the timing parameter, D-to-Q delay. Several assumptions are made during the power measurement of all flip-flops: (1) the data activity factor (average number of output transitions per clock cycle) is 0.25; (2) low-to-high and high-to-low data transitions are equally likely. The cell layout areas are estimated by an internal tool at Intel, with a worst case error of 5% compared to real layouts. The SERs are obtained from an internal simulator at Intel.

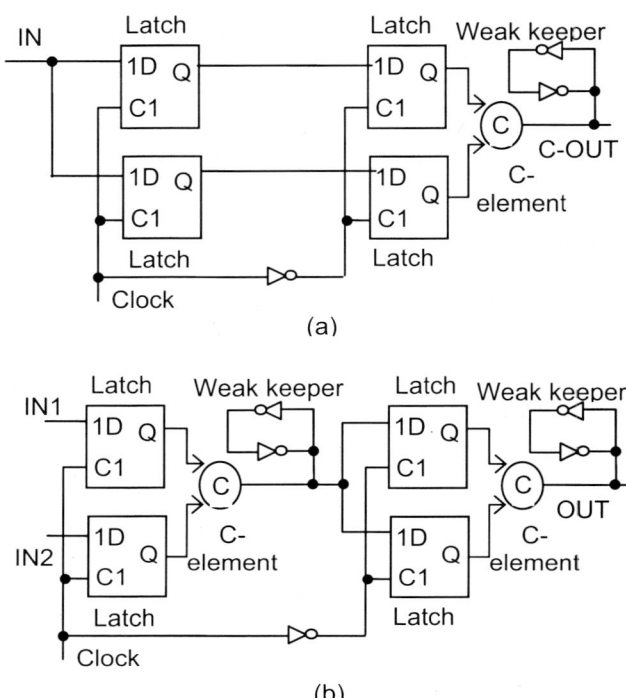

Figure 7. BISER for flip-flop based designs: (a): Flip-flop design for correcting soft errors in flip-flops; (b) Flip-flop design for correcting soft errors in flip-flops and combinational logic.

The selective node engineering technique, which increases the capacitances of selective nodes of a circuit, is an effective approach for designs requiring 30-50% undetected soft error rate reduction. For circuit hardening and BISER techniques, power overheads are derived based on an Alpha processor model with 10-fold chip-level soft error rate reduction [Zhang 06]. The power and area overheads are significantly lower for the BISER technique because it reuses already existent design-for-testability and debug resources. Moreover, the BISER technique allows insertion of an economy mode which enables reuse of the same core design for various applications with soft error protection and power trade-offs.

For the BISER technique, the power overhead is between 7-11%. In comparison, hardware duplication and time redundancy techniques such as multi-threading for error detection and Software Implemented Hardware Fault Tolerance (SIHFT) have very significant power overheads. For chip-level duplication, the power overhead is expected to be greater than 100%. For more fine-grained duplication (e.g., [Spainhower 99]), the power overhead is lower. (We estimated the power overhead to be similar to area overhead in the absence of published data). These numbers are greater than even a worst-case scenario in which all flip-flops (rather than the subset of important flip-flops) are protected with BISER resulting in 14-22% power overhead. Moreover, time redundancy techniques have very significant performance overheads (40-200%) [Mukherjee 02, Oh 02a], and are mainly applicable for designs with well-defined architectures such as microprocessors.

Tables 1 and 2 imply that the BISER technique is most cost-effective for soft error protection. One major advantage of the BISER based error blocking technique is that it doesn't require any error recovery mechanisms and does not incur significant costs associated with the design and validation of recovery mechanisms.

6   Conclusion

The BISER technique is an efficient and practical way to design systems with built-in soft error correction. Comparative analysis with existing techniques demonstrates that the BISER technique combines the major benefits of circuit-level error correction and architectural techniques such as time redundancy and error detection, while avoiding their drawbacks. This is possible because the characteristics of soft errors are utilized by the BISER technique instead of general error models used by techniques such as duplication. This may limit the use of the BISER technique since all error sources may not have characteristics similar to radiation-induced soft errors.

Acknowledgment

We thank K. Ganesh, V. Zia, P. Shipley, J. Yang, S. Walstra, A. Vo, and J. Maiz from Intel Corporation for discussion and assistance during the course of this research. Prof. Subhasish Mitra is partially supported by DARPA/MARCO Gigascale Systems Research Center (GSRC).

Table 1. Comparative analysis of various soft error protection techniques: (a) Quantitative analysis; (b) Qualitative Analysis.

(a)

	BISER [Mitra 05a, 05b, 06, Zhang 06]	Transistor sizing [Karnik 02, Zhou 06]	Circuit hardening [Calin 96]	Hardware duplication [Bartlett 04]	Time redundancy [Mukherjee 02, Oh 02a, 02b, 02c, Saxena 00]
SDC reduction	**Latch: 20X Comb. Logic: 12-64X**	1.5X	Latch: 20X Comb. Logic: None	Almost all	Almost all
DUE reduction	**Latch: 20X Comb. Logic: 12-64X**	1.5X	Latch: 20X Comb. Logic: None	Increased DUE	Increase SUE
Power penalty (resilient mode)	**7–11%**	3%	12–18%	40–100%	> 40%
Power penalty (economy mode)	**1.5%**	3%	12–18%	Very little	Very little
Speed penalty	**Latch correction: 0–1% Comb. Logic correction: ~ 5%**	0–10.4%	0–1%	Very small	50%
Area penalty	**Die size increase not expected**	Die size increase not expected	Die size increase not expected	40–100%	Die size increase not expected
Recovery design & validation efforts	**None**	None	None	Significant	Significant
Configurability	**Yes**	No	No	Yes	Yes
Applicability	**General**	General	General	General	Processor designs
Flip-flop and comb. Logic protection	**Both**	Both	Latches & flip-flops only	Both	Both

(b)

	BISER [Mitra 05a, 05b, 06, Zhang 06]	Transistor sizing [Karnik 02, Zhou 06]	Circuit hardening [Calin 96]	Hardware duplication [Bartlett 04]	Time redundancy [Mukherjee 02, Oh 02a, 02b, 02c, Saxena 00]
SDC reduction	**Latch: A Comb. Logic: A**	C	Latch: A Comb. Logic: F	A+	A+
DUE reduction	**Latch: A Comb. Logic: A**	C	Latch: A Comb. Logic: F	D	D
Power penalty (resilient mode)	**B**	A	B	D	C
Power overhead (economy mode)	**A-**	B	C	A	A+
Speed penalty	**Latch correction: A Comb. Logic correction: B**	B	A	A	D
Area penalty	**A**	A	A	C	A
Recovery design & validation efforts	**A+**	A+	A+	D	D
Configurability	**A+**	D	D	A+	A+
Applicability	**A+**	A+	A+	A+	C
Flip-flop and comb. Logic protection	**A+**	A+	D	A+	A+

References

[Bartlett 04] Bartlett, W., and L. Spainhower, "Commercial Fault Tolerance: A Tale of Two Systems," *IEEE Trans. Dependable and Secure Computing*, Vol. 1, No. 1, pp. 87–96, 2004.

[Baumann 05] R.C. Baumann, "Radiation-induced soft errors in advanced semiconductor technologies", *IEEE Transactions on Device and Materials Reliability*, Vol. 5, No. 3, pp. 305–316, 316, 2005.

[Calin 96] Calin, T., M. Nicolaidis, and R. Velaco, "Upset Hardened Memory Design for Submicron CMOS Technology," *IEEE Trans. Nucl. Sci.*, Vol. 43, pp. 2874–2878, Dec. 1996.

[Hazucha 00] P. Hazucha, C. Svensson, "Impact of CMOS technology scaling on the atmospheric neutron soft error rate", *IEEE Transactions on Nuclear Science*, Vol. 47, No. 6, pp. 2586–2594, 2000.

[Hennessy 02] Hennessy J., and D. Patterson, *Computer Architecture: A Quantitative Approach*, 3rd ed., Morgan Kaufmann, 2002.

[Karnik 02] T. Karnik, *et al.*, "Selective Node Engineering for Chip-level Soft Error Rate Improvement," *Proc. VLSI Circuits Symp.*, pp. 204–205, 2002.

[Mitra 05a] Mitra, S., N. Seifert, M. Zhang, Q. Shi and K.S. Kim, "Robust System Design with Built-In Soft Error Resilience," *IEEE Computer*, Vol. 38, No. 2, pp. 43–52, Feb. 2005.

[Mitra 05b] Mitra, S., M. Zhang, T.M. Mak, N. Seifert, V. Zia and K.S. Kim, "Logic Soft Errors: A Major Barrier to Robust Platform Design," *Proc. Intl. Test Conf.*, 2005.

[Mitra 06] Mitra, S., M. Zhang, N. Seifert, B. Gill, S. Waqas and K.S. Kim, "Combinational Logic Soft Error Correction," *Proc. Intl. Test Conf.*, 2006, to appear.

[Mukherjee 02] Mukherjee, S., M. Kontz, and S. Reinhardt, "Detailed Design and Evaluation of Redundant Multithreading Alternatives," *Proc. Intl. Symp. Computer Architecture*, 2002.

[Mukherjee 03] Mukherjee S., *et al.*, "A Systematic Methodology to Compute the Architectural Vulnerability Factors for a High-Performance Microprocessor," MICRO, 2003.

[Nguyen 03] Nguyen, H.T., and Y. Yagil, "A Systematic Approach to SER Estimation and Solutions", *Proc. Intl. Reliability Physics Symp.*, pp. 60–70, 2003.

[Oh 02a] Oh, N., P.P. Shirvani and E.J. McCluskey, "Error Detection by Duplicated Instructions in Super-Scalar Processors," *IEEE Trans. Reliability*, Vol. 51, No. 1, pp. 63–75, March 2002.

[Oh 02b] Oh, N., P.P. Shirvani and E.J. McCluskey, "Control-Flow Checking by Software Signatures," *IEEE Trans. Reliability*, Vol. 51, No. 1, pp. 111–122, March 2002.

[Oh 02c] Oh, N., S. Mitra and E.J. McCluskey, "ED4I: Error Detection by Diverse Data and Duplicated Instructions," *IEEE Trans. Computers, Special Issue on Fault-Tolerant Embedded Systems*, Vol. 51, No. 2, pp. 180–199, Feb. 2002.

[Saxena 00] Saxena, N.R., S. Fernandez Gomez, W.J. Huang, S. Mitra, S.Y. Yu and E.J. McCluskey, "Dependable Computing and On-line Testing in Adaptive and Reconfigurable Systems," *IEEE Design and Test of Computers*, pp. 29–41, Jan-Mar 2000.

[Seifert 04] Seifert N., and N. Tam, "Timing Vulnerability Factors of Sequentials", *IEEE Trans. Device and Materials Reliability*, Vol. 4, No. 3, pp. 516–522, September 2004.

[Seifert 06] N. Seifert, *et al.*, "Radiation-induced Soft Error Rates of Advanced CMOS Bulk Devices," *IEEE International Reliability Physics Symposium*, pp. 217–225, 2006.

[Spainhower 99] Spainhower, L., and T.A. Gregg, "S/390 Parallel Enterprise Server G5 Fault Tolerance," *IBM Journal Res. and Dev.*, Vol. 43, pp. 863–873, Sept./Nov., 1999.

[Walstra 05] S.V. Walstra and Changhong Dai, "Circuit-level modeling of soft errors in integrated integrated circuits", *IEEE Transactions on Device and Materials Reliability*, Vol. 5, No. 3, 2005, pp. 358–364.

[Wang 04] N. Wang, *et al.*, "Characterizing the Effects of Transient Faults on a High-Performance Processor Pipeline", *Intl. Conf. Dependable Systems and Networks*, pp. 61–70, 2004.

[Zhang 06] M. Zhang, *et al.*, "Sequential Element Design with Built-In Soft Error Resilience," *IEEE Trans. VLSI*, Vol. 14, No. 12, pp. 1368–1378, 2006.

[Zhou 06] Q. Zhou, and K. Mohanram, "Gate sizing to radiation harden combinational logic," *IEEE Trans. CAD*, Vol. 25, No. 1, pp. 155–166, Jan. 2006.

Library Compatible
Variational Delay Computation

Luis Guerra e Silva[1], Zhenhai Zhu[2], Joel R. Phillips[2], and L. Miguel Silveira[1]

[1] Cadence Laboratories / INESC-ID
IST / Tech. U. Lisbon
Lisbon, Portugal
{lgs,lms}@inesc-id.pt
[2] Cadence Berkeley Laboratories
Cadence Design Systems
San Jose, CA 95134, U.S.A.
{zhzhu,jrp}@cadence.com

Abstract. With technology steadily progressing into nanometer dimensions, precise control over all aspects of the fabrication process becomes an area of increasing concern. Process variations have immediate impact on circuit performance and behavior and standard design and signoff methodologies have to account for such variability. In this context, timing verification, already a challenging task due to the sheer complexity of todays designs, becomes an increasingly difficult problem. Statistical static timing analysis has been proposed as a solution to this problem, but most of the work has focused in the development of timing engines for computing delay propagation. Such tools rely on the availability of delay formulas accounting for both cell and interconnect delay that take into account unpredictable variability effects. In this paper, we concentrate on the impact of interconnect on delay and propose an extension to the standard modeling strategies that is variation-aware and compatible with such statistical engines. Our approach, based on a specific type of perturbation analysis, allows for the analytical computation of the quantities needed for statistical delay propagation. We also show how perturbation analysis can be performed when only the standard delay table lookup models are available for the standard cells. This makes the proposed approach compatible with existing timing analysis frameworks. Results from applying our proposed modeling strategy to computing delays and slews in several instances accurately match similar results obtained using electrical level simulation.

1 Introduction

The impact of process variation on circuit performance is an area of increasing concern, both in the semiconductor industry, as well as academic research. In the research community, considerable work has been devoted to the development of statistical static timing analysis [1, 2]. Nowadays, designers spend a considerable amount of their verification budget trying to make sure that their circuits will

Please use the following format when citing this chapter:

Guerra e Silva, L., Zhu, Z., Phillips, J.R. and Silveira, L.M., 2007, in IFIP International Federation for Information Processing, Volume 249, VLSI-SoC: Research Trends in VLSI and Systems on Chip, eds. De Micheli, G., Mir, S., Reis, R., (Boston: Springer), pp. 157–176

work under all possible settings. To achieve this, they target the worst possible scenarios by considering so-called pessimistic conditions, and design in order to ensure that such corner cases are accounted for. This analysis is usually based on assuming worst-case conditions on all possible variations simultaneously. Such an scenario is pessimistic and may lead to considerable over-design.

Improving this situation requires tools that are better suited to handle realistic variations and the complex inter-relations that exist between those variations. Not only should those tools directly make use of realistic process information, thus making them better suited to model the unpredictability of process parameter variations, but they should be able to implicitly determine how such variations affect the circuit behavior. Such a formulation makes it possible to compute on a single analysis the circuit behavior not only due to a given parameter setting, but to a variety of settings. The recent development and availability of statistical timers that are based on a parametric description of delay in terms of random process variables is an example of movement in this direction [1]. Other approaches targeting direct determination of the worst parameter settings with respect to delay also follow the same trend [3].

A timing analyzer consists of several component pieces. In a statistical context, the most well-studied part of the timing engine is the timing graph traversal, which manages the calculation of arrival times and slews at the level of abstraction of a timing graph. An equally important, if more mundane, component is the delay calculation engine. The delay calculator takes as input the cell and interconnect models and produces a delay expression in a form that can be consumed by the graph engine. This paper is concerned with a portion of the delay calculation step, the impact of interconnect on delay. We explore how commonly used interconnect modeling strategies can be extended to be compatible with the most recent generation of statistical timing analysis tools [1]. Specifically, we wish to produce cell and interconnect delay as affine functions of process parameters. We assume that one of several recently proposed approaches for interconnect reduction under process variation is available to generate tractably sized reduced order models [4–6]. The key technology in our approach is a specific type of perturbation analysis. While digital circuits are strongly nonlinear with respect to the circuit inputs, cell delays are often close to linear *with respect to process parameters*. In this paper we adapt the general development of *linear time-varying* (LTV) perturbation theory [7, 8] for extraction of variation-aware delay models to the specific needs of delay calculation for precharacterized standard cells. LTV perturbation theory has been widely used in RF analysis with great success [9] and is at the heart of many interesting new developments. The advantage of this type of approach over, for example, differencing repeated delay calculation runs is that it is essentially an analytic method. Differencing type approaches can suffer from severe robustness problems that make them difficult to use reliably. In addition, our technique can potentially be made very fast, handling parametric models with ten to twenty parameters at minimal penalty relative to a non-variational calculation.

The outline of this paper is as follows: in Section 2, we review the basics of delay computation, the general mechanics of the procedure including cell and interconnect delay, assuming no variations are taken into account. Then, in Section 3, we introduce the general perturbation formulation and discuss the specific specialization of the more general technique to cell-level interconnect-related delay. In Section 4, we also discuss how perturbation analysis can be performed when only delay table lookup models are available for the standard cells. A key point is that *analytic* expressions for delay sensitivities can be obtained without having to have closed-form expressions for the cell delay elements (however, see [10] for such closed-form expressions). Finally in Section 5 we discuss the utilization of adjoint methods to accelerate the computation of timing models when large numbers of parameters are present. Results of using our proposed approach are shown in Section 6 and conclusions are drawn in Section 7.

2   Nominal Delay Calculation

Timing verification is an enabling methodology for optimizing performance and ensuring that circuits satisfy certain timing and frequency requirements. To that end, timing verifiers determine approximate but safe estimates of the worst-case delay through a circuit: for every input and output signal, there are many possible paths through the circuit, each path consisting of a set of interconnected network cells. Timing verification deals with the identification and analysis of the critical paths, also known as the longest delay paths in the circuit. In addition to finding critical-path delays, timing verifiers can also be used to do miscellaneous static analysis, like finding high-speed components off the critical path that can be slowed down to save power and several other relevant tasks. However, the most common usage is indeed to determine the worst case paths. Computing the delay along a path requires the computation of the delay of every cell along that path, plus the added delay due to interconnect between the cells. In this section we review the standard computation of cell and interconnect delay.

2.1   Mechanics of Delay Computation

Timing analysis constitutes the foundation of any timing verification methodology. The typical timing analysis methodology consists in *arrival time* computation, which is concerned with computing the time instants at which signal transitions reach "interesting" nodes in the circuit, often corresponding to primary outputs or register inputs, where specific timing constraints must be enforced.

Two main approaches have been proposed for timing analysis: block-based and path-based. In the block-based approach, characterized by linear runtime, arrival times are pushed through the circuit in a levelized fashion, performing sum operations with cell or interconnect delays and min/max operations to compute the arrival times in the outputs of multi-input cells, assuming that the earliest/latest input transition determines the output transition. The alternative path-based approach consists in individually computing the delay of each

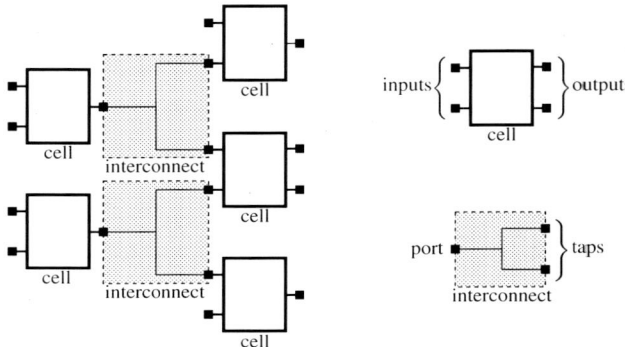

Fig. 1. Typical partition of a digital IC topology for timing analysis.

path in the circuit by adding the delay of all the cells and interconnect along that path. Even though more accurate, this approach is computationally much more expensive than the former, since the number of paths is known to grow exponentially with the number of nodes. Clearly any timing analysis approach requires the computation of cell and interconnect delays.

For timing analysis purposes, the digital IC topology is usually partitioned into cells and interconnect nets, as illustrated in Figure 1. Primary inputs and outputs are usually represented by the corresponding *pads*, which are a particular type of cells. Cell input and output pins are connected by interconnect nets. Each interconnect net can be seen as distributing the signal injected in its input, designated by *port*, to each of its outputs, designated by *taps*, that are connected to cell input pins. For typical cell and interconnect delay models, the slew of the input signal(s) is a required parameter. Accordingly, the slew of the output signal(s) is a result produced by the model. Therefore, once the circuit is properly partitioned and all the cell and interconnect delay models are in place, the task of the delay computation engine is to forward propagate the slews and invoke the appropriate delay models that will compute delays and output slews given the input slews and output loads.

2.2   Cell Delay and Cell Loading

Mainly for historical reasons, the most common modeling strategy for cell library characterization is based in delay look-up tables (LUTs) sometimes referred to as *dot-lib (.lib)* tables. This is a simplified model where delay and power information is maintained in the form of a few parameters. In this simplified model the timing behavior of a cell is usually characterized by a set of lookup tables that, for each input/output pin pair, describe the delay and output slew of the cell as a function of the input slew and output load. Such a model is illustrated in Figure 2 where the standard definitions are also used, namely input and output slews are defined as $s = t_H - t_L$, where t_L and t_H are the time instants at which the respective voltage waveform reaches some pre-defined values, V_L and V_H, related to the

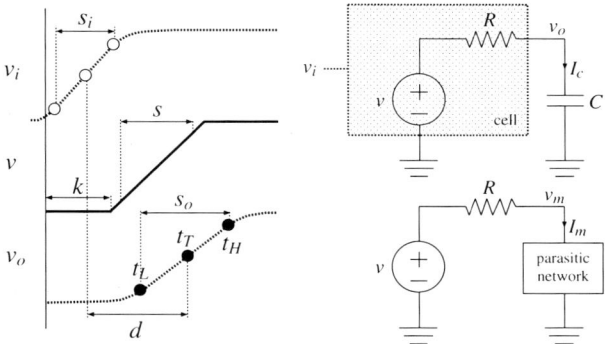

Fig. 2. Voltage source based cell model, loaded by the effective capacitance (top right), and by the parasitic network (bottom right) and corresponding waveforms (left).

definition of noise margins. In a similar manner, delay is defined as the time it takes the output of a cell to reach its transition midpoint, from the time the cell input waveform reached its own midpoint. Cell characterization is performed by simulating the cell behavior as a function of input slew and loading capacitances. These results are then stored in look-up tables as mentioned, which are accessed to determine delay and slew in specific instances.

The outlined delay modeling strategy assumes a voltage source model for the cell characterization, as illustrated in Figure 2, since delay and slew values implicitly characterize the output voltage waveforms of the cell. However, in recent years, current source models are gaining more prominence, since they are more effective in handling complex interconnect loading effects. Even though throughout this paper we assume voltage source delay models, the proposed techniques can also be directly applied when using current source delay models.

In Figure 2, the output load is assumed to be a single lumped capacitance that somehow models the capacitive effects introduced by the interconnect and by the input pins of the cells connected to same net. In reality, however, the interconnect attached to the driver cell is a complex RC network that in deep submicron processes is very poorly modeled by a lumped capacitance. The loading effect of interconnect on the cell, i.e. the impact of downstream interconnect on the cell delay itself, cannot be accurately obtained simply by looking at the total capacitance on the net. To try to account for the effects of complex interconnect, while still preserving table-based cell models, the concept of *effective capacitance* [11, 10] has been widely adopted. For the remainder of this paper we will consider that the C shown in Figure 2 is such an effective capacitance.

The idea behind the effective capacitance consists in determining the value of C that in a certain sense approximates as accurately as possible the behavior of the original parasitic network. In Figure 2. the output stage of a cell (or more accurately, of an output pin of a cell) is modeled by a voltage source, producing a voltage ramp v, with slew s, and a series resistor, with resistance R, that models the output resistance of the pin. The figure depicts the output stage of a cell loaded by the effective capacitance C (top right), and by the original parasitic RC network, obtained by layout extraction (bottom right). In the following, without loss of generality, in order to simplify the description, we restrict ourselves to the case of rising output waveforms for non-inverting cells. Clearly any other case can be derived in a similar manner.

The simple RC circuit on the top of Figure 2 is an approximated model of the output stage of a cell connected to an effective capacitance, that is itself an approximation of the interconnect load. For a given input slew s_i and a given effective capacitance C, we can compute the estimated cell delay d and the estimated output slew s_o, by a table lookup in the timing characterization of the cell. Using this information, we can compute the three time instants at which the waveform of the output voltage v_o should cross V_L, V_T and V_H, respectively,

$$t_L = \frac{s_i}{V_H - V_L} V_T + d - \frac{s_o}{V_H - V_L}(V_T - V_L) \tag{1}$$

$$t_T = \frac{s_i}{V_H - V_L} V_T + d \tag{2}$$

$$t_H = \frac{s_i}{V_H - V_L} V_T + d + \frac{s_o}{V_H - V_L}(V_H - V_T) \tag{3}$$

Assuming the voltage v to be a rising ramp of slew s, shifted in time by k,

$$v(s, k, t) = \begin{cases} 0 & if \ 0 \le t < k \\ \frac{V_H - V_L}{s} t & if \ k \le t < k + \frac{s V_{DD}}{V_H - V_L} \\ V_{DD} & if \ t \ge k + \frac{s V_{DD}}{V_H - V_L} \end{cases} \tag{4}$$

the output voltage, v_o, produced by the simple RC circuit presented in Figure 2, with time constant $\tau = RC$, is given by,

$$v_o(s, k, \tau, t) = \begin{cases} 0 & if \ 0 \le t < k \\ \frac{V_H - V_L}{s}\left(-\tau + t - k + \tau e^{-\frac{t-k}{\tau}}\right) & if \ k \le t < k + \frac{s V_{DD}}{V_H - V_L} \\ V_{DD} - \frac{V_H - V_L}{s}\left(e^{\frac{s V_{DD}}{V_H - V_L}} - 1\right)\tau e^{-\frac{t-k}{\tau}} & if \ t \ge k + \frac{s V_{DD}}{V_H - V_L} \end{cases} \tag{5}$$

In order to simplify our notation, in the following we will assume,

$$\phi = \langle s, k, R, C \rangle. \tag{6}$$

Using Eqn. (5), we can compute a waveform for v (e.g. s and k) and a resistance R, such that the waveform of the response v_o crosses (t_L, V_L), (t_T, V_T) and (t_H, V_H), thus matching the tabulated behavior of the cell and its output response. This problem can be stated by the following three equations,

$$v_o(t_L, \phi) = V_L \tag{7}$$

$$v_o(t_T, \phi) = V_T \tag{8}$$

$$v_o(t_H, \phi) = V_H \tag{9}$$

The waveform of v can be seen as the "ideal" output voltage of the cell, under a zero output load. We should not lose track of the fact that our goal is to determine an appropriate value for the effective capacitance C. The previous derivations assumed that such a value was somehow known. However, all that is required is that C should approximate the behavior of the original parasitic network as accurately as possible. Several criteria [12] can be used when defining what effective capacitance provides a good approximation of the behavior of the original parasitic network. In this work we consider that the effective capacitance that better approximates the behavior of the original parasitic network is the one that draws the same average current, over the transition period (e.g. when the output voltage switches from V_L to V_H). Formally,

$$\langle I_c \rangle = \langle I_m \rangle \Leftrightarrow \frac{1}{s_o} \int_{t_L}^{t_H} I_c \, dt = \frac{1}{t'_H - t'_L} \int_{t'_L}^{t'_H} I_m \, dt \tag{10}$$

where $v_m(t'_L) = V_L$ and $v_m(t'_H) = V_H$. An analytical expression for $\langle I_c \rangle$ can be derived. On the other hand, $\langle I_m \rangle$ must be computed by numerically integrating the port current, obtained by interconnect simulation, as detailed in Section 2.3.

From Eqns. (7), (8), (9), and (10) we can compute the value of ϕ that both matches the output waveform v_o with the tabulated timing information at t_L, t_T and t_H, and also that matches the average current drawn by the original parasitic network and the effective capacitance. Since Eqns. (7), (8), (9) and (10) contain nonlinear terms, an implicit iterative method must be used to solve them. We have used Newton's method in this work. Once the value of the effective capacitance C is known, we can compute the delay d and output slew s_o of the cell by a simple lookup in the timing characterization of the cell. This completely characterizes the cell output waveform within the constraints of the simple model. Such a waveform constitutes the input to the interconnect model.

2.3 Interconnect Delay

Assuming that the cell output voltage waveform has been computed, signals are then propagated along the path through an interconnect net. The input of such nets, the port, is tied to the output of a cell, and the net outputs, the taps, connect to the inputs of several other cells. At the timing level, the difference in the timing of the transition at the cell output (port) and next cell inputs (taps) we refer to as intrinsic interconnect delay. There are various methods of computing the interconnect delay ranging from closed-form expressions, descendants of the Elmore delay formula, to numerical solution of the underlying interconnect equations. In this work we assume that the circuit equations of the cell driver plus interconnect network are solved numerically, either via direct integration or an equivalent process like recursive convolution. Likewise the slew at the output nodes must be computed to be used in the analysis of the following cell.

The general state-space representation of a parasitic RC network (either in its original of reduced form) is

$$C \frac{d}{dt} x(t) + G x(t) = u(t) \tag{11}$$

$$y(t) = N^T x(t) \tag{12}$$

where $x \in \mathbb{R}^n$ is the vector of circuit state variables, u is the input excitation, y is the output response, C and G are the matrices describing the reactive (capacitances) and dissipative (conductances) parts of the circuit and N selects the output response.

Assuming a cell characterization in terms of voltage source models, as illustrated in Figure 2, the input excitation is the voltage waveform, v_m, and the output response are the voltage waveforms in the taps, v_{tap}. Therefore, we have,

$$u(t) = Bv_m \tag{13}$$

$$v_{tap}(t) = L^T x(t) \tag{14}$$

where B is a matrix describing the node where the input voltage is injected, and L is an incidence-type matrix describing which voltage nodes are monitored (taps). In the particular case of voltage source models, the current drawn by the parasitic network, I_m, is also relevant, both for computing the effective capacitance and the input voltage waveform. Hence, an additional equation should be added,

$$I_m(t) = M^T x(t) \tag{15}$$

where M selects the output current out of the state vector x.

3   Variation-Aware Methodology

3.1   General Perturbation Formulation

In this section, we will discuss the parametric analysis of the intrinsic interconnect delay itself. The impact of the interconnect parameters on the cell delay (i.e. variation in cell loading effects) is taken up in the next section.

The starting point of our analysis is the general formulation of time-varying linear perturbation theory (see [8] for details). We assume the existence of a set of nonlinear differential-algebraic equations whose topology is fixed, but whose constitutive relations depend on a continuous way on a set of parameters. Without loss of generality the basic circuit equations can be written as

$$\frac{d}{dt}q(x, \lambda) + i(x, \lambda) = u(t) \tag{16}$$

where x again represents the circuit state variables, for example, node voltages, $q \in \mathbb{R}^n$, the dynamic quantities such as stored charge, $i \in \mathbb{R}^n$, the static quantities such as device currents, t, time, and $u(t) \in \mathbb{R}^n$, the independent inputs such as current and voltage sources. In departure from the usual case, we introduce a p-element parameter vector $\lambda \in \mathbb{R}^p$. These parameters represent properties of the circuit, such as wire width or thickness, that induce variation in the circuit behavior through the q and i functions.

The perturbation approach to modeling the parameter variation treats the parameters as fluctuations $\Delta\lambda$ around a nominal value λ_0, and assumes the circuit response x can be treated similarly, i.e.

$$\lambda = \lambda_0 + \Delta\lambda \tag{17}$$
$$x(t) = x_0(t) + \Delta x(t). \tag{18}$$

Expanding i and q as function of x, λ and keeping the first order variations, we get

$$q(x, \lambda) = q(x_0, \lambda_0) + \frac{\partial q}{\partial \lambda}\Delta\lambda + \frac{\partial q}{\partial x}\Delta x \tag{19}$$

$$i(x, \lambda) = i(x_0, \lambda_0) + \frac{\partial i}{\partial \lambda}\Delta\lambda + \frac{\partial i}{\partial x}\Delta x. \tag{20}$$

Assuming a solution to the nominal case, $x_0(t)$ is obtained, that is

$$\frac{d}{dt}q(x_0, \lambda_0) + i(x_0, \lambda_0) = u(t) \tag{21}$$

then substituting the perturbation expansions (19) and (20) into Eqn. (16) and using (21) to eliminate the nominal-case terms, we obtain the equations for the first-order perturbation expansion as

$$\frac{d}{dt}\left[\frac{\partial q}{\partial x}\Delta x\right] + \frac{\partial i}{\partial x}\Delta x = -\left[\frac{d}{dt}(\frac{\partial q}{\partial \lambda})\Delta\lambda + \frac{\partial i}{\partial \lambda}\Delta\lambda\right] \tag{22}$$

The simplest way to compute waveform sensitivities from Eqn. (22) is by solving it once for each parameter in turn, as

$$\text{for each } k: \quad \frac{d}{dt}\left[\frac{\partial q}{\partial x}\frac{\partial x}{\partial \lambda_k}\right] + \frac{\partial i}{\partial x}\frac{\partial x}{\partial \lambda_k} = -\left[\frac{d}{dt}(\frac{\partial q}{\partial \lambda_k}) + \frac{\partial i}{\partial \lambda_k}\right]. \tag{23}$$

This gives the final expression

$$x(t, \lambda) = x_0(t) + \sum_{k=1}^{p}\frac{\partial x}{\partial \lambda_k}(t)\Delta\lambda_k. \tag{24}$$

Once the sensitivities in the waveforms are known, the next step is to translate to sensitivity of delay. As discussed, delay can be computed as $d = t_2 - t_1$ where t_2, t_1 are the crossing times of the two waveforms of interest. The sensitivity in a crossing time can be related to the sensitivity of the waveform value $x(t)$ at that point via the slew, $\partial x/\partial t$. Suppose there is a small change ΔT in the crossing time of a given waveform. With a linear model, the corresponding change in the voltage is

$$\Delta X = \frac{\partial x}{\partial t}\Delta T. \tag{25}$$

Conversely, if the perturbation in the waveform ΔX can be computed, the change in crossing time is given by

$$\Delta T = \frac{\Delta X}{\frac{\partial x}{\partial t}}. \tag{26}$$

Therefore we can compute the sensitivity of the delay as

$$\frac{\partial d}{\partial \lambda_k} = \frac{\left.\frac{\partial x}{\partial \lambda_k}\right|_{t_2}}{\left.\frac{\partial x}{\partial t}\right|_{t_2,\lambda_0}} - \frac{\left.\frac{\partial x}{\partial \lambda_k}\right|_{t_1}}{\left.\frac{\partial x}{\partial t}\right|_{t_1,\lambda_0}} \tag{27}$$

Note that for this computation, the waveform sensitivity is only needed at a few points in time, a fact that can be used to speedup computations (see Section 5).

This is the formulation for a general first-order perturbation analysis. In the following we restrict ourselves to the problem at hand, namely modeling the linear interconnect sub-circuits assuming variations in parameters affecting the interconnect elements.

3.2   Specialization to Interconnect

Our concern in this document is with the special case of interconnect parameters, so simplifications of the general theory are possible. On-chip cell-level interconnect models are usually written in terms of capacitances and resistances, or equivalently, capacitances and conductances. Inductance is typically neglected at this level and for the sake of simplicity we will proceed likewise; it is however easy to see that the derivation is quite similar when inductance is involved. Therefore, in this case,

$$q(x,\lambda) = C(\lambda)x \qquad i(x,\lambda) = G(\lambda)x \tag{28}$$

so that

$$\frac{\partial k}{\partial \lambda_k} = \frac{\partial G}{\partial \lambda_k}x \qquad \frac{\partial q}{\partial \lambda_k} = \frac{\partial C}{\partial \lambda_k}x \tag{29}$$

Let us then assume, for now, that for every element in the parasitic network (resistor or capacitor), a linear variational model is available. Such a model contains the nominal values for the elements and also the sensitivities to each parameter. Therefore, the conductance and the capacitance matrices have the form:

$$G = G_0 + \sum_{k=1}^{p}(G_k\Delta\lambda_k), \qquad C = C_0 + \sum_{k=1}^{p}(C_k\Delta\lambda_k) \tag{30}$$

where G_0 and C_0 are the nominal values of the elements in the interconnect network and the sensitivities $\frac{\partial G}{\partial \lambda_k}$ and $\frac{\partial C}{\partial \lambda_k}$ to each parameter λ_k are given by

$$\frac{\partial G}{\partial \lambda_k} = G_k, \qquad \frac{\partial C}{\partial \lambda_k} = C_k. \tag{31}$$

The nominal value corresponds to the solution of the equations with each $\Delta\lambda_k = 0$, that is $\lambda = \lambda_0$. Assuming the variational formulation for G presented in Eqn. (30), and for x presented in Eqn. (18) we obtain, for instance for $i(x,\lambda)$:

$$i(x,\lambda) = \left[G_0 + \sum_{k=1}^{p}(G_k\Delta\lambda_k)\right](x_0 + \Delta x) \tag{32}$$

Simplifying and eliminating the (non-linear) cross-product terms, we obtain:

$$i(x, \lambda) \approx G_0 x_0 + G_0 \Delta x + \sum_{k=1}^{p} (G_k x_0 \Delta \lambda_k) \tag{33}$$

implying that:

$$i_0 \equiv i(x_0, 0) = G_0 x_0, \quad \frac{\partial i}{\partial x} = G_0, \quad \frac{\partial i}{\partial \lambda_k} = G_k x_0. \tag{34}$$

An identical procedure can be applied to $q(x, \lambda)$ leading, as expected, to:

$$q(x, \lambda) \approx C_0 x_0 + C_0 \Delta x + \sum_{k=1}^{p} (C_k x_0 \Delta \lambda_k) \tag{35}$$

and therefore, that:

$$q_0 \equiv q(x_0, 0) = C_0 x_0, \quad \frac{\partial q}{\partial x} = C_0, \quad \frac{\partial q}{\partial \lambda_k} = C_k x_0 \tag{36}$$

Eqns. (21) and (22) which describe the general perturbation analysis framework, can therefore, in the specialization of parameter-varying interconnect, be written as:

$$C_0 \frac{d}{dt} x_0(t) + G_0 x_0(t) = u(t) \tag{37}$$

$$C_0 \frac{d}{dt} [\Delta x] + G_0 \Delta x = - \sum_{k=1}^{p} \left[\frac{d}{dt} (C_k x_0(t)) \Delta \lambda_k + G_k \Delta \lambda_k \right] \tag{38}$$

The delay modeling problem is completed by adding the notion of inputs and outputs to form state-space models. In the case of cell-level interconnect, the inputs are represented by drivers, the output stages of cells. If the cell library is characterized using current source models, then the input is a fixed current source,

$$u(t) = B i_{drv}(t) \tag{39}$$

where B is simply an incidence matrix indicating at which node each driver is connected to. Similarly, if the cell library is characterized using voltage source models (as in the case under study), we have

$$u(t) = B v_{drv}(t) \tag{40}$$

as in Eqn. (13), where $v_{drv} = v_m$. Other models may be used, like nonlinear current source models [13, 14].

Recalling Eqn. (14), the full set of equations is now

$$C_0 \frac{d}{dt} x_0(t) + G_0 x_0(t) = u(t) \tag{41}$$

$$v_{0,tap}(t) = L^T x_0(t) \tag{42}$$

$$C_0 \frac{d}{dt} [\Delta x] + G_0 \Delta x = - \sum_{k=1}^{p} \left[\frac{d}{dt} (C_k x_0(t)) \Delta \lambda_k + G_k \Delta \lambda_k \right] \tag{43}$$

$$\Delta v_{tap} = L^T \Delta x \tag{44}$$

These equations can be written more compactly if we define

$$s_k(t) = -\left[C_k\frac{d}{dt}x_0(t) + G_kx_0(t)\right] \tag{45}$$

where $x_0(t)$ is the nominal solution computed above. s_k can be interpreted as the "equivalent source" that will allow determination of the sensitivity to the kth interconnect parameter. With this definition, the final, complete set of equations is then rewritten as

$$C_0\frac{d}{dt}x_0(t) + G_0x_0(t) = u(t) \tag{46}$$

$$v_{0,tap} = L^Tx_0(t) \tag{47}$$

$$C_0\frac{d}{dt}[\Delta x] + G_0\Delta x = \sum_{k=1}^{p}s_k(t)\Delta\lambda_k \tag{48}$$

$$\Delta v_{tap} = L^T\Delta x \tag{49}$$

3.3   Interconnect Sensitivity Calculation

The process of sensitivity calculation can now be concisely stated. First, solve Eqns. (46) and (47) to get the nominal case responses. Then, for each parameter k, solve

$$C_0\frac{d}{dt}\left[\frac{\partial x}{\partial\lambda_k}\right] + G_0\left[\frac{\partial x}{\partial\lambda_k}\right] = s_k(t) \tag{50}$$

$$\frac{\partial v_{tap}}{\partial\lambda_k} = L^T\left[\frac{\partial x}{\partial\lambda_k}\right] \tag{51}$$

to get the sensitivity of the response waveforms. From the sensitivity waveforms, the delay sensitivity can be computed using Eqn. (27) at the appropriate time-points. Of course, in practice, it is useful to diagonalize the state-space model above, i.e. to put the C_0, G_0 matrices into pole-residue form, as numerical solution of the multiple systems is much more efficient.

4   Cell Delay Sensitivity Calculation

In the preceding section, we have seen how to perform variation-aware delay computation, by computing the sensitivities of the response waveforms in inter-connect blocks. However, it is also necessary to show that similar sensitivities can be computed at the output of cells, in particular assuming that cell delay computation is still based on delay table models.

To show this, we refer back to the derivation in Section 2 and in particular to Eqns. (7), (8), (9) and (10). If we perform an expansion around a nominal point ϕ_0, keeping the first order variations, and eliminating the nominal-case terms, we obtain,

$$\Delta v_o(t_L, \Delta\phi) = 0 \tag{52}$$

$$\Delta v_o(t_T, \Delta\phi) = 0 \tag{53}$$

$$\Delta v_o(t_H, \Delta\phi) = 0 \tag{54}$$

$$\langle\Delta I_c\rangle(t_L, t_H, \Delta\phi) = \langle\Delta I_m\rangle \tag{55}$$

Noticing the dependence of t_L, t_T and t_H, on d and s_o, and their dependence on s_i and C, we obtain the generic equation,

$$\frac{\partial v_o}{\partial s}\Delta s + \frac{\partial v_o}{\partial k}\Delta k + \frac{\partial v_o}{\partial R}\Delta R$$
$$+ \left(\frac{\partial v_o}{\partial C} + \frac{\partial v_o}{\partial t_X}\frac{dt_X}{dC}\right)\Delta C + \frac{\partial v_o}{\partial t_X}\frac{dt_X}{ds_i}\Delta s_i = 0 \tag{56}$$

where

$$\frac{dt_X}{dC} = \frac{\partial t_X}{\partial s_o}\frac{\partial s_o}{\partial C} + \frac{\partial t_X}{\partial d}\frac{\partial d}{\partial C}, \qquad \frac{dt_X}{ds_i} = \frac{\partial t_X}{\partial s_i} + \frac{\partial t_X}{\partial s_o}\frac{\partial s_o}{\partial s_i} + \frac{\partial t_X}{\partial d}\frac{\partial d}{\partial s_i}. \tag{57}$$

t_X can be replaced by t_L, t_T or t_H to obtain Eqns. (52), (53), and (54), and all derivatives are computed at time t_X. For Eqn. (55) a similar expansion can be performed,

$$\left(\frac{\partial\langle I_c\rangle}{\partial s} - \frac{\partial\langle I_m\rangle}{\partial s}\right)\Delta s + \frac{\partial\langle I_c\rangle}{\partial k}\Delta k$$
$$+ \left(\frac{\partial\langle I_c\rangle}{\partial R} - \frac{\partial\langle I_m\rangle}{\partial R}\right)\Delta R + \frac{d\langle I_c\rangle}{dC}\Delta C + \frac{d\langle I_c\rangle}{ds_i}\Delta s_i = \langle\Delta I_m\rangle \tag{58}$$

where

$$\frac{d\langle I_c\rangle}{dC} = \frac{\partial\langle I_c\rangle}{\partial C} + \frac{\partial\langle I_c\rangle}{\partial t_L}\frac{dt_L}{dC} + \frac{\partial\langle I_c\rangle}{\partial t_H}\frac{dt_H}{dC} \tag{59}$$

$$\frac{d\langle I_c\rangle}{ds_i} = \frac{\partial\langle I_c\rangle}{\partial t_L}\frac{dt_L}{ds_i} + \frac{\partial\langle I_c\rangle}{\partial t_H}\frac{dt_H}{ds_i} \tag{60}$$

Δs_i and $\langle\Delta I_m\rangle$ are related to the parameter variation vector, $\Delta\lambda$, by the following expressions,

$$\Delta s_i = \frac{\partial s_i}{\partial\lambda}\Delta\lambda \tag{61}$$

$$\langle\Delta I_m\rangle = \frac{\partial\langle I_m\rangle}{\partial\lambda}\Delta\lambda \tag{62}$$

where $\frac{\partial s_i}{\partial\lambda}$ and $\frac{\partial\langle I_m\rangle}{\partial\lambda}$ are the sensitivity vectors. Resorting to Eqns. (56), (58), (61), and (62), we can now represent Eqns. (52), (53), (54), and (55) in matrix form as,

$$J\Delta\phi = \left(Q\frac{\partial s_i}{\partial\lambda} + W\frac{\partial\langle I_m\rangle}{\partial\lambda}\right)\Delta\lambda \tag{63}$$

where J, Q and W are given by

$$J = \begin{bmatrix} \frac{\partial v_o}{\partial s}\big|_{t_L} & \frac{\partial v_o}{\partial k}\big|_{t_L} & \frac{\partial v_o}{\partial R}\big|_{t_L} & \frac{\partial v_o}{\partial C}\big|_{t_L} + \frac{\partial v_o}{\partial t_L}\frac{dt_L}{dC} \\ \frac{\partial v_o}{\partial s}\big|_{t_T} & \frac{\partial v_o}{\partial k}\big|_{t_T} & \frac{\partial v_o}{\partial R}\big|_{t_T} & \frac{\partial v_o}{\partial C}\big|_{t_T} + \frac{\partial v_o}{\partial t_T}\frac{dt_T}{dC} \\ \frac{\partial v_o}{\partial s}\big|_{t_H} & \frac{\partial v_o}{\partial k}\big|_{t_H} & \frac{\partial v_o}{\partial R}\big|_{t_H} & \frac{\partial v_o}{\partial C}\big|_{t_H} + \frac{\partial v_o}{\partial t_H}\frac{dt_H}{dC} \\ \frac{\partial\langle I_c\rangle}{\partial s} - \frac{\partial\langle I_m\rangle}{\partial s} & \frac{\partial\langle I_c\rangle}{\partial k} & \frac{\partial\langle I_c\rangle}{\partial R} - \frac{\partial\langle I_m\rangle}{\partial R} & \frac{d\langle I_c\rangle}{dC} \end{bmatrix} \tag{64}$$

$$Q = \begin{bmatrix} -\frac{\partial v_o}{\partial t_L}\frac{dt_L}{ds_i} \\ -\frac{\partial v_o}{\partial t_T}\frac{dt_T}{ds_i} \\ -\frac{\partial v_o}{\partial t_H}\frac{dt_H}{ds_i} \\ -\frac{d\langle I_c \rangle}{ds_i} \end{bmatrix} \quad , \quad W = \begin{bmatrix} 0 \\ 0 \\ 0 \\ 1 \end{bmatrix} \tag{65}$$

$\frac{\partial s_i}{\partial \lambda}$ results from the variational timing analysis on the interconnect of the input net, as described in Section 3. $\frac{\partial \langle I_m \rangle}{\partial \lambda}$ can be computed by integrating the sensitivities of the port current, I_m, for the transition period and dividing by its width. All the derivatives in J, Q and W can either be computed analytically or by accessing the timing characterization of the cell.

If $N_C = [0\ 0\ 0\ 1]$ is a vector that "selects" the capacitance row of $\Delta\phi$,

$$\Delta C = N_C \Delta\phi = N_C J^{-1}\left(Q\frac{\partial s_i}{\partial \lambda} + W\frac{\partial \langle I_m \rangle}{\partial \lambda}\right)\Delta\lambda \tag{66}$$

Acknowledging the dependence of the delay d and the output slew s_o on the input slew s_i and the capacitance C, the following expressions can be derived,

$$\Delta d = \frac{\partial d}{\partial s_i}\Delta s_i + \frac{\partial d}{\partial C}\Delta C \tag{67}$$

$$\Delta s_o = \frac{\partial s_o}{\partial s_i}\Delta s_i + \frac{\partial s_o}{\partial C}\Delta C \tag{68}$$

where $\frac{\partial d}{\partial s_i}$, $\frac{\partial d}{\partial C}$, $\frac{\partial s_o}{\partial s_i}$ and $\frac{\partial s_o}{\partial C}$ can be computed by direct analysis of the lookup table that contains the timing characterization of the cell. Substituting Eqns. (61) and (66) in Eqns. (67) and (68), we can derive the sensitivities of the delay and output slew to the parameters,

$$\frac{\partial d}{\partial \lambda} = \frac{\partial d}{\partial s_i}\frac{\partial s_i}{\partial \lambda} + \frac{\partial d}{\partial C}N_C J^{-1}\left(Q\frac{\partial s_i}{\partial \lambda} + W\frac{\partial \langle I_m \rangle}{\partial \lambda}\right) \tag{69}$$

$$\frac{\partial s_o}{\partial \lambda} = \frac{\partial s_o}{\partial s_i}\frac{\partial s_i}{\partial \lambda} + \frac{\partial s_o}{\partial C}N_C J^{-1}\left(Q\frac{\partial s_i}{\partial \lambda} + W\frac{\partial \langle I_m \rangle}{\partial \lambda}\right) \tag{70}$$

5   Optimizations for Large Numbers of Parameters

In this section, we discuss how using adjoint methods [15] can accelerate the computation of the timing models when large numbers of parameters are present. Since the computation time is nearly independent of the number of parameters, the sensitivity to a larger number of parameters can be done simultaneously. Thus, if only a few crossing times are of interest, the computation is very cheap on an information-gained basis. When device mismatch is of interest, the sensitivities to multiple model parameters, for every device in a circuit, are needed. Mismatch could be caused by purely stochastic mechanisms, such as dopant fluctuations in MOSFET channels. Systematic effects such as optical proximity printing errors may also lead to device-by-device parameter variations.

Let us suppose the nominal system in Eqn. (46) has been discretized into M time-points and an operating point $x_0(t)$ has been computed. For a given timepoint t_k, $k = 1 \ldots M$, let us introduce the capacitance and conductance matrices C, G as follows:

$$C_k \equiv \left. \frac{\partial q}{\partial x} \right|_{x(t_k)} \qquad G_k \equiv \left. \frac{\partial i}{\partial x} \right|_{x(t_k)} \tag{71}$$

Similarly, define the "source" functions s as

$$s_k^{(q),(l)} \equiv -\frac{d}{dt} \left[\frac{\partial q}{\partial \lambda_l} \right]_{x(t_k)} \quad, \quad s_k^{(i),(l)} \equiv - \left. \frac{\partial i}{\partial \lambda_l} \right|_{x(t_k)} \tag{72}$$

$$s_k^{(l)} = s_k^{(q),(l)} + s_k^{(i),(l)}. \tag{73}$$

We "pack" the time-varying quantities into matrices and vectors with a block structure. If there are N equations in (22) and M timepoints, then the vectors

$$\mathcal{X} = \begin{bmatrix} \Delta x_1 \\ \Delta x_2 \\ \ldots \\ \Delta x_M \end{bmatrix}, \quad s^{(l)} = \begin{bmatrix} s_1^{(l)} \\ s_2^{(l)} \\ \ldots \\ s_M^{(l)} \end{bmatrix} \tag{74}$$

have M sections, each section a vector of N entries. The vector $\mathcal{X}$ represents the waveforms of perturbations due to parameter fluctuation. The vectors $s^{(l)}$ will be used to form the p columns (one for each parameter) of the matrix $\mathcal{S}$,

$$\mathcal{S} = \begin{bmatrix} s^{(1)} & s^{(2)} & \ldots & s^{(p)} \end{bmatrix} \tag{75}$$

Likewise the matrix

$$\mathcal{G} = \begin{bmatrix} G_1 & & & \\ & G_2 & & \\ & & \ddots & \\ & & & G_M \end{bmatrix} \tag{76}$$

has $M \times M$ blocks, each block an $N \times N$ matrix.

After time-discretization, a composite capacitance matrix $\mathcal{C}$ may also be formed. The precise structure of this matrix depends on the discretization scheme used. For example, for a backward Euler discretization with timesteps $h_1, \ldots, h_M$, the matrix $\mathcal{C}$ becomes

$$\mathcal{C} = \begin{bmatrix} \frac{C_1}{h_1} & & & \\ -\frac{C_1}{h_2} & \frac{C_2}{h_2} & & \\ & & \ddots & \\ & & -\frac{C_{(M-1)}}{h_M} & \frac{C_M}{h_M} \end{bmatrix} \tag{77}$$

Eq. (22) can be written as one composite matrix equation[3]

$$\mathcal{C}\mathcal{X} + \mathcal{G}\mathcal{X} = \mathcal{S}\Delta\lambda. \tag{78}$$

To extract the sensitivity of the waveforms to the parameter λ_k, we solve

$$(\mathcal{C} + \mathcal{G})\mathcal{X}_k = \mathcal{S}e_k \tag{79}$$

where e_k denotes the kth unit vector (all zero, except entry k, where it is unity).

For the delay computation, the sensitivity of the waveform at a specific time-point j and node k is needed. Construct the (block-structured) vector

$$E = \begin{bmatrix} E_1 \\ E_2 \\ \dots \\ E_M \end{bmatrix} \tag{80}$$

with the vectors E_l given by

$$E_l = e_k, l = j \quad E_l = 0, l \neq j. \tag{81}$$

Then the required sensitivity a_k is $a_k = E^T \mathcal{X}_k$.

Note that $(\mathcal{C} + \mathcal{G})$ is block-lower-triangular. This means that operations with $(\mathcal{C} + \mathcal{G})^{-1}$ are cheap to compute. Of course, the matrices $\mathcal{C}, \mathcal{G}$ are never written down explicitly, we only perform implicit operations as as multiplying $(\mathcal{C} + \mathcal{G})^{-1}$ times a vector. Clearly, to extract the full set of sensitivity information, we must perform p solves – one for each parameter. This is acceptable if p is small, but problematic if p is large. On the other hand, for one solve, we obtain the sensitivities of the waveforms at *all* nodes and all *all* timepoints. The computational complexity is $O(pNM)$ for the solution.

The idea of adjoint analysis is to obtain the sensitivity of a voltage waveform at a *single* timepoint and *single* node, to perturbations of all parameters simultaneously, at all timepoints. With the above notation, the notation of the procedure is simple. First we solve

$$(\mathcal{C} + \mathcal{G})^T \mathcal{U} = E. \tag{82}$$

Denoting the vector of sensitivities $\eta = [a_1, a_2, \dots, a_p]$, we have

$$\eta = \mathcal{U}^T \mathcal{S}. \tag{83}$$

If the sensitivities for multiple timepoints or nodes are to be computed, there is one solve of Eq (82) for each such observation point. The computation of $\mathcal{S}$ is done once, and shared across all solves. If t is the number of such terminal points, the computational complexity is $O(tNM)$ for matrix solution. Compared to the direct computation, savings is possible if $t < p$.

---

[3]We have omitted uncertainty in the initial condition, which will contribute an additional term to $s^{(q),(l)}$ above.

We have not yet discussed the computation time for constructing the matrix S. At worst, this is $O(pDM)$ where D is the number of devices. However, usually either the number of parameters is small, p is $O(1)$, or each device depends on only a small number of parameters. In either case, the complexity becomes $O(DM)$ $O(NM)$ if the implementation is done so as to exploit such structure.

6   Experimental Results

A realistic circuit block was synthesized and mapped to an industrial 90nm technology. As process parameters, we considered the widths and thicknesses of the six metal layers needed to route the block. During parasitic extraction of the design, we computed the nominal values and sensitivities of each parasitic element (resistors and grounded capacitors), relative to each one of the 12 parameters.

In order to validate the interconnect delay and slew computations, we selected from the design 3671 nets, including nets in the internal logic, nets in the clock tree and nets in the pad wiring. For each of these nets, we computed the parametric delay and slew expressions for each of its taps (resulting in 13870 taps among all nets), while the port was excited by a rising voltage ramp. To assess the accuracy of the proposed methodology, the delay and slew sensitivities were compared to transistor-level simulations performed using the circuit simulator SPECTRE. In Figure 3 we present scatter plots of the sensitivities computed by both methods, for 4 parameters. In Figure 4 we present histograms of the relative errors for other 4 parameters. Both figures clearly show that the computed sensitivities accurately match those obtained by simulation.

In order to validate the cell delay and output slew computations we proceeded as follows. For a given standard cell of that same 90nm technology, and using Spice-level models, we generated a dotlib-style lookup table of size 7x7, for delay and output slew, as a function of input slew and load. Using these tables, and applying the proposed methodology, we computed the delay and output slew sensitivities for one of the cell instances in the previously mentioned design, considering its loading net obtained from extraction. Using the methodology proposed in Section 4 we generated the sensitivities of delay and output slew to all 12 parameters. Next, varying the parameter values, a similar set of sensitivities was also computed with SPECTRE, using accurate Spice-level models for the cell. The delay and output slew sensitivity values obtained using the proposed method were then assessed by computing its relative error versus the SPECTRE-generated data. These relative errors are shown in Figure 5 (left plot). As can be observed, the errors are in general small, usually in the low percentage range. The only exception to this rule is the pathological case of the slew sensitivity to parameter #2, whose absolute value is small, the smallest of all the sensitivities and near machine precision. In order to investigate this behavior, we introduced a variation in the input slew depending on parameter #2, so that the delay and output slew sensitivity values to this parameter would become larger. As a result we observed that when this happened the relative error dropped to the normal range, as shown in Figure 5 (right plot). Considering that the size

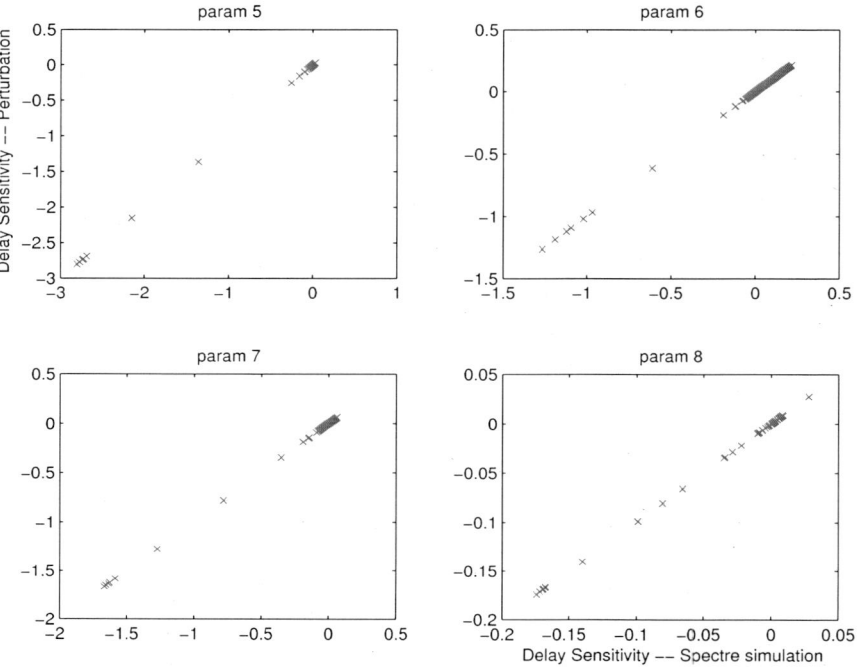

Fig. 3. Computed delay sensitivities vs. transistor-level simulation.

of the dotlib-style lookup table used was only 7x7 (typical value), providing a rough approximation of the behavior of the cell, and that the parasitic network was also approximated by a single lumped capacitance, we believe that the accuracy of the computed values is fairly good. Better accuracy should be obtained by using larger lookup tables, or by extending the proposed model for handling tables depending on other parameters.

7   Conclusions

In this paper we have developed an analytic delay calculation methodology suitable for use in a statistical static timing methodology. Our approach, based on a specific type of perturbation analysis, allows for the analytical computation of the quantities needed for statistical delay propagation. We also showed how perturbation analysis can be performed when only the standard cell delay table lookup models are available. The techniques proposed are robust and show good correlation with transistor level calculations. Furthermore, they can be directly applied when cell characterization is based either in voltage or current source models. Future work will show how to develop models that include nonlinear contributions from the process parameters.

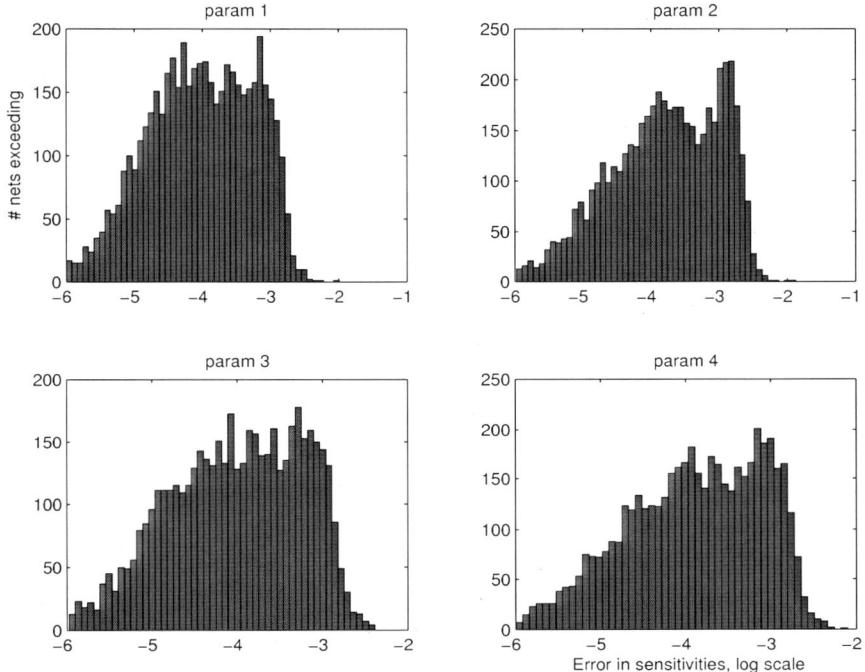

Fig. 4. Histograms of errors in computed delay sensitivities.

Acknowledgments

The authors want to thank Roberto Passerone for providing example circuits, and Vinod Kariat and Igor Keller for suggestions and discussions.

References

1. C. Visweswariah, K. Ravindran, K. Kalafala, S. G. Walker, and S. Narayan. First-Order Incremental Block-Based Statistical Timing Analysis. In *Proceedings of the Design Automation Conference*, pages 331–336, San Diego, CA, June 2004.
2. H. Chang and S. S. Sapatnekar. Statistical Timing Analysis Considering Spatial Correlations using a Single Pert-like Traversal. In *Proceedings of the International Conference on Computer Aided-Design*, pages 621–625, San Jose, CA, November 2003.
3. Luis Guerra e Silva, L. Miguel Silveira, and Joel R. Phillips. Efficient Computation of the Worst-Delay Corner. In *Proceedings of Design, Automation and Test in Europe, Exhibition and Conference*, Nice, France, April 2007.
4. J. Wang, P. Ghanta, and S. Vrudhula. Stochastic Analysis of Interconnect Performance in the Presence of Process Variations. In *Proceedings of the International Conference on Computer Aided-Design*, pages 880–886, San Jose, CA, November 2004.

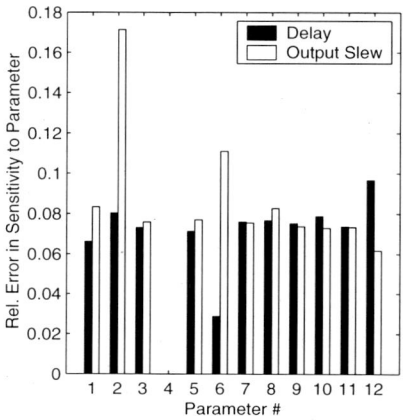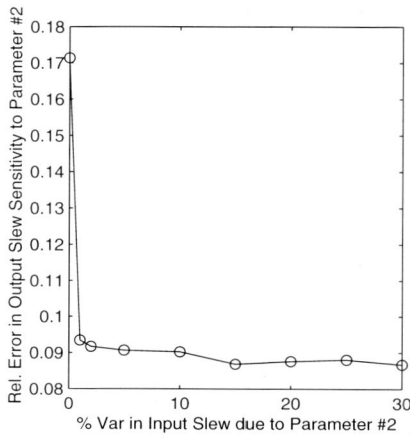

Fig. 5. Relative errors in computed cell delay and output slew sensitivities.

5. Joel R. Phillips. Variational Interconnect Analysis Via PMTBR. In *Proceedings of the International Conference on Computer Aided-Design*, pages 872–879, San Jose, CA, November 2004.

6. X. Li, P. Li, and L. Pileggi. Parameterized interconnect order reduction with Explicit-and-Implicit multi-Parameter moment matching for Inter/Intra-Die variations. In *Proceedings of the International Conference on Computer Aided-Design*, pages 806–812, San Jose, CA, November 2005.

7. Z. Wang, R. Murgai, and J. Roychowdhury. ADAMIN: Automated, accurate macomodelling of digital aggressors for power and ground supply noise prediction. *IEEE Transaction on CAD*, 24:56–64, January 2005.

8. Joel R. Phillips. Model Computation for Statistical Static Timing Analysis, May 2006. Cadence Internal Report.

9. R. Telichevesky, J. White, and K. Kundert. Efficient AC and Noise Analysis of Two-Tone RF Circuits. In *Proceedings of the Design Automation Conference*, June 1996.

10. Sani R. Nassif and Zhuo Li. A More Effective C_{EFF}. In *Proceedinge of the Sixth International Symposium on Quality of Electronic Design*, pages 654 – 661, San Jose, CA, March 2005.

11. J. Qian, S. Pullela, , and L. Pillage. Modeling the Effective Capacitance for the RC Interconnect of CMOS Gates. *IEEE Trans. on VLSI*, 13:1526 – 1535, 1994.

12. Florentin Dartu, Noel Menezes, and Lawrence T. Pileggi. Performance Computation for Precharacterized CMOS Gates with RC Loads. *IEEE Trans. on CAD*, 15(5):544 – 553, May 1996.

13. Igor Keller, Nishath Verghese, and Kenneth Tseng. A Robust Cell-Level Crosstalk Delay Change Analysis. In *Proceedings of the International Conference on Computer Aided-Design*, San Jose, CA, November 2004.

14. J.F. Croix and D.F Wong. Blade and Razor: Cell and Interconnect Delay Analysis using Current-Based Models. In *Proceedings of the Design Automation Conference*, pages 386 – 389, Anaheim, CA, June 2003.

15. S.W. Director and R. A. Rohrer. The Generalized Adjoint Network and Network Sensitivities. *IEEE Trans. on Circuit Theory*, CT-16(3):318–323, August 1969.

A Power-Efficient Methodology for Mapping Applications on Multi-Processor System-On-Chip Architectures

Giovanni Beltrame, Donatella Sciuto, and Cristina Silvano

Politecnico di Milano, DEI, Milano 20133, Italy

Abstract. This work introduces an application mapping methodology and case study for multi-processor on-chip architectures. Starting from the description of an application in standard sequential code (e.g. in C), first the application is profiled, parallelized when possible, then its components are moved to hardware implementation when necessary to satisfy performance and power constraints. After mapping, with the use of hardware objects to handle concurrency, the application power consumption can be further optimized by a task-based scheduler for the remaining software part, without the need for operating system support. The key contributions of this work are: a methodology for high-level hardware/software partitioning that allows the designer to use the same code for both hardware and software models for simulation, providing nevertheless preliminary estimations for timing and power consumption; and a task-based scheduling algorithm that does not require operating system support. The methodology has been applied to the co-exploration of an industrial case study: an MPEG4 VGA real-time encoder.

1 Introduction

Technological advances have made multiprocessor implementations of embedded systems a viable alternative to traditional single-processor and pure-hardware designs. Such multiprocessor designs offer high levels of performance, flexibility and, at the same time, promise low-cost and power-efficient implementations. One of the most promising approaches to design such systems is the Multiprocessor System-on-a-Chip (MPSoC) paradigm. A typical MPSoC system consists of a number of processing elements (PEs), which can be programmable processors or fixed application-specific co-processors, and storage elements (SEs) connected to PEs via an on-chip communication architecture. As a result, MPSoC architectures represent heterogeneous systems that offer flexible parallel processing resources for implementation of bandwidth-demanding multimedia applications.

However, MPSoC platforms introduce several design challenges associated with their parallel and heterogeneous architecture. Platform-based design [1] faces the problem of defining a configurable microarchitecture platform, onto which an application can be mapped through a well defined parallel programming model, specified via an application-program interface (API) platform. The mapping of an application to a microarchitecture platform starts from a complex system specification and goes through a possible extensive design space exploration phase. In this context, the reuse of a large

Please use the following format when citing this chapter:

Beltrame, G., Sciuto, D. and Silvano, C., 2007, in IFIP International Federation for Information Processing, Volume 249, VLSI-SoC: Research Trends in VLSI and Systems on Chip, eds. De Micheli, G., Mir, S., Reis, R., (Boston: Springer), pp. 177–196.

base of existing software to perform the exploration of different possible implementations constitutes an important concern. The existing software commonly written in C/C++ language with a single-processor architecture in mind cannot be directly reused in a multiprocessor environment, especially if it consists of a heterogeneous mix of different software and hardware components. The existing software needs to be adapted to the parallel capabilities of the architecture. Furthermore, to enable fast and flexible exploration of the possible application-to-architecture mappings, it is necessary to automate the hardware-software partitioning of the application. Therefore, there is a need for a disciplined approach based on a unified parallel modelling paradigm that would enable a smooth translation of existing sequentially-coded software algorithms into their parallel models suitable for the design space exploration of MPSoC platforms.

In this work, we show how it is possible to map an application to the MultiFlex [2] platform, exploiting the features offered by the combined use of the DSOC and the SMP programming model to provide a novel way of performing initial partitioning and exploration. The use of the DSOC programming model together with the transaction-level modeling (TLM) [3] infrastructure allows easy moving of a component from hardware to software and vice-versa.

Once the application has been partitioned and mapped to the target platform according to its performance and power constraints, it is possible to further optimize its power consumption with the use of a proper Dynamic Power Management System (DPMS). In this work we extend our previous work [4] with a task-based DPMS scheduler, that optimizes power consumption without affecting the system's performance.

The rest of this paper is structured as follows: Section 2 describes current approaches to the parallel mapping problem; Section 3 outlines the proposed mapping flow; Section 4 introduces power optimization to the overall architecture; Section 5 describes an industrial case study to which the proposed methodology was applied; and finally, Section 6 draws some concluding remarks.

2   Related work

The present paper focuses on a mapping methodology of applications onto MPSoC platforms in order to identify the best trade-off of power and performance behavior of the application on a given platform configuration. The next step is task allocation for dynamic power consumption optimization. Most literature is either focused on programming models to solve the mapping of software applications onto specific platforms or on the scheduling for dynamic power management on multiprocessors. So far, no paper considered yet the entire design flow from a comprehensive perspective.

2.1   Designing MPSoCs

The architectural changes introduced in the emerging MPSoCs have a direct consequence in how software engineers program. This fact has already been acknowledged by several researchers, who have proposed preliminary solutions. Most of them agree on the importance of new high-level programmer views of SoC.

A number of programming models focused on multiprocessor SoCs have been presented, such as the MESCAL approach [1], which has served as base for further different programming models. Nevertheless, most of them are application or domain specific.

A more general approach composed of two SoC parallel programming models has been introduced in [5]. The Distributed System Object Component (DSOC) model and the Symmetric Multi-Processing (SMP) model are inspired by leading-edge approaches for large system development, but adapted and constrained for the SoC domain.

Various actor-oriented frameworks are proposed to capture arbitrary Models of Computation (MoC) for the purpose of system level modeling and tool supported paths to exploration, implementation and/or verification [6]. The modeling strategy presented in this paper can be implemented on top of any of these MoC generic frameworks. We selected SystemC mainly because of the broad user acceptance and commercial tool support. Complementary to our top-down refinement flow, the Component Based Design paradigm [7] advocates the bottom-up platform composition from a parameterizable IP library, containing off-the-shelf processing elements, communication fabrics and hardware dependent software layers. This approach is clearly advantageous for the rapid exploration and implementation of the general purpose portion of the application, whereas our approach is focused on application specific architectures executing the data-processing part.

The highest possible abstraction level for design space exploration and application mapping is static performance analysis [8, 9]. Other approaches are closer related to simulation frameworks for top-down exploration and refinement like ARTEMIS [10] and StepNP [5]. Some recent works present simulation frameworks for mapping applications based on SystemC [11], and mapping and scheduling of applications on parallel architectures [12, 13].

2.2 Power optimization

Dynamic Power Management (DPM) is a design methodology that dynamically reconfigures an electronic system to provide the requested services and performance levels with a minimum number of active components or a minimum load on such components.

Dynamic Voltage/Frequency Scaling (DVFS) requires processors to adapt their voltage and frequency at run-time, according to some control actions. The work in [14] introduce architectural and implementation issues together with energy saving bounds concerning DVFS techniques.

A significant amount of research on DVFS scheduling and algorithms have been proposed on both single and multi-processor systems. DVFS has been implemented in several contemporary microprocessors as Intel XScale, AMD Mobile K6 Plus and Transmeta Crusoe [15]. These can be classified as compile-time and run-time policies [15]. Run-time policies have drawn more attention because of the ability to reduce energy consumption in response to workload variations. A run-time DVFS policy consists of two elements [15]:

- *Scaling points*: these are the positions where voltage/frequency scaling occurs. They can be signaled by timer interrupts, cache misses, etc. The time frame enclosed by two scaling points is referred as a *scaling unit*.

– *Scaling criteria*: it is the policy that determines the voltage/frequency level of the next scaling point.

Depending on scaling points, DVFS policies can be classified as interval-based policies (timer interrupts) [16, 17], micro-architecture-based policies (cache misses and performance counters),and task-based policies (task arrivals and completions) [18]. Other techniques have also been introduced for multiprocessor embedded systems, such as [19], focusing mostly on scheduling algorithms.

3   Mapping and Exploration Design Flow

The proposed mapping flow allows the designer to co-explore the application design space including both the architecture model and the application source code, as shown in Figure 1. This flow is targeted to highly parallel applications, like, for instance, multimedia ones. This flow starts from an executable specification of an application, in a

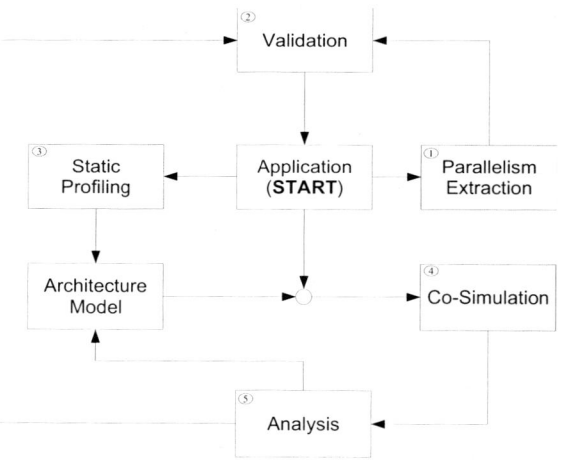

Fig. 1. The proposed mapping and exploration flow

language that can be compiled and executed on the target micro-architecture platform. The mapping phase starts by considering a fully software implementation. The application undergoes the following steps:

– *Static Profiling*: the source code of the application is profiled natively on a workstation and the total computation effort needed by the application is estimated, given performance constraints. It is worth noting that this provides the *lower bound* to the number of processors, in case of a fully software solution.
– *Parallelism Extraction*: since the target architecture is an MPSoC, the intrinsic parallelism must be extracted from the application to exploit the available system resources. The general problem of automatic parallelization of the code is out of

the scope of this paper. We consider code that has been already parallelized for the MultiFlex programming models. This step identifies data dependencies at very coarse-grain, e.g. image macroblocks for an MPEG4 encoding algorithm.

- *Validation*: any modification to the reference source code is validated against the reference data, to avoid losing the original program behavior due to some neglected dependencies.
- *Architecture Modeling*: static profiling provides the lower bound in terms of computational resources to run the application and to meet the constraints, hence we need to model the architecture components considering these basic requirements. As an example, the static profiling states the minimum number of processors for a fully software implementation.
- *Co-simulation*: the architecture model and the software are co-simulated, extracting both performance and power measures.
- *Analysis*: simulation results are collected and analyzed to modify both the application (e.g. to increase the parallelism) and the architecture model, by modifying the initial hardware and software partitioning, for example by moving some parts of the application in hardware or by varying some platform architectural parameters.

The mapping continues to cycle through these steps until the constraints for the application are fully met. These steps are detailed in the following.

3.1   Static Profiling

Statically profiling an application consists of determining its computational requirements. The idea is to define a *lower bound* to the resources needed by the application. In this way, it is possible to configure the micro-architecture platform so that it satisfies the lower bound. Static profiling can be done natively on any machine that supports a profiling tool-set, like `gprof`or `iprof`. `iprof` identifies the number of instructions needed to execute the native code, while `gprof` provides analysis on the call graph. Merging the two outputs allows the designer to identify the computational needs of the application and how those need are distributed in the code. As an example, an MPEG4 encoder developed at STMicroelectronics (described in detail in Section 5), when profiled on a x86 architecture, shows a requirement of 4.081 GIPS (Giga Instructions Per Second). It is worth noting that this requirement is strictly valid only on the same Instruction Set Architecture (ISA) on which the application was profiled. More complex (or simpler) ISAs may run the code using less (or more) instructions.

Considering a micro-architecture platform where the target processor is an ARM9 core running at 200MHz, the bound for a fully software implementation is 21 ARM CPUs. In fact, ARM9 has at most a CPI (cycles-per-instruction) equal to 1:

$$c \times f \times CPI = 21 \times 200 \cdot 10^6 \times 1$$
$$= 4200 GIPS$$

with c and f the number of processors and their frequency respectively. Even supposing that the micro-architecture platform has enough computational power, a sequential application cannot run on all the processing elements and therefore the designer needs to modify it to exploit all possible inherent parallelism.

3.2 Parallelization

Parallelization requires the identification of sections of code working on independent sections of the application's input data. This is not trivial even for applications that exhibit an "embarrassing" level of parallelism like multimedia audio and video encoding. The kind of parallelization needed for distributing tasks to the processing elements of an MPSoC is coarse- to medium-grained. Although there is a large amount of research devoted to the automatic parallelization of code [20, 21], coarse-grain parallelizing compilers are still not widespread in the industry. It is more common to directly apply programming models to sequential code, leaving the designer to identify data-dependencies in the code. As an example, a routing application like an IPv4 forwarder, can be parallelized in such a way that every packet is manipulated by a separate thread. The use of the MultiFlex SMP approach reduces the effort [2] because it provides a well-established parallel API, that is entirely similar to multi-threaded programming when using a traditional UNIX-like operating system, in particular POSIX threading. In addition, the Concurrency Engine (a hardware SMP accelerator that is part of the Multi-Flex approach) provides load balancing functionalities, leaving the designer to the only effort of determining the data dependencies of the target application. The proposed flow

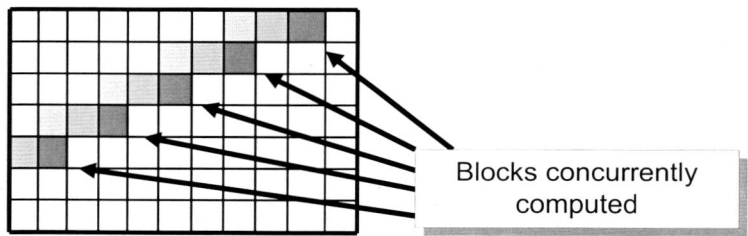

Fig. 2. Parallelization of a motion estimation algorithm over the macroblocks of a frame

involves the application of the MultiFlex SMP programming model, identifying data dependencies manually. As an example, considering STM's MPEG4 encoder, independent data blocks in the motion estimation algorithm are represented by a "knight's move" on the chessboard formed by dividing a frame to be encoded in macroblocks (squares of 2x2 blocks of 8x8 pixels), as typical in JPEG and MPEG compression [22]. The parallelized application starts a thread for each independent macroblock to be encoded, as shown in Figure 2.

3.3 Validation

After some parts of the application have been parallelized, the resulting code has to be verified again in order to prove that the functional requirements of the application still hold. Applying regression tests to simulated code can be excessively time consuming and may seriously slow down the design process. To overcome such problem, we exploit the similarities of the MultiFlex programming model to Pthreads. It is conceivable to compile natively the same code that would run on the micro-architecture platform, given that the API is the same. Using Pthreads it is possible to explore the parallelization of the application natively, obtaining the exact same behavior that would take place on the simulated platform. The only attention to be given is to use only those Pthread's concurrency control structures that have been implemented in the Concurrency Engine (such as semaphores, monitors, conditions). Validating the parallel design implies enforcing proper synchronization through the use of those structures avoiding any sort of deadlock or race condition. The time saving introduced by this solution is conspicuous. As an example, a simulation run over 30 frames of the MPEG4 encoder takes about 1s natively, while it requires minutes (or even hours, depending on the accuracy level) when it is executed, on the same workstation, with an instruction-set simulator (ISS).

3.4   Simulation

Once the first run of mapping has been applied, the architecture model and the software can be co-simulated. Co-simulation is performed using a transaction-level simulator (in our case, StepNP [5]), first using timed functional models. The first run of simulation provides data concerning the actual performance of the system, giving bounds to delay and energy consumption. The results of the simulation provide data to proceed in the exploration of the platform configuration. Examples of these results might be low processor utilization, which mean that channel latency is excessive and has to be accounted for, as an example using hardware multi-threading [23]. Whenever the application meets the required constraints, simulation can be performed at a lower abstraction level, going into progressive refinements that lead to the actual implementation of the system.

It is unlikely for a complex, high speed, application, that a full software solution gives acceptable performance. In this case, some parts of the system may be implemented in hardware, raising the performance but, at the same time, raising the platform's design cost and lowering its flexibility.

3.5   Hardware-software partitioning

One of the main advantages of the proposed flow is the ease of HW/SW partitioning through the use of DSOC. In fact DSOC does not make a distinction, from the point of view of the user, between hardware and software components. The DSOC ORB routes requests and the MP engine takes care of the marshaling and unmarshaling activities. In fact, it is possible to create transaction-level models of the application functions using the exact same code that is used for software models, as shown in Figure 3.

During the static profiling phase, the critical kernels of the applications have been identified. These are implemented as DSOC objects, defining their function signatures as appropriate interfaces using SIDL. The SIDL compiler generates skeletons and stubs needed for communication between DSOC clients and servers. Using a transaction-level DSOC object adapter, it is possible to connect any DSOC object to a communication channel. Since the code used by DSOC objects can be either compiled for the simulated processors or natively, it is kept into a separate library compiled in both forms. Being StepNP (and MP4Free) based on SystemC, which produces native executable simulators, the natively-compiled library can be linked directly to the simulator, and accessed through the stubs. This allows the creation of untimed functional models of each component modeled as DSOC. Models can be turned into timed functional ones adding appropriate `wait()` statements in the SystemC wrapper. It is worth noting how this approach is not specific to a type of function or hardware component, but it is completely general: it is sufficient to define a SIDL interface to connect a new object.

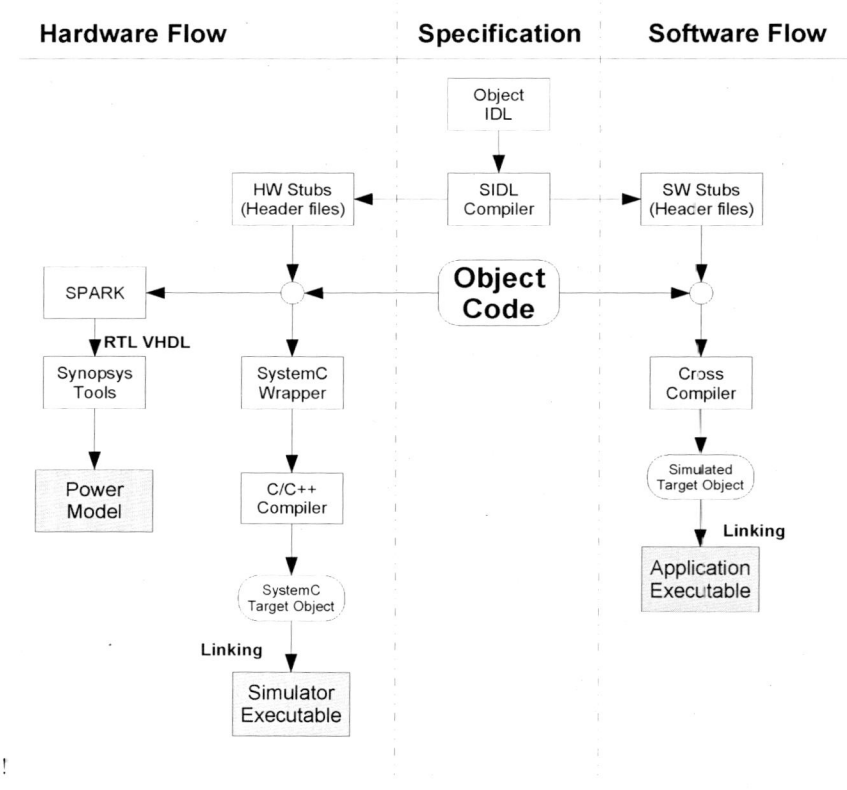

tb!

Fig. 3. Exploiting DSOC to re-use code for both hardware and software object models

The proposed methodology, however, suffers from a minor limitation. Since the simulated application and SystemC have different address spaces, it is not possible to use pointers when passing parameters to hardware components. This means that is is not possible for a hardware component to access memory via DMA, unless the SystemC wrapper is sophisticated enough to support address space conversion.

Concerning power consumption, the timed functional models associated with each DSOC object have no power model. To overcome this limitation, we roughly estimated the energy consumption per access to the components using SPARK [24]. SPARK is a behavioral C synthesizer, that produces RT VHDL code. The VHDL code is synthesized as well in Synopsys Design Analyzer and estimated with Power Compiler. The result gives a first estimation of the energy cost per access to the component, as shown in Figure 3.

Exploiting DSOC, SystemC and TLM, the designer can perform the HW/SW partitioning of the system as a matter of turning some switches and running simulations, greatly simplifying high-level design space exploration. The methodology has been applied to a set of critical kernels, as outlined in the following.

4   Power-Aware Scheduling on MPSoCs

After the application has been partitioned and mapped to the MPSoC platform, in this work we also propose a power-aware task-driven scheduling algorithm that, with respect to previous approaches, works at run-time without the need of an operating system on a multi-processor system. The basic idea is to exploit the MultiFlex [2] hardware-assisted programming models and task structures to gather the necessary information for the scaling criteria to assign voltages to processors. The scheduler is included in the Concurrency Engine as a hardware component, and, if only hardware threads are used, does not require any operating system support except that needed to access the Concurrency Engine (CE). The scheduler acts transparently, identifying task events (start, finish, resource availability) snooping the requests to the CE core. The proposed voltage scheduler, described in the following, has three key advantages:

1. It is a task-driven scheduler but nevertheless does not require any operating system support.
2. It is based on very simple, constant time algorithms
3. Its granularity is very coarse, trying to get as close as possible to the optimal value of one voltage setting per deadline [25]

Due to the structure of most common SMP programming models, i.e. a main thread forking parallel worker threads, it is easier than in the general case to predict the next value of the average load of a processor for the next scheduling unit, due to the predictable behavior of the system. The scheduling points are the fork and join operations (therefore the classification of the policy as event-driven) defined by a predictive approach, outlined in the following. We define the *main thread* the hardware thread executing the main flow of a program, and *worker thread* every other thread. *Tasks* are functions or programs executed by worker threads; *jobs* are instances of a task, that is, they are a mapping between a task and a working data set. A fork operation consists of creating a set of jobs that have to be executed by worker threads. According to the MultiFlex model, only hardware threads are allowed in the system, and the CE maps jobs to threads according to its own internal scheduling algorithm.

For each scheduling unit, we determine the best load obtained for each processor, then use them to compute the predicted value of the next scheduling point. As a prediction scheme, we use an exponentially smoothed moving average, obtained as a weighted average of the best loads of the given task. This prediction scheme works well for applications whose tasks roughly keep the same behavior in time. The assumption is not a limitation if tasks (i.e. functions) don't have a timing behavior that is strongly dependent on data. Nevertheless, it is possible to use another and perhaps more effective prediction scheme. The average load of a processor is determined with its internal instruction counter and is a number less than or equal to one. The idea is to scale the voltage/frequency of all the processors to keep the average load as close as possible to one. Load average tends to be lower than one due to wait states caused by channel latency, contention, etc..

The targets of the MultiFlex approach are multimedia and network applications: these applications require a huge amount of computational resources, but they are also quite linear and repetitive. This linearity allows collecting statistics on the current program in order to manage future iterations of the same task. According to the programming model, the sequence of fork-join of an application is computed sequentially, and tasks are forked to one or more threads. The main thread is stalled until all the worker threads have completed their task. For each task we keep track of the worst execution time τ as an exponentially smoothed moving average. At every scheduling point, the voltage/frequency of the processor is set according to the predicted worst load average for the next set of jobs. Since most DVFS processors can modify their voltage settings in discrete intervals, we approximate the setting to the voltage/frequency tuple $s_i = (V_{dd}, f)$ such that

$$f \geq f_{max} \cdot l_w \qquad (1)$$

where f_{max} is the maximum frequency value and l_w is the predicted average load for the next scheduling unit. This means that the frequency will be higher or equal to the frequency needed to complete the task without incurring in performance penalties, according to the predicted load. The higher the accuracy of the prediction, the better the DVFS result. This approach can be extended to perform static voltage scheduling, if the execution time of each task is known a priori. Finally, we can assume that the scheduling points are far enough in time to reduce the scheduling points in such a way that the transition cost is not affecting the overall power consumption of the application. This has been proved valid for an MPEG4 encoder application: fork-join sections generate a high number (up to 10^3) of threads, and each task requires a time $t \gg t_{tran}$, making the overhead negligible.

With the classic SMP approach, scaling voltage of active processors may still incur in energy waste due to idle processing elements. This may happen in two conditions:

1. The inherent parallelism of the application for a task does not allow forking enough threads to cover all the processing elements of the target platform. This means that some processors will be idle during the execution, until the next join.
2. Job distribution is not uniform during a join phase: some jobs finish earlier than others and leave their processor idle.

To avoid this energy expenditure, the DVFS subsystem of the CE applies DPM methodologies, turning off the unused hardware if the conditions arise and this does not affect the performance of the system due to restart delays.

In the following, for the sake of simplicity, we will consider mono-threaded processors, i.e. processors that can execute only one thread at a time. However, the approach still holds for multi-threaded processing elements: a processor executing n threads is considered as n single-threaded processors that belong to the same voltage cluster and can be turned off if and only if all n threads are idle.

4.1   The Concurrency Engine Scheduler

To maximize the effectiveness of the approach, the DVFS subsystem of the Concurrency Engine has to interface with its multi-processor task scheduler. Our approach is designed to interface with both static (fixed for all the tasks) and dynamic scheduling approaches, but in this work, we will focus on the standard dynamic scheduler of the Concurrency Engine. The original implementation of the CE scheduler is a simple FIFO: every new job is mapped in sequence to available worker threads. Every time a job is completed, if there are some unmapped jobs, the first in the queue is assigned to the newly freed resource.

To manage the state of the processors (idle, on, off), the CE monitors the execution of every job. Whenever a task is about to be completed, some processors become idle, but turning them off immediately is not necessarily the most effective strategy. In fact, it is possible that their restart is too slow for the next job to be scheduled without delays, hindering the system's global performance. In the original CE implementation, after the last job is assigned, all the processors are active, finishing their jobs before deadline τ, given by the application constraints (e.g. 30 fps for a video encoder). At the time τ all processors are ready for the next fork, but in the interval between the completion of their job and τ, they are idle. We define the *critical time* $t_{c,i}$ of a task i, the minimum time required to turn off and restart a processor before the beginning of the next task. Finally, we define the *shutdown time* t_d and the *start time* t_s as the time needed to turn off a processor and to turn on one that was previously off, respectively. The time $t_{s,i}$ is the time during task i after which it is not possible to turn on a processor before the next task.

4.2   Scheduling with on/off management

Since the CE possesses all the information needed to schedule tasks over free resources, we can add the proper signals to switch on and off each processor independently. Each processor needs a non-null time to change its state (in either direction), so the CE must check if there is enough time available to shutdown and restart a processor before the beginning of the next task, considering also the voltage schedule of the current task.

Since the DVFS subsystem predicts the duration of a task τ, the CE has an estimation of the time needed to execute the task. Supposing that the restart time is known as $t_{rs} = t_d + t_s$, where t_d is the shutdown time and t_s is the start time. Therefore:

$$t_{c,i} = \tau - t_{rs} \qquad (2)$$

The improvement when compared to the standard CE scheduler is little: only processors that can be restarted before τ are turned off, that is processors are stopped only if the number of remaining jobs is lower than the number of available processors, and there is no guarantee that there will be any in all tasks. The constraint that all the processors have to be active at time τ, before starting a new job assignment, can be relaxed if not all the processors are needed for the next task. This can be discovered by the CE after a first conservative run where all tasks are supposed to be needing all computational resources, or can alternatively be defined at design time. In this case, shutdown and restart operations can be allowed over task boundaries (i.e. when there are no more waiting jobs to be mapped), leading to three cases:

- Case 1: $\tau_{i+1} - \tau_i > t_{rs}$
  In this case it is possible to shutdown a processor in any moment between t_0 and τ_0. Considering the example in Figure 4, supposing that the task ending in τ_2 needs only 3 resources, processors $P2$ and $P5$ can be shut down and their restart can be scheduled at $t_{c,2}$ without incurring in any penalty. This case provides the maximum

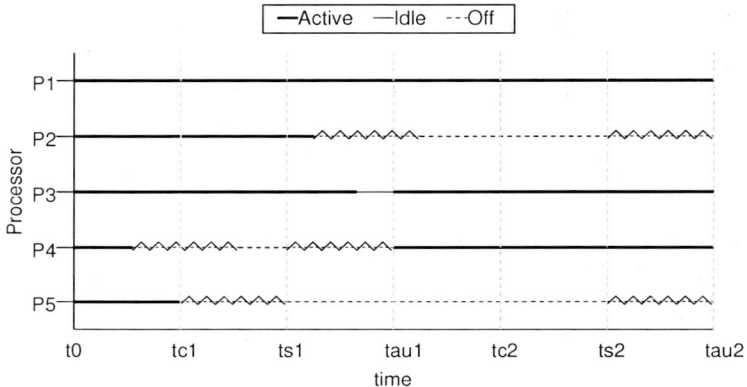

Fig. 4. Shutdown over task boundaries with $\tau_{i+1} - \tau_i > t_{rs}$

power saving, because it enforces loose constraints on the processors' shutdown process.
- Case 2: $t_{start} < \tau_{i+1} - \tau_i < t_{rs}$
  In this case, it is only possible to start (and not restart) a processor before the next deadline
- Case 3: $\tau_{i+1} - \tau_i < t_s$
  In this case, it is not possible to stop and restart any processor before τ_2

Scheduling with task reordering It is also possible to reorder job mapping to processors so that at the task boundary, the number of required resources k_{next} is less than the number of processing elements k. In fact, the programming model implies that each job is independent from the others, as each job is working on a different data set. Roughly, each job will take the same execution time σ when executed repeatedly using the same resources. If there are k processing elements and there are n jobs to be completed, there are always k active jobs until the last job is scheduled, and the total execution time is:

$$\tau \approx (n \bmod k) \times \sigma \tag{3}$$

We call each set of k jobs an *iteration* of the task. If the number of jobs n is a multiple of the number of processors, then it is not possible to perform reordering among iterations. However, if $k \bmod n \neq 0$, it is conceivable to reorder the jobs mapping among iterations, to reduce the fragmentation of idle times at task boundaries.

Given the number of processors k and the number of jobs n, supposing initially, that $n > k$, the number of mapped jobs for all the iterations of the task is described by the regular expression: $[k] + \lambda$, where

$$\lambda = \begin{cases} k & \text{if } n \bmod k = 0 \\ n \bmod k & \text{otherwise} \end{cases}$$

As an example consider $n = 16, k = 5, \lambda = 1$, in this case $[k] + \lambda = \text{"5551"}$. Therefore, the job distribution over the iterations is the sequence 5-5-5-1, 5 jobs for the first 3 iterations and 1 for the last. Let us rewrite the expression as $k'k^*\lambda$, where $k' \in [1, k]$. Keeping constant $k' + \lambda$, it is possible to rearrange the job assignment to best fit the scheduling of shutdown and restart operations in the first and the last iteration. We define the tuple (k'_i, λ_j) the value of λ and k' for the ith and jth task, respectively.

As an example, assume that $k = 5$, then $\lambda = \{1, 2, 3, 4, 5\}$, and $k' = 5$; with these values, all the possible pairs of (k', λ) and their rearrangements are: $(5, 1) \Rightarrow (3, 3), (5, 2) \Rightarrow (4, 3), (5, 3) \Rightarrow (5, 3), (5, 4) \Rightarrow (5, 4), (5, 5) \Rightarrow (5, 5)$.

This arrangement reduces the probability that $\lambda_j > k'_{j+1}$. In fact, if $\lambda_j > k'_{j+1}$, some processor may be idle at the beginning of the first iteration of task $j + 1$. Considering the example and two consecutive tasks, there are 9 different (λ_j, k'_{j+1}) tuples, and only 3 have $\lambda_j > k'_{j+1}$. In the worst case, 2 processors every 4 join operations will be in idle state for a time smaller than t_{rs}. Removing the hypothesis that $n > k$, in the worst case, a processor can be in an idle state for an iteration. This reordering table is computed at design time and hard-coded in the CE, and depends on the number of processing elements available in the target SoC. The CE FIFO scheduler uses the reordering table for scheduling whenever it receives a fork command by the main thread.

If we consider that, in general, the execution time needed by a task of more than a few assembly instructions is greater than t_{rs}, it is possible to merge the three cases presented in Section 4.2 into a single algorithm. The algorithm merges the last and the first iteration of two consecutive tasks, considering two separate deadlines. The behavior is shown in Figure 5, and the algorithm follows:

```
 1:  scheduled_active = 0;
 2:  for all  P_j  do
 3:      if  P_j completes his job before t_c,i  then
 4:          shutdown P_j;
 5:          if length(scheduled_active) ¡ k_next  then
 6:              schedule P_j for restart at t_s,i
 7:              scheduled_active++;
 8:          else
 9:              schedule P_j for restart at t_s,i+1
10:          end if
11:      else if  P_j completes his job in [t_c,i, t_c,i+1] ∧ k_a ≥ k_next  then
12:          shutdown P_j;
13:          schedule P_j for restart at t_s,i+1;
14:      else
```

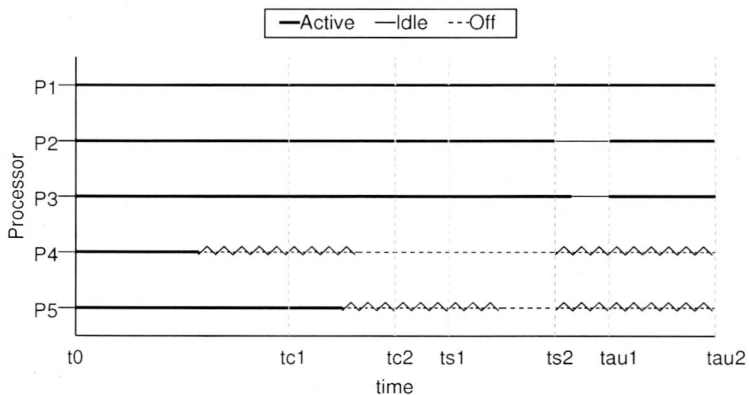

Fig. 5. Behavior of the generalized DPMS

15:        scheduled_active++;
16:    **end if**
17: **end for**

It worth noting that if $\tau_{i+1} - \tau i > t_{rs}$, we obtain Case 1, and if $t_s < \tau_{i+1} - \tau i < t_{rs}$, the result is an improved arrangement of Case 2, that were detailed in Section 4.2. Theoretically, it is possible that $\tau_i + \tau_{i+1}$ is lower than the time required to restart a processor and only in such case, the scheduler considers also τ_{i+2} and so on. However, it is reasonable to assume that the time required to execute a task is significantly larger than the time needed to restart a processor, and therefore only two deadlines have to be considered when scheduling shutdowns and restarts. In fact, considering as an example an MPEG4 algorithm parallelized for the purpose, each task requires on average $7ms$ while an ARM10 processor running at 200MHz requires $\sim 10\mu s$to awake from shutdown mode.

The introduced algorithms have linear complexity in the number of computational resources, making them viable for a fast hardware implementation.

5   Case Study: an MPEG4 encoder

This section describes how the methodology was applied to map an MPEG4 onto the MultiFlex platform. This platform is constituted of a variable number of ARM processors with a variable number of hardware threads, a two-level cache structure and the STBus interconnection network [12].

The application, specified in C, has been initially profiled statically, using gprof and iprof (GNU open source tools) on a Linux machine. Profiling defined a lower bound on the number of processors needed for execution: the computing power needed by the applications amounts to 4.08 GIPS. A full software solution would require a minimum of 21 ARM CPUs running at 200MHz (each one providing at most 200MIPS). Table 2 shows the results for the 9 functions that take most of the execution time during the encoding of a frame. These functions represent only a very small portion of the application code (approximately 6% of 8086 lines of code) but they cover 82.81% of all computational resources needed for execution.

Table 1. Static profiling results of the MPEG4 application on Linux

Function	Execution Time [%]	Lines of Code	Fraction of source [%]
BSAD	27.98	90	1.11
BQ	19.17	100	1.21
BDCT	10.36	80	1.11
BZIGZAG	6.22	5	0.06
BIDCT	5.70	110	1.2
BADD	4.66	15	0.18
BDIFF	3.63	17	0.21
BQI	2.59	37	0.45
BSUM	2.59	10	0.12
TOTAL	82.81	465	5.65

As a first design choice, these blocks were selected for a possible hardware implementation in a coprocessor or MPEG4 accelerator: these functions are present in all versions of the MPEG algorithm [22], and moving them to hardware blocks does not hinder the overall flexibility of the system. These blocks correspond to 83% of all computation time but less than 6% of all the application lines of code. These functions were modeled as DSOC servers, with the application software accessing either the hardware or software versions of the models.

The remaining 17% of the application computation is executed as software. The profiling of the distributed application shows that 800 MIPS are required to run the application on the ARM processors. The data access bandwidth of these processors is 1.7 GB/s.

Concerning the hw/sw partitioning of the application, we applied DSOC programming model allowing to easily switching from software to hardware an vice-versa, using timed functional models. The code was compiled and executed on StepNP in an Instruction Set Simulator (ISS) when simulated as software and it was instrumented for timing analysis, compiled natively, and executed in SystemC space when simulated as hardware.

Adding power estimation required to build models for the hardware components. To simplify the modeling, we used SPARK [24] and synthesized RTL directly from the source code, and we used Synopsys Power Compiler to derive a power model based on the access to the devices using STM technology libraries at $0.18 \mu m$, as shown in Table 2. The use of SPARK provides good results since the MPEG4 high-computation kernels are very simple functions.

Switching progressively each kernel to hardware, starting from a full-software solution and from the most computationally-intensive kernel, brings to the results shown in Figure 6. It is remarkable how turning BDCT into hardware constitutes a significant

Table 2. SPARK results for the MPEG4 critical kernels

Function	Cell Pow. [μW]	Interconnect Pow. [μW]	Total [μW]	Leakage [μW]
BSAD	17,028	16,063	33,091	20,494
BQ	6,426	1,572	7,999	5,903
BDCT	0,156	0,009	0,165	0,060
BZIGZAG	5,133	1,716	6,850	4,837
BIDCT	23,179	14,037	37,216	24,761
BADD	16,069	5,247	21,316	15,170
BDIFF	38,160	14,114	52,274	29,481
BQI	4,308	1,430	5,739	7,057
BSUM	0,329	0,781	1,110	3,035

energy saving, while it does not have the same effect on performance. Only turning both BDCT and IDCT into hardware has the effect of raising the performance. This means that the DCT is one of the major bottlenecks of the system, and therefore justifying the hardware design choice. The frame rate increases as more components are turned to hardware, but this is not sufficient to reach the 30 frames per second constraint, and more optimizations have to be done in terms of parallelism exploitation.

To exploit the MPSoC architecture, the MPEG4 application has then been split into parallel sections working on independent data. This phase has been optimized manually for the target MPEG4 application. The inner loops were parallelized using fork-join constructs. Data dependencies were carefully analyzed and verified after parallelization.

The architecture has been explored with the proposed flow, and the graph of Figure 7 summarizes the overall performance results, expressed in frames per second (fps) achieved, for a range of architecture parameters. These include the number of processors (2 to 5) and the number of threads per processor (2 to 8). The upper curve represents

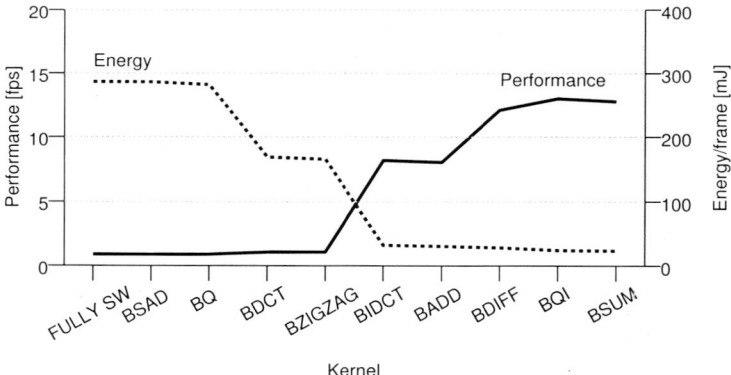

Fig. 6. Performance and power efficiency of the MPSoC with varying hardware/software partitioning

the theoretical upper bound for a perfect parallelization (i.e. results for a single processor accessing local memory, and then simply multiplied by the number of processors). This theoretical result does not include any inter-processor communication code and assumes zero bus latency. The best result makes use of 5 processors and 8 hardware threads per processor. In this case, 28.5 fps is achieved, or 86% of the theoretical best result of 33 fps. The system was simulated with and increasing number of processors, after parallelization, showing a result close to the theoretical upper bound when using a no-wait-state channel and a result very close to the latter when using STBus.

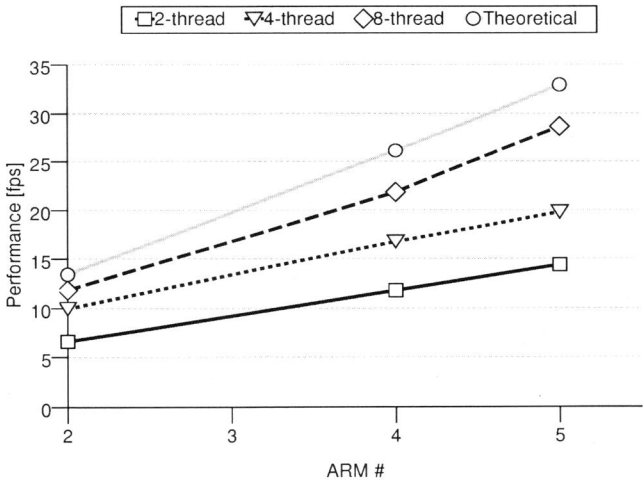

Fig. 7. Performance of the MPEG4 application with different configurations

5.1   Power optimization

Application mapping results are very effective for the given platform, as the overall average load for each processor is roughly 85%. Therefore, we cannot expect large savings in power consumption, because the available resource usage is only 15% away from maximum usage. Nevertheless, it is still possible to gain a 10% power saving applying the methodologies outlined in this work. For less efficient mappings, these savings might be even larger.

The DVFS scheduling algorithm was tested on the platform and the resulting voltage schedule is shown in Figure 8. The algorithm managed to increase the average load per processor from 85% to 96%, and after the first three frames (during which the MPEG4 pipeline was filling up) the voltage stabilizes to the best feasible value without affecting the algorithm performance, maintaining the average load to its best value.

Concerning the processor state management, the results are shown in Figure 9: the number of cycles spent by the processors in idle state (but fully powered) is reduced by

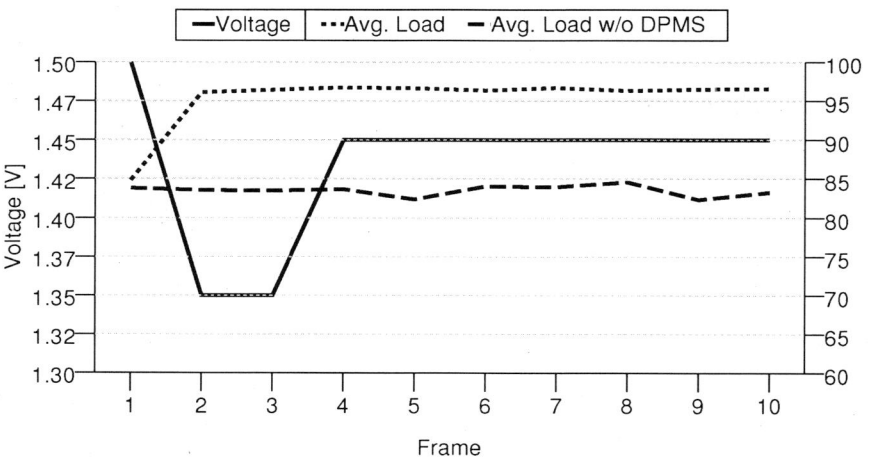

Fig. 8. Voltage scheduling performed by the algorithm and energy results

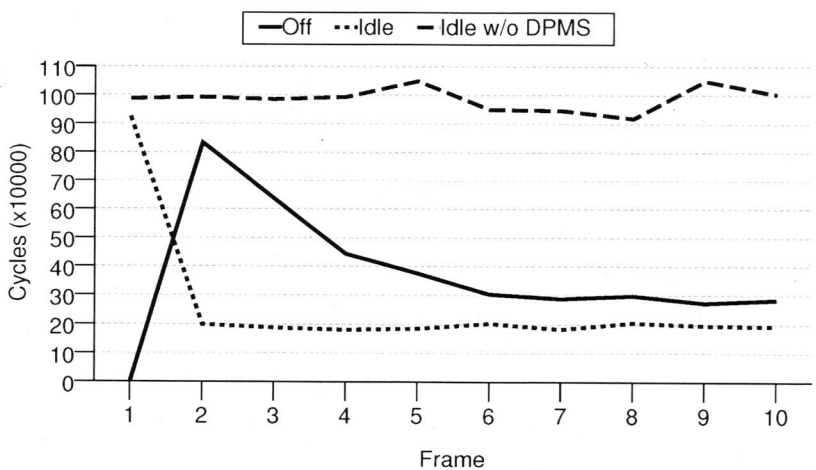

Fig. 9. Average cycles in different states of the processors

80%, and processors spend more time in low-power mode (the Off state in the figure). The remainder number of idle cycles is spent while switching from one state to the other.

Due to the high efficiency of the mapping an the state swticthing cost for the arm processor, the overall energy saving is roughly 10%, which is inline with the increase in average load per processor.

6   Concluding Remarks

This paper presented a mapping methodology for applications on the MultiFlex platform. In addition, this work presents a fast partitioning exploration scheme that takes advantage of the DSOC programming model. These methodologies have been applied to a multimedia case study: an industrial MPEG4 encoder, showing the validity of the approach. Future works include the integration of the methodology with automatic exploration algorithms and automatic parallelization of the application code. Concerning power consumption, this work also introduces a novel low-power voltage scheduler and a dynamic power management system for the MultiFlex system. This DPMS has three key advantages: it is task-driven without needing any operating system support, its algorithms are linear in the number of computational resources, and the scheduling granularity is very coarse compared to the target application structure. Future work will add multiple sleep states for the processor cores (with different wake-up times) and compare scheduling results with optimal values.

References

1. Keutzer, K., Newton, A., Rabaey, J., Sangiovanni-Vincentelli, A.: System-level design: Orthogonalization of concerns and platform-based design. IEEE Transactions On Computer-Aided Design of Integrated Circuits and Systems **19** (2000) 1523–1543
2. Paulin, P.G., Pilkington, C., Langevin, M., Bensoudane, E., Nicolescu, G.: Parallel programming models for a multi-processor SoC platform applied to high-speed traffic management. In: CODES+ISSS'04: Proceedings of the Conference. (2004) 48–53
3. Cai, L., Gajski, D.: Transaction level modeling: an overview. In: CODES+ISSS'03: Proceedings of the Conference. (2003) 19–24
4. Beltrame, G., Sciuto, D., Silvano, C., Paulin, P., Bensoudane, E.: An application mapping methodology and case study for multi-processor on-chip architectures. In: VLSI-SoC'06: Proceedings of the Conference. (2006)
5. Paulin, P.G., Pilkington, C., Bensoudane, E.: StepNP: A system-level exploration platform for network processors. IEEE Design and Test of Computers **1** (2002) 2–11
6. Balarin, F., Watanabe, Y., Hsieh, H., Lavagno, L., Passerone, C., Sangiovanni-Vincentelli, A.: Metropolis: An integrated electronic system design environment. IEEE Computer **34** (2003) 45–52
7. Dziri, M.A., Cesrio, W., Wagner, F.R., Jerraya, A.A.: Unified component integration flow for multi-processor soc design and validation. In: DATE'04: Proceeding of the Conference. (2004)
8. Jersak, M., Henia, R., Ernst, R.: Context-aware performance analysis for efficient embedded system design. In: DATE'04: Proceedings of the Conference. (2004)

9. Pop, T., Eles, P., Peng, Z.: Holistic scheduling and analysis of mixed time/event-triggered distributed embedded systems. In: CODES'02: Proceedings of the Symposium. (2002) 187–192

10. Pimentel, A.D., Lieverse, P., van der Wolf, P., Hertzberger, L., Deprettere, E.F.: Exploring embedded-systems architectures with Artemis. IEEE Computer **34**(11) (2001) 57–63

11. Kempf, T., Doerper, M., Leupers, R., Ascheid, G., Meyr, H., Kogel, T., Vanthournout, B.: A modular simulation framework for spatial and temporal task mapping onto multi-processor soc platforms. In: DATE'05: Proceedings of the Conference. (2005) 876–881

12. Paulin, P.G.: Automatic mapping of parallel applications onto multi-processor platforms: a multimedia application. In: Digital System Design, Euromicro Symposium. (2004) 2–4

13. Pazos, N., Maxiaguine, A., Ienne, P., Leblebici, Y.: Parallel modelling paradigm in multimedia applications: Mapping and scheduling onto a multi-processor system-on-chip platform. In: Proceedings of the International Global Signal Processing Conference, Santa Clara, California (2004)

14. Zhai, B., Blaauw, D., Sylvester, D., Flautner, K.: Theoretical and practical limits of dynamic voltage scaling. In: DAC '04: Proceedings of Conference. (2004) 868–873

15. Xie, F., Martonosi, M., Malik, S.: Efficient behavior-driven runtime dynamic voltage scaling policies. In: CODES+ISSS '05: Proceedings of the Conference. (2005) 105–110

16. Lorch, J.R., Smith, A.J.: PACE: A new approach to dynamic voltage scaling. IEEE Transactions on Computers **53** (2004) 856–869

17. Choi, K., Soma, R., Pedram, M.: Fine-grained dynamic voltage and frequency scaling for precise energy and performance trade-off based on the ratio of off-chip access to on-chip computation times. In: DATE '04: Proceedings of the Conference. (2004)

18. Andrei, A., Schmitz, M., Eles, P., Peng, Z., Al-Hashimi, B.M.: Overhead-conscious voltage selection for dynamic and leakage energy reduction of time-constrained systems. In: DATE '04: Proceedings of the Conference. (2004)

19. Kadayif, I., Kandemir, M., Vijaykrishnan, N., Irwin, M., Kolcu, I.: Exploiting processor workload heterogeneity for reducing energy. In: DATE'04: Proceedings of the Conference. (2004)

20. Bacon, D.F., Graham, S.L., Sharp, O.J.: Compiler transformations for high-performance computing. ACM Computing Surveys **26** (1994) 345–420

21. Banerjee, U., Eigenmann, R., Nicolau, A., Padua, D.A.: Automatic program parallelization. Proceedings of the IEEE **81** (1993) 211–243

22. Murray, D.J., VanRyper, W.: Encyclopedia of Graphics File Formats. O'Reilly Associates (1996)

23. Beltrame, G., Palermo, G., Sciuto, D., Silvano, C.: Plug-in of power models in the StepNP exploration platform: Analysis of power-performance trade-offs. In: CASES'04: Proceedings of the Conference. (2004) 85–92

24. Gupta, S., Dutt, N., Gupta, R., Nicolau, A.: SPARK: A high-level synthesis framework for applying parallelizing compiler transformations. In: VLSID'03: Proceedings of the Conference. (2003)

25. Ishihara, T., Yasuura, H.: Voltage scheduling problem for dynamically variable voltage processors. In: ISLPED'98: Proceedings of the Symposium. (1998) 197–202

Frequency and Speed Setting for Energy Conservation in Autonomous Mobile Robots

Jeff Brateman[1], Changjiu Xian[2], and Yung-Hsiang Lu[3]

[1] Purdue University, West Lafayette, Indiana brateman@purdue.edu
[2] Purdue University, West Lafayette, Indiana cjx@purdue.edu
[3] Purdue University, West Lafayette, Indiana yunglu@purdue.edu

Abstract. Autonomous mobile robots have been achieving significant improvement in recent years. Intelligent mobile robots may detect hazardous materials or survivors after a disaster. Mobile robots usually carry limited energy (mostly rechargeable batteries) so energy conservation is crucial. In a mobile robot, the processor and the motors are two major energy consumers. While a robot is moving, it has to detect an obstacle before a collision. This results in a real-time constraint: the processor has to distinguish an obstacle within the traveled time interval. This constraint requires that the processor run at a high frequency. Alternatively, the robot's motors can slow down to enlarge the time interval. This paper presents a new approach to simultaneously adjust the processor's frequency and the motors' speed to conserve energy and meet the real-time constraint. We formulate the problem as non-linear optimization and solve the problem using a genetic algorithm for both continuous and discrete cost functions. Our experiments demonstrate that more energy can be saved by adjusting both the frequency and the speed simultaneously.

1 Introduction

Autonomous mobile robots provide great potential in transportation, entertainment, environment sensing, search, rescue, reconnaissance, hazard detection, and carpet cleaning [6] [7]. Mobile robots usually carry limited energy, such as rechargeable batteries, so energy conservation is crucial. Makimoto et al. [12] predicted that robots would be a major challenge for future low-power designs. A robot requires many different sensors to detect the environment. Among all sensing technology, stereovision is widely used for determining the distances of obstacles [10] [15]. In a mobile robot, the processor and the motors are two major energy consumers [13]. In this paper, we consider a robot with only one motor, but the method can be generalized to multiple motors.

Even though dynamic voltage scaling (DVS) and energy conservation for mobile robots have been studied [1] [3] [9] [11] [13] [20] [23] [25] [26] [27], the close interaction between computation and motion remains unexplored. This paper presents a probabilistic approach for energy reduction in a mobile robot. We consider a mobile robot moving across an environment with static (i.e. not

Please use the following format when citing this chapter:

Brateman, J., Xian, C. and Lu, Y.-H., 2007, in IFIP International Federation for Information Processing, Volume 249, VLSI-SoC: Research Trends in VLSI and Systems on Chip, eds. De Micheli, G., Mir, S., and Reis, R., (Boston: Springer), pp. 197–216

moving) obstacles, using stereovision to calculate the distance to each obstacle. We assume that each obstacle represents a pass/stop signal, and the minimum distance between signals is a known constant. The robot must recognize the actual distance to the signal before crossing the minimum distance to avoid any chance of failure. The computation cycles needed to recognize the distance to the signals follow a probability distribution. Our method controls both the robot's processor frequency (and voltage) and the motor's speed to reduce the total energy consumption. Our method can save up to 15% additional energy when it is compared with existing solutions that adjust the frequencies only and use constant motor's speeds.

A robot is a real-time system. The processor has to determine the distance of an obstacle before the robot collides with the obstacle. The robot can stop during the distance calculation. However, to conserve energy the robot should be moving while performing this calculation. Many studies have been conducted on energy conservation for real-time systems [9] [18] [25] [27]. Existing studies assume that the deadlines are *externally* determined. For example, a video player has to provide 30 frames per second to prevent jitters. This 33 ms deadline for each frame is given by human's visual perception and cannot be changed by the video player. In contrast, in a mobile robot the deadline is not pre-determined for vision. If the obstacle is static, the robot can slow down or even stop to postpone the deadline before an impending collision. Hence, the deadline is determined by the interaction between the robot's processor and its motor. This paper studies energy conservation in a real-time system in which deadlines can be *internally* adjusted. Our earlier work [3] presents the system using only discrete frequencies and discrete motor speeds, and finds the optimal schedule through an exhaustive search method. We extend this work by generating a schedule using a genetic algorithm. We show this method can obtain a near optimal schedule using both discrete and continuous frequencies and speeds.

2 Related Work

2.1 Probability-Based Voltage Scaling

Some studies have been conducted for dynamic voltage scaling (DVS) by considering the probability distributions of tasks' cycle demands [9] [11] [25] [27]. When different instances of a task's execution cycles follow a known probability distribution, the processor can start at a low frequency (and voltage). If one instance requires fewer cycles, energy is saved because of the lower voltage. If the instance requires more cycles, the processor's frequency gradually rises to ensure that the instance can finish before the deadline. This approach is called *accelerating frequencies*. Lorch et al. [11] use accelerating frequencies for a single task and treat concurrent tasks as a single joint workload. Accelerating frequencies are also used for multiple tasks based on their worst-case execution cycles

(WCEC) [9]. Yuan et al. [27] combine accelerating frequencies with soft real-time constraints for multimedia applications. Xu et al. [25] study accelerating scheduling in systems with discrete frequencies.

Suppose a task demands at most W cycles and the distribution of the cycles is expressed by the cumulative distribution function (CDF). The probability that the jth cycle is needed is $P(j) = 1 - CDF(j-1)$. Note that P is non-increasing because CDF is non-decreasing. Since a task may demand millions of cycles, it is impractical to store the distribution in individual cycles. Thus, we partition $[0, W]$ into n bins and each bin contains b cycles ($b = \lceil \frac{W}{n} \rceil$). The CDF is then a function of the bins. The probability that the jth bin is needed is $P(j) = 1 - CDF(j-1)$. The frequency assigned to the jth bin is f_j and the execution time for this bin is $\frac{b}{f_j}$. The processor's power is proportional to $v^2 f$ and $v \propto f$ (here v is the voltage). The energy for this bin is $(v_j^2 f_j) \times \frac{b}{f_j} \propto b f_j^2$. The expected energy consumption for this bin is proportional to the product of the energy and the probability: $b f_j^2 P(j)$. Suppose the task is released at time zero and the deadline is t. The goal is to find a schedule $\{f_1, f_2, ..., f_n\}$ to minimize the total expected energy. This is formulated as follows.

$$\text{minimize} \sum_{1 \leq j \leq n} b f_j^2 P(j) \tag{1}$$

$$\text{subject to} \sum_{1 \leq j \leq n} \frac{b}{f_j} \leq t \tag{2}$$

Based on earlier studies [11] [25] [27], the optimal schedules can be obtained by assigning f_j:

$$f_j = \frac{\sum_{i=1}^{n} b \sqrt[3]{P(i)}}{t \sqrt[3]{P(j)}} \tag{3}$$

2.2 Energy Conservation for Mobile Robots

Batteries are often used to provide power for mobile robots; however, batteries are heavy and have limited energy capacity. A Honda humanoid robot can walk for only 30 minutes with a battery pack [1]. Rybski et al. [20] show that power consumption is one of the major issues in robot design. Sun et al. [23] present an algorithm for finding the energy-efficient paths on terrains. Yamasaki et al. [26] present an energy-efficient walk generation algorithm for a humanoid robot. A case study [13] shows that motor power is less than 50% of the total power in a mobile robot. Hence, the power for electronic components cannot be ignored. In recent years, small robots have been studied for sensing [2] [5] [7] [21].

2.3 Image Correspondence for Stereovision

Robots can detect their surroundings, including distances to objects with two cameras and stereovision. Several advances have made stereovision both precise

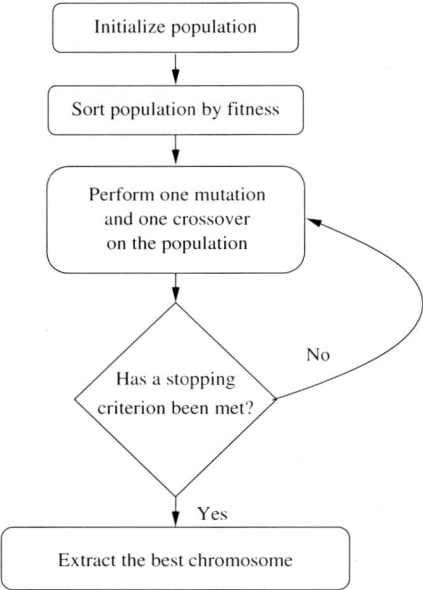

Fig. 1. A simplified view of the GENITOR algorithm.

and accurate [10]. Redert et al. [19] show the advances made for those seeking high-accuracy, high-resolution 3D scene acquisition. Stereovision has been used in mobile robots for both navigation, and terrain mapping [15] [16].

2.4 Genetic Algorithm

Genetic algorithms have been used in many practical applications [4] for problems where optimal schedules take more than polynomial time to find. GENITOR [24] is a steady-state genetic algorithm that has been shown to perform well for several problem domains [17] [22] such as resource allocation, job shop scheduling, and neural networks. A simplified view of the GENITOR algorithm is shown in Figure 1. To generate a better schedule using the GENITOR algorithm, several steps are performed. First, an initial population is generated, either through simple heuristics or random generations. The population consists of many chromosomes, or schedules in the search space. Next, the algorithm performs evolution until a stopping criterion is reached, such as reaching a maximum number of iterations or a homogeneous population.

In every iteration, one mutation and one crossover operations are performed. If the chromosome generated by a mutation or a crossover is better than the worst chromosome in the population, the new chromosome is inserted into the sorted population and the worst chromosome is removed. The fitness function is the criteria which allow a chromosome to be ranked better than another. The probability of selecting a chromosome for the mutations and crossovers is given by the linear bias function defined in [24]. To achieve the linear bias effects, the chromosomes remain sorted by their evaluation of the fitness function.

2.5 Paper Contributions

This paper makes the following contributions: (a) We consider a real-time system in which the deadline is determined by the interaction between two components: processor and motor. (b) The overall energy consumption is modeled as an optimization problem. (c) A probabilistic solution is presented to find the processor's frequency and the motor's speed. (d) We then use a genetic algorithm to find a sub-optimal schedule quickly. (e) We consider continuous processor frequencies and continuous motor speeds, and we use the genetic algorithm to obtain an energy-efficient schedule.

3 Problem Formulation

This section formulates the problem to conserve the energy of a mobile robot by adjusting the robot's processor frequency and the motor's speed. We first use a motivating example to illustrate the important concept and then formulate the problem as a probabilistic non-linear optimization problem. Next, we discuss the properties of the formulation presented in Section 3.2. We describe how to solve the optimization problem using discrete frequencies and discrete speeds with an exhaustive search. Then we use a genetic algorithm to find energy-efficient schedules for either discrete or continuous frequencies and speeds.

3.1 Motivating Example

Suppose the total power of a robot's motor is $s^2 + s + 1$ at speed s meters per second. Here, the constant 1 is used to model the DC loss of the motor. The processor's power consumption is $f^3 + 1$ at frequency f MHz and a constant leakage power of 1. Suppose the robot has to travel along a road. The road contains signs indicating whether the robot can pass or has to stop. The signs do not change (unlike traffic lights) and the minimum distance between two adjacent signs is 100 meters. Even though the distance between signs may be larger than 100 meters, the robot must recognize the sign by the time it has traveled the minimum distance to guarantee success, as shown in Figure 2. If the robot fails to recognize the sign in time, the robot may collide with the sign and fail.

We define the optimal speed as the speed to consume the minimum energy per unit distance. Suppose the minimum distance between signs is D. The time to cross this distance is $\frac{D}{s}$. The total energy consumption is $(s^2 + s + 1)\frac{D}{s}$ and the energy per unit distance is $\frac{s^2+s+1}{s} = s + 1 + \frac{1}{s}$. Thus, the optimal speed is 1

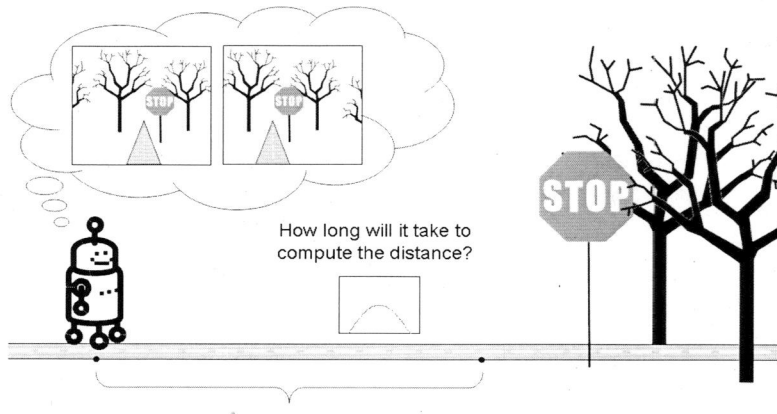

D = minimum distance to travel

Fig. 2. Problem formulation showing that a robot must travel a minimum distance while completing a task (stereovision distance calculation) with uncertain execution time.

meter per second. If the robot moves at this speed, it takes 100 seconds to cross the minimum distance. The worst-case execution cycle is 150 million cycles and the processor has to operate at 1.5 MHz to ensure recognizing every sign before the robot reaches the sign. The total energy consumed by the motor at $s = 1$ is $3 \times 100 = 300$ J. The total energy consumed by the processor at $f = 1.5$ MHz is $(f^3 + 1)\frac{cycles}{f} = 4.38 \times 100 = 438$ J. The overall energy is 738 J to cross the minimum distance between two signs.

If we consider the power of the motor and the processor simultaneously, we can reformulate the problem as follows. The time to cross the distance is $\frac{100}{s}$ at speed s. The processor has to operate at $1.5s$ MHz to meet the deadline. The total energy is $\frac{100}{s} \times \{(s^2 + s + 1) + [(1.5s)^3 + 1]\}$. The minimum energy value occurs when $s \approx 0.62$ and the overall energy consumption is 614 J, or a 17% reduction from 738 J. This shows the importance of considering both frequency and speed simultaneously.

We consider a further extension of this example. The computation cycles vary due to the scene complexity surrounding the signs: among all signs, 30% require only 50 million cycles, 40% for 100 million cycles, and the remaining 30% for 150 million cycles. The probability can be expressed in the following way. The first 50 million cycles are always needed so the probability is 100%. The second 50 million cycles are needed with probability 70%. Finally, the last 50 million cycles are needed with a probability of only 30%. With this additional information, we can compute the *expected*, rather than the worst-case, energy consumption. We want to lower the expected energy, but still finish detecting the sign in the worst case. If the motor's speed is a constant at 1 m/s, the deadline is 100 seconds. We can adopt the strategy with accelerating frequencies

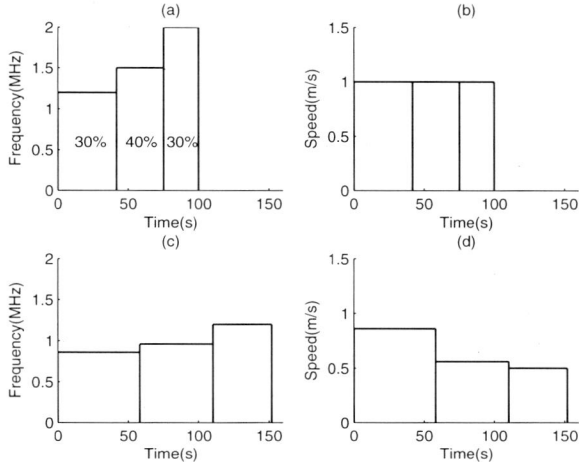

Fig. 3. Processor and motor scaling schedule assuming a constant motor speed in (a) and (b). Processor and motor scaling schedule if the motor speed is allowed to change in (c) and (d).

explained in Section 2.1 as shown in Figures 3 (a) and (b). The overall system saves energy in average cases because most tasks need only 50 or 100 million cycles. Meanwhile, the system still meets the deadline in the worst cases by using a higher frequency when needed. This, however, results in an energy consumption of 611 J, less than 1% reduction from 614 J. We can consider accelerating frequencies for the processor and simultaneously *decelerating speeds* for the motor and save more energy, as shown in Figures 3 (c) and (d). By decreasing the motor's speed, the processor's frequency does not have to rise significantly, and its expected energy is reduced substantially. This approach can further reduce the expected energy to 529 J in this example, or 14% additional savings. The following sections will explain how to determine the frequency and the speed simultaneously to achieve better energy savings.

3.2 Constrained Optimization Problem

The minimum distance between two signs is a known constant, D. The maximum number of cycles needed for recognition is W and is divided into n bins. Each bin has $b = \lceil \frac{W}{n} \rceil$ cycles. We use $P(i)$ to represent the probability that the ith ($1 \leq i \leq n$) bin of cycles is needed. As defined in Section 2.1, $P(i) = 1 - CDF(i-1)$ and $P(i) \geq P(i+1)$. The processor operates at frequency f_i for the ith bin. When the processor is computing for the ith bin, the robot moves at speed s_i. The execution time for the ith bin is $\frac{b}{f_i}$. The distance traveled during this time is $d_i = s_i \frac{b}{f_i}$. The timing constraint is that the processor has to finish the computation of all bins before the robot crosses the distance of D. In other words, the sum of d_i cannot exceed D:

$$\sum_{i=1}^{n} d_i \leq D \Rightarrow \sum_{i=1}^{n} \frac{bs_i}{f_i} \leq D \tag{4}$$

Let $\alpha(f_i)$ be the power consumption of the processor at frequency f_i when voltage scaling is also applied. When the processor finishes the task, the processor's frequency can be reduced to zero. In this case, the processor consumes static power $\alpha(0)$. Let $\beta(s_i)$ be the power consumption of the motor at speed s_i. The *expected* energy for crossing the distance is the sum of the processor energy and the sum of the motor energy over all bins. The energy consumed can be divided into two parts: (i) when the processor is still computing, and (ii) when all computation has finished.

When the ith bin is being computed, the processor consumes power $\alpha(f_i)$ and the motor consumes power $\beta(s_i)$. The duration of this bin is $\frac{b}{f_i}$, and this occurs with probability $P(i)$. Therefore, the expected energy is

$$\sum_{i=1}^{n} \frac{P(i)b}{f_i}(\alpha(f_i) + \beta(s_i)) \tag{5}$$

To compute the energy in (ii), we have to first determine the distance the robot has traveled while the processor is computing. The total expected distance traveled is $\sum_{i=1}^{n} \frac{bP(i)s_i}{f_i}$ and the remaining distance is $D - \sum_{i=1}^{n} \frac{bP(i)s_i}{f_i}$. When the robot is traveling through this remaining distance, the processor is turned off and consumes idle power $\alpha(0)$. Let s_o be the speed for the remaining distance. The time to cross the remaining distance is $\frac{1}{s_o}(D - \sum_{i=1}^{n} \frac{bP(i)s_i}{f_i})$. Hence, the total expected energy is

$$\frac{1}{s_o}(D - \sum_{i=1}^{n} \frac{bP(i)s_i}{f_i})[\alpha(0) + \beta(s_o)] \tag{6}$$

The optimization problem is to find the values of f_i and s_i ($1 \leq i \leq n$) and s_o for minimizing the sum of (5) and (6).

$$\min \sum_{i=1}^{n} \frac{P(i)b}{f_i}(\alpha(f_i) + \beta(s_i)) + \frac{1}{s_o}(D - \sum_{i=1}^{n} \frac{bP(i)s_i}{f_i})[\alpha(0) + \beta(s_o)] \tag{7}$$

with the constraint in (4). This is a problem of constrained optimization.

3.3 Frequency and Speed Scheduling

The above formulation has three sets of variables: the processor's frequency f_i, the motor's speed s_i, and time. The time intervals have been discretized; hence, the frequency and the speed can change only at the boundaries of bins. Each bin takes b clock cycles on the processor. We use $P(i)$ to express the probability that the ith bin is needed. In our formulation, the time intervals are not divided into equal durations (measured by seconds). Instead, the duration of the ith interval is determined by the ratio of b and f_i in order to simplify the expression in (7). It is possible to generalize the formulation and use continuous time so that (a) The duration of a constant frequency is not determined by the value of this frequency. (b) The frequency and the speed do not have to change at the same time. If we use continuous time to model the problem, the frequency and the speed are expressed as $f(t)$ and $s(t)$ respectively. The search space becomes substantially larger and it is difficult to find optimal schedules. Hence, in the rest of this paper, we use discrete time by allowing the frequency and the speed to change only at the boundaries of bins.

Our solution uses accelerating frequencies (i.e. $f_i \leq f_{i+1}$, $1 \leq i \leq n-1$) and decelerating speeds (i.e. $s_i \geq s_{i+1}$, $1 \leq i \leq n-1$). To find the initial values for f_1 and s_1, we examine the schedulability of the problem using the constraint of inequality (4). The initial value of f_1 is the lowest frequency to satisfy (4) when all s_i's are assigned the lowest speed. Similarly, the initial value of s_1 is the highest speed to satisfy (4) when all f_i's are assigned the highest frequency. If f_1 exceeds the highest available frequency or s_1 is below the minimum available speed, no schedule can be found. After finding the initial values for f_1 and s_1, we enumerate all feasible schedules and find the schedule that provides the minimum expected energy and meets the constraint in (4). For a small value of n, it takes only several minutes on a modern computer to find the optimal schedule. This schedule can be computed off-line, and loaded into the robot so that it can change to the correct speed and frequency while the task is still executing. As n increases, the time to find this schedule becomes more important, as there are many more combinations of f_i and s_i. This becomes a problem of scalability with the number of bins.

In most cases, it is impractical to wait hours to generate a schedule for different values of D. This is especially true for dynamic environments and unknown operating environment. Therefore, it is preferable to find a schedule quickly even though it may not be optimal.

3.4 Optimization using Genetic Algorithm

To determine a schedule in a reasonable amount of time, a genetic algorithm is used. Even though the genetic algorithm does not guarantee to reach the optimal schedule, we will show that the schedule produced still saves energy and approaches the optimal schedule. The technique is one adopted from [24].

A chromosome contains all the frequency and speed assignments for the mobile robot for each bin. Each chromosome contains a value for f_i and s_i ($1 \leq i \leq n$) and has $2n$ parameters, where n is the number of bins in the problem. In the discrete case, f_i and s_i are restricted to a limited set of discrete frequencies and discrete speeds, predetermined before the algorithm is run. We are minimizing the expected energy, therefore the fitness function is Equation 7.

The initial population is composed of arbitrary chromosomes. We select the first chromosome to have all frequencies set to the maximum available frequency, and all speeds set to their minimum speeds. If this schedule is not feasible, i.e., the processor still cannot finish executing the task before the robot arrives at its destination, then no schedule can be found. The remainder of the chromosomes are randomly generated.

For mutation, either one or two parameters are selected at random to change. A parameter is one of any f_i or s_i. Allowing two randomly selected parameters to change produces better schedules than changing only one parameter. In our algorithm, half of the multations change two parameters in a schedule simultaneously. Changing only one parameter in a schedule may result in being trapped in a local minimum because the mutated chromosomes are either infeasible (robot no longer meets its deadline) or increase the power consumption. To perform the crossover operation, each parameter is ordered, and a cutoff point is determined at random. This is shown in Figure 4. Any parameters before the cutoff remain the same as their parents and any chromosomes parameters after the cutoff are swapped from one parent to the other. This creates two potentially better child chromosomes.

4 Simulations

4.1 Overview

We consider both discrete and continuous frequency and speed settings for our experiments. The experiments were conducted over several workloads, using both an exhaustive search method and a genetic algorithm. Our simulations show up to 15% energy savings over those methods that scale processor frequencies only.

4.2 Hardware Models

Table 1. XScale's frequency/voltage and power.

Frequency(MHz)	150	400	600	800	1000
Voltage(V)	0.75	1.0	1.3	1.6	1.8
Power(mW)		80	170	400	900	1600

We use the voltage and frequency settings of the Intel XScale processor [25]. For the discrete experiments, we use five discrete frequency settings. Their associated power consumption is shown in Table 1. For the continuous frequency

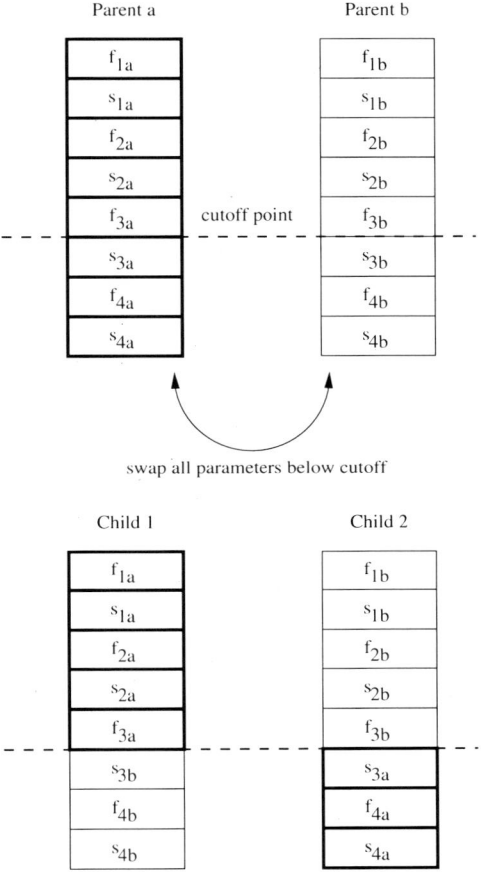

Fig. 4. Crossover operation of a sample chromosome with 4 bins and 8 parameters.

settings, we use a third-order polynomial power model based on the discrete values in Table 1 and allow the frequency to vary anywhere between 150MHz and 1GHz. The motor power is from the measurements performed by Mei et al. [14] shown in Figure 5. We limit the motor's speed between 0.5 m/s and 5 m/s with 0.5 m/s as the step size, for the discrete case. For the continuous motor speeds, we limit the motor speeds to a range of 0.5 m/s and 5 m/s. All of our calculations assume the minimum distance to travel is 500 meters for D's value.

All experiments were performed using Matlab 7.1 running on Windows XP SP2. The hardware consisted of an Intel Pentium 4 CPU running at 3.4 GHz with 1 GB of RAM. These values are important for the execution time of the genetic algorithm and the exhaustive search, as seen later in Figure 12.

We compare our approach with three other methods. The first uses a constant frequency and a constant speed. The frequency and the speed are selected

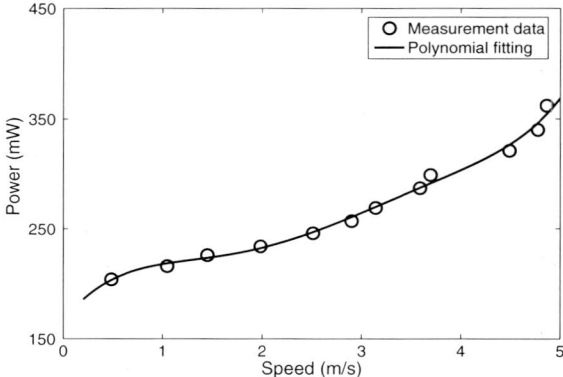

Fig. 5. Power efficiency of a robot at different speeds.

from the discrete settings such that they minimize the total energy consumption and satisfy the constraint. In the synthetic distributions, a search finds the optimal energy consumption schedule that meets the constraint to be a frequency of 400 MHz and a speed of 1.5 m/s. The second uses a constant speed and accelerating frequencies. The third uses a constant frequency and decelerating motor speeds. The processor frequency is set to the middle frequency 600 MHz. The fourth uses both accelerating frequencies and decelerating speeds; this is the method proposed in this paper.

4.3 Workloads

We use two types of workloads: synthesized workloads with different distribution functions, and a distribution function generated from captured stereo images.

The synthetic benchmarks have distributions of uniform, Gaussian, and exponential functions. These synthetic workloads have worst-case execution cycles (WCEC) of 100 billion cycles. For the uniform distribution, the actual number of needed cycles is between 0 and WCEC. For the Gaussian distribution, the mean is half WCEC and the standard deviation is a quarter WCEC. For the exponential distribution, the mean is a quarter WCEC. The distributions are normalized after removing the negative cycles and the cycles above WCEC. We varied the mean and the standard deviation (STD) of the synthetic workloads to show how different values affect the energy savings of our schedule. The energy savings calculations are done using the genetic algorithm with continuous frequencies and continuous speeds. Each run is performed over a range of means, where each mean is calculated as a percentage of the original $WCEC$. To find average energy savings, three calculations are performed for each workload. The uniform and Gaussian workloads are generated by selecting a value of the standard deviation. The exponential workload is generated only over variable means.

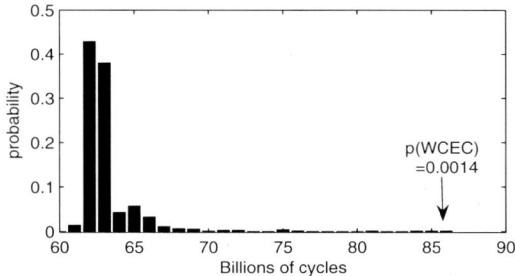

Fig. 6. Probability distribution of stereovision computations.

We generated the image workload from pairs of stereovision images taken from the image database of the city of West Lafayette and Indianapolis in the state of Indiana [8]. Pairs of stereo images are compared, and distances for several objects are returned.

4.4 Experimental Results

The experiments compare our method with several workloads. We analyze an image processing algorithm to obtain the distribution of execution cycles, and how our method performs on the workload. The genetic algorithm experiments show how a schedule can be obtained in a reasonable time, even if it is not optimal. We then show how altering synthetic workloads affects the energy savings.

Figure 6 shows the distribution of the needed cycles for running the correspondence programs on 700 pairs of images. Note that there is great potential for energy savings as the probability of the WCEC (85.7 billion cycles) is only 0.14%. We can see that the majority of tasks execute in around 62 billion cycles.

Figure 7 shows the relative energy consumption of the four methods for the four benchmarks, using the exhaustive search method. All numbers are normalized related to the first method with a constant frequency and a constant speed. As can be seen in this figure, our method can save 20% to 50% energy compared with the first method in the four benchmarks. Compared with the second and the third methods, our method can save an additional 7% to 15% energy. These results are generated using 10 bins.

An exponential distribution shows the greatest potential for savings as compared with the constant frequency and the constant motor speed schedule. In an exponential distribution, the task finishes quickly more often, and has a low probability of finishing near the WCEC. We can see the potential for reducing the expected energy as opposed to WCEC scheduling. In the stereovision distribution, the energy savings is not as high as the exponential distribution because no task finishes before $\frac{2}{3} \times WCEC$ cycles, but our method still saves 20% energy over the constant frequency and constant motor speed scheduling.

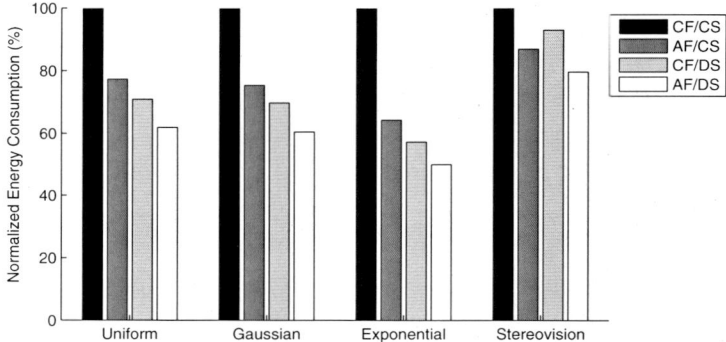

Fig. 7. Normalized energy consumption from the set of the three synthetic tasks with uniform, Gaussian, and exponential distributions, and the stereovision benchmark, respectively. The four methods use either constant frequencies (CF) or accelerating frequencies (AF), and constant speeds (CS) or decelerating speeds (DS).

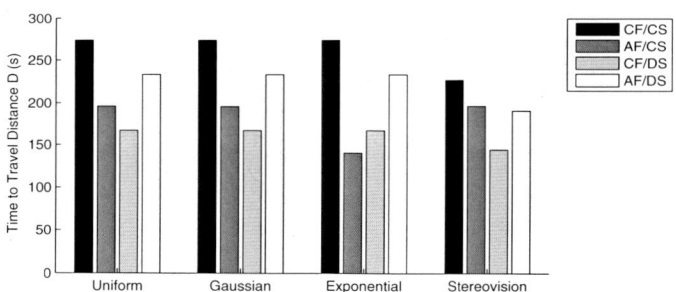

Fig. 8. Time to finish each of the three synthetic task and the stereovision benchmark for a task of WCEC cycles.

One advantage of using our method over a constant frequency and constant speed schedule is that our method will not necessarily increase the worst case travel time of the robot to the minimum distance D. For this analysis, we assume that the task takes the maximum number of cycles to execute, namely, its WCEC. Figure 8 shows the time required to travel the minimum distance. We see that in all cases, dividing the frequency and speed schedule into 10 bins allows the robot to tune its speed better, so that the robot takes less time traveling the minimum distance than the constant frequency and constant speed schedule allows.

Figure 9 shows the energy consumption for a growing number of bins, using discrete parameters. This figure indicates that energy consumption decreases as the number of bins grows because more frequencies and speeds can be used.

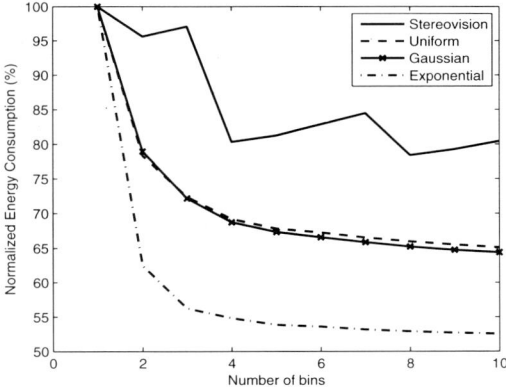

Fig. 9. Energy consumption of the optimal schedule of each benchmark over a different number of bins using discrete parameters and the exhaustive search method.

It should be noted that with the stereovision distribution, the energy actually increases in some cases. This is due to the division of the PDF into a relatively small number of bins. Some of the areas with high probability are divided in some sizes of n, resulting in an increased expected energy. However, energy is still reduced from the extreme case of one bin. Because of the small number of bins used to compute the frequency and the speed schedule, our method can be applied to practical systems, even though the method has exponential computation time.

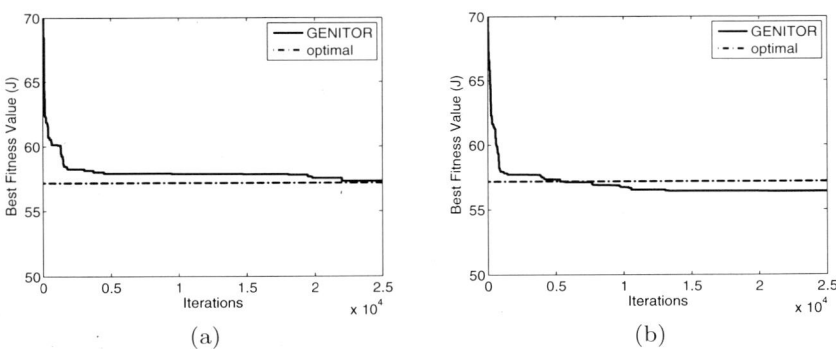

Fig. 10. GENITOR improvement versus the number of iterations compared with the optimal energy consumption for the exponential workload with *discrete* (a) and *continuous* (b) frequencies and speeds.

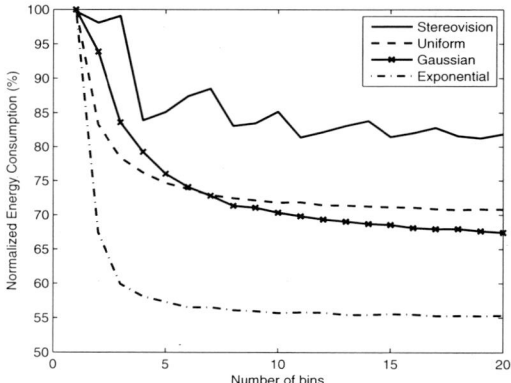

Fig. 11. Energy consumption of each benchmark over different number of bins using continuous parameters and a genetic algorithm to find a schedule.

Genetic Algorithm The genetic algorithm had a population of 50 chromosomes, starting with 50 random schedules. We can see from Figure 10 (a) that the energy consumed approaches the exhaustive search optimal schedule after only a few thousand iterations. For the exponential distribution, the energy consumed by the schedule generated by the GENITOR algorithm was within 0.24% of the energy consumed by the optimal schedule. Other distributions' simulations perform in a similar manner, and all schedules were computed in about 4 minutes.

We can see in Figure 10 (b) the results of running the genetic algorithm using continuous parameters compared with the optimal schedule using discrete parameters. The result is 14% more energy-efficient than the discrete optimal schedule. This figure shows the advantage of using continuous parameters over discrete parameters.

We show the effects of increasing the number of bins in Figure 11. Increasing the number of bins increases the number of parameters that can be adjusted. These schedules were calculated using continuous parameters, as these were shown to provide better schedules than discrete parameters. This figure can be compared to Figure 9, where each graph is calculated using discrete parameters and normalized with the original one bin discrete parameter schedule. The figure also indicates that the energy consumption begins to approach diminishing returns as the number of bins exceeds 15. In other words, a large number of bins cannot provide a significant amount of additional savings.

We use a genetic algorithm over an exhaustive search to reduce the time for finding energy-efficient schedules. In Figure 12, we show the execution time of the exhaustive search compared with the execution time of GENITOR. No times were recorded for 1 or 2 bins exhaustive search because the execution

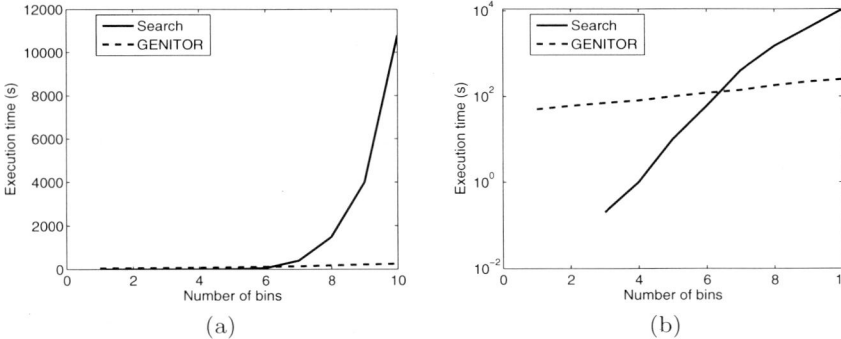

Fig. 12. Execution time of the exhaustive search method and GENITOR algorithm as the number of bins increases over a linear scale (a) and a log scale (b), using discrete processor frequencies and motor speeds.

time was negligible. We can see that using a genetic algorithm will begin to save hours as the number of bins increases.

Variable Synthetic Workloads The uniform distribution results are shown in Figure 13 (a). The figure shows the increased energy consumption as the mean increases. For a small STD value, the task executing almost always executes its worst case execution cycles, while the workloads with a large STD have constant energy consumption over the selected ranges of the mean. This occurs when probabilities that are assigned to bins below 0% or above 100% get clipped, and are normalized so they sum to one. The result is the appearance of constant energy savings.

Figure 13 (b) shows the Gaussian distribution results. For each STD, we see that the energy consumption increases as the mean increases. This increase becomes more significant as the STD increases. With a large STD, the distributions approach a uniform distribution for each mean, therefore the energy savings remains constant. We also see the crossing point in the middle because a distribution with a large STD performs as well as a schedule with a small STD with a mean of 50% WCEC.

The exponential distribution results are shown in Figure 13 (c). The figure indicates the increasing energy consumption as the mean increases, but takes on a different shape than the other workloads. This is because even with the increased mean, the majority of tasks will complete early. The rate of increase with mean is small, and our method still saves energy in the worst case.

5 Conclusions

This paper presents a method to simultaneously scale processor frequencies and motor speeds for autonomous robots with hard deadlines. However, each

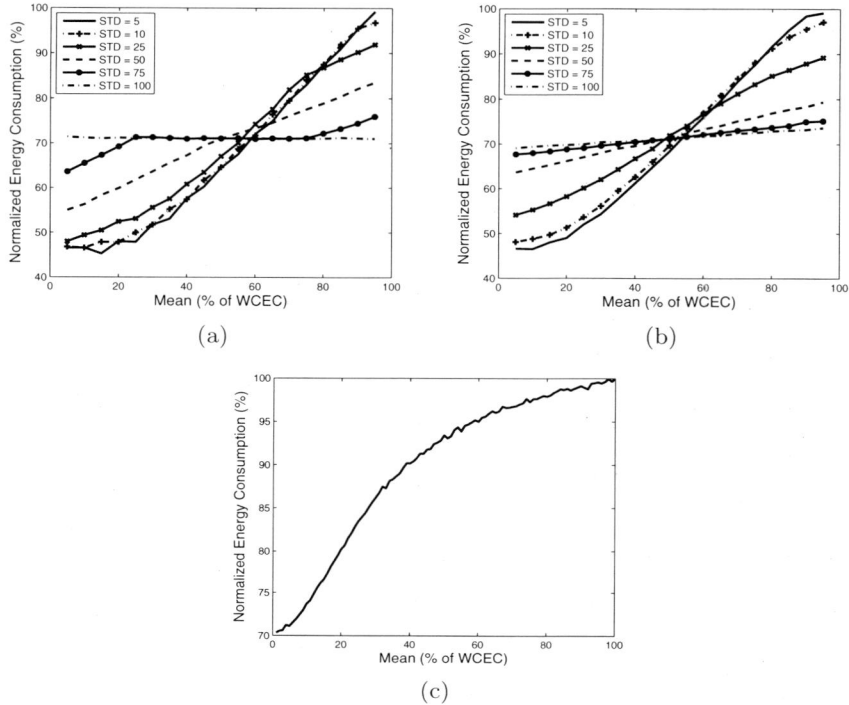

Fig. 13. Energy consumption over a range of means of uniform (a), Gaussian (b), and exponential (c) synthetic workloads.

deadline is not a time deadline, rather it is a distance deadline. This problem is formulated as an optimization problem. An exhaustive search method is presented to find the optimal solution among discrete processor frequencies and motor speeds. A genetic algorithm is used to find a near-optimal solution in less time than the exhaustive search. The genetic algorithm is modified so that it can handle continuous processor frequencies and motor speeds.

A probability distribution of the number of cycles required for stereovision distance calculation is used for our simulations, along with three synthetic distributions. Our experimental results show that we achieve energy savings from 7% to 15% more than only scaling the processor frequency. These results can be achieved through the calculation of an optimal schedule off-line. We can save more energy if continuous processor frequencies and motor speeds are available using the genetic algorithm. We also show that the genetic algorithm can be used for greater energy savings with increasing numbers of bins.

6 Acknowledgments

This work is supported in part by National Science Foundation IIS-0329061, CNS-0347466, and Purdue Research Foundation. Any opinions, findings, and conclusions or recommendations expressed in this paper are those of the authors and do not necessarily reflect the views of the sponsors.

References

1. Aylett, R.: Robots: Bringing Intelligent Machines To Life. Barrons (2002)
2. Birch, M.C., Quinn, R.D., Hahm, G., Phillips, S.M., Drennan, B., Fife, A., Verma, H., Beer, R.D.: Design of a Cricket Microrobot. In: International Conference on Robotics and Automation, pp. 1109–1114 (2000)
3. Brateman, J., Xian, C., Lu, Y.H.: Energy-efficient scheduling for autonomous mobile robots. In: IFIP International Conference on Very Large Scale Integration (VLSI-SoC), pp. 361–366 (2006)
4. Chaiyaratana, N., Zalzala, A.: Recent developments in evolutionary and genetic algorithms: Theory and applications. In: International Conference on Genetic Algorithms In Engineering Systems:Innovations And Applications, pp. 270–277 (1997)
5. Colot, A., Caprari, G., Siegwart, R.: InsBot: Design of An Autonomous Mini Mobile Robot Able To Interact With Cockroaches. In: International Conference on Robotics and Automation, pp. 2418–2423 (2004)
6. Davids, A.: Urban Search and Rescue Robots: From Tragedy To Technology. IEEE Intelligent Systems **17**(2), 81–83 (2002)
7. Drenner, A., Burt, I., Dahlin, T., Kratochvil, B., McMillen, C., Nelson, B., Papanikolopoulos, N., Rybski, P.E., Stubbs, K., Waletzko, D., Yesin, K.B.: Mobility Enhancements to the Scout Robot Platform. In: International Conference on Robotics and Automation, pp. 1069–1074 (2002)
8. Gautam, S., Sarkis, G., Tjandranegara, E., Zelkowitz, E., Lu, Y.H., Delp, E.J.: Multimedia for Mobile Users: Image Enhanced Navigation. In: Multimedia Content Analysis, Management, and Retrieval, IST/SPIE Symposium on Electronic Imaging (2006)
9. Gruian, F.: Hard Real-Time Scheduling for Low-Energy Using Stochastic Data and DVS Processors. In: International Symposium on Low Power Electronics and Design, pp. 46–51 (2001)
10. Lenz, R.K., Tsai, R.Y.: Techniques for Calibration of the Scale Factor and Image Center for High Accuracy 3-D Machine Vision Metrology. IEEE Transactions onPattern Analysis and Machine Intelligence **10**(5), 713–720 (1988)
11. Lorch, J.R., Smith, A.J.: Improving Dynamic Voltage Scaling Algorithms with PACE. In: ACM SIGMETRICS International Conference on Measurement and Modeling of Computer Systems, pp. 50–61 (2001)
12. Makimoto, T., Sakai, Y.: Evolution of Low Power Electronics and Its Future Applications. In: International Symposium on Low Power Electronics and Design, pp. 2–5 (2003)
13. Mei, Y., Lu, Y.H., Hu, Y.C., Lee, C.G.: A Case Study of Mobile Robot's Energy Consumption and Conservation Techniques. In: International Conference on Advanced Robotics, pp. 492–497 (2005)

14. Mei, Y., Lu, Y.H., Hu, Y.C., Lee, C.S.G.: Energy-Efficient Motion Planning for Mobile Robots. In: International Conference on Robotics and Automation, pp. 4344–4349 (2004)
15. Murray, D., Jennings, C.: Stereo Vision Based Mapping and Navigation for Mobile Robots. In: IEEE International Conference on Robotics and Automation, vol. 2, pp. 1694–1699 (1997)
16. Negishi, Y., Miura, J., Shirai, Y.: Vision-Based Mobile Robot Speed Control Using a Probabilistic Occupancy Map. In: IEEE International Conference on Multisensor Fusion and Integration for Intelligent Systems, pp. 64–69 (2003)
17. Oltikar, M., Brateman, J., White, J., Martin, J., Knapp, K., Maciejewski, A.A., Siegel, H.J.: Robust resource allocation in weather data processing systems. In: International Conference on Parallel Processing Workshops, pp. 445–454 (2006)
18. Pillai, P., Shin, K.G.: Real-time Dynamic Voltage Scaling for Low-power Embedded Operating Systems. In: ACM Symposium on Operating Systems Principles, pp. 89–102 (2001)
19. Redert, A., Hendriks, E., Biemond, J.: Correspondence Estimation in Image Pairs. IEEE Signal Processing Magazine 16(3), 29–46 (1999)
20. Rybski, P.E., Papanikolopoulos, N.P., Stoeter, S.A., Krantz, D.G., Yesin, K.B., Gini, M., Voyles, R., Hougen, D.F., Nelson, B., Erickson, M.D.: Enlisting Rangers and Scouts for Reconnaissance and Surveillance. IEEE Robotics and Automation Magazine 7(4), 14–24 (2000)
21. Sibley, G.T., Rahimi, M.H., Sukhatme, G.S.: Robomote: A Tiny Mobile Robot Platform for Large-scale Ad-hoc Sensor Networks. In: International Conference on Robotics and Automation, pp. 1143–1148 (2002)
22. Sugavanam, P.V., Siegel, H.J., Maciejewski, A.A., Ali, S.A., Al-Otaibi, M., Aydin, M., Guru, K., Horiuchi, A., Krishnamurthy, Y.G., Lee, P., Mehta, A.M., Oltikar, M., Pichel, R., Pippin, A.J., Raskey, M., Shestak, V., Zhang, J.: Processor allocation for tasks that is robust against errors in computation time estimates. In: International Parallel and Distributed Processing Symposium (2005)
23. Sun, Z., Reif, J.: On Energy-Minimizing Paths on Terrains for A Mobile Robot. In: International Conference on Robotics and Automation, pp. 3782–3788 (2003)
24. Whitley, L.D.: The genitor algorithm and selection pressure: Why rank-based allocation of reproductive trials is best. In: International Conference on Genetic Algorithms, pp. 116–123 (1989)
25. Xu, R., Xi, C., Mehlem, R., Moss, D.: Practical PACE for Embedded Systems. In: ACM International Conference On Embedded Software, pp. 54–63 (2004)
26. Yamasaki, F., Hosoda, K., Asada, M.: An Energy Consumption Based Control for Humanoid Walking. In: IEEE/RSJ International Conference on Intelligent Robots and System, pp. 2473–2477 (2002)
27. Yuan, W., Nahrstedt, K.: Energy-Efficient Soft Real-Time CPU Scheduling for Mobile Multimedia Systems. In: ACM Symposium on Operating Systems Principles, pp. 149–163 (2003)

Configurable On-Line Global Energy Optimization in Multi-Core Embedded Systems Using Principles of Analog Computation

Zeynep Toprak Deniz, Yusuf Leblebici, and Eric Vittoz

Ecole Polytechnique Fédérale de Lausanne (EPFL)
Lausanne, Switzerland
{zeynep.toprak,yusuf.leblebici,eric.vittoz}@epfl.ch

Abstract. This work presents the design of an on-line energy optimizer unit, which is capable of dynamically adjusting power supply voltages and operating frequencies of multiple processing elements (PE), tailored to the instantaneous workload information and is fully adaptive to variations in process and temperature. The circuit design borrows some of the basic principles of analog computation to continuously optimize the system-wide energy dissipation of multiple cores. The analogy between the energy minimization problem under timing constraints in a general task graph and the power minimization problem under Kirchhoff's current law (KCL) constraints in an equivalent resistive network is exploited. To our best knowledge, this is the first study of its kind to demonstrate an on-line solution to complex, multi-variable energy optimization problem which allows dynamic adjustment of individual operating frequencies and supply voltages of multiple processing elements.

1   Introduction

The continuing exponential growth of complexity in VLSI systems is largely supported by the advances in silicon processing technology, which enable integration of ever more complex functions on a single chip. Future Systems-on-Chip (SoC) are generally envisioned as high-performance embedded systems composed of a heterogeneous network of processing elements (meaning non-identical elements in functionality, size, performance and even the design methodology), providing integrated solutions to challenging design problems in the mobile telecommunication, consumer electronics and multimedia domains. Application demands and the continuous trend towards mobile, distributed systems have also made battery-powered portable electronic systems very popular and virtually ubiquitous. Nowadays such systems are widely used in many applications, such as mobile computing, information appliances as well as various industrial, medical and military applications. Hence, energy dissipation and energy/performance trade-offs have emerged as major factors in determining the weight, the size and the life-time (autonomy) of portable devices. Thus, the ultimate energy management goal in such complex systems is to reduce "system-level" or global

Please use the following format when citing this chapter:

Deniz, Z.T., Leblebici, Y. and Vittoz, E., 2007, in IFIP International Federation for Information Processing, Volume 249, VLSI-SoC: Research Trends in VLSI and Systems on Chip, eds. De Micheli, G., Mir, S., Reis, R., (Boston: Springer), pp. 217–240

energy consumption, rather than concentrating on local minimization. A number of system-level energy optimization techniques have been presented in the literature recently [1, 2].

The significance of the problem of energy optimization in multi-core systems where the individual energy demands of various processing elements (PE) are governed by instantaneous workload requirements is underlined by the increasing prominence of multi-core systems that must operate under strict energy budget constraints, in mobile applications. A range of solutions have been proposed over the last few years, which are mostly based on static, off-line calculation of a limited set of operating points in the form of optimum voltage and frequency assignments, that are subsequently chosen according to actual demands. These observations lead to the conclusion that implementation of sophisticated energy management techniques will be necessary in SoC/NoC (Network-on-Chip) architectures that consist of multiple functional units, where each unit is experiencing a non-uniform workload during operation time. In such systems, fine grained energy management is generally implemented as Dynamic Voltage Scaling (DVS), which refers to varying the operation speed of a processor by changing the clock frequency along with the supply voltage. DVS is usually implemented as an open-loop technique whereby the single digital core is characterized for throughput at a given clock speed and at a given voltage with ample margin allowed for temperature, power supply and fabrication variations necessitating extensive characterizations, to build a hard-coded table of speed versus supply voltage that insures performance criteria for each wafer [3]-[6].

All components in a SoC/NoC, executing a specific application, are expected to have varying energy requirements in time domain which could be easily derived from their instantaneous workload estimations. Based on this "local" information, the optimum energy allocations for all sub-blocks of a complex SoC/NoC could be computed locally in time domain, for any given task. This approach is widely used in the literature to reduce the energy consumption of individual PEs [7, 8]. However, these "locally optimum" allocations may not always coincide with the global optimum for the overall system, especially taking into account the interaction of the various system components. Thus, a system-wide (global), continuous-time optimization approach would be expected to yield better results from a system point of view (see Fig. 1). However, system-level energy optimization under performance constraints is a challenging problem. The concept of system stability needs to be considered when several components adopt dynamic policies to control energy consumption and performance [9]. Possible oscillations in power/performance space that could be caused by applied energy management policies are undesirable, and should be avoided. Hence, it is preferable to implement very efficient on-line dynamic power management techniques, in a centralized fashion, guaranteeing globally optimum results and system-level stability.

Another important issue that needs to be addressed is the power consumption of the optimizer block, i.e. the optimizer, itself. This issue has not been carefully validated until now, and has been largely neglected in the literature. However,

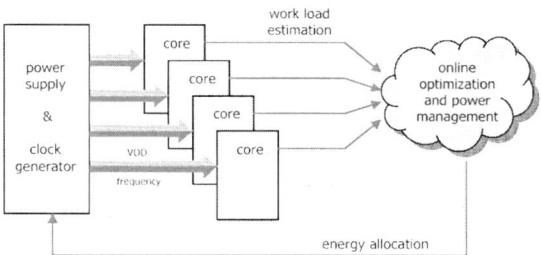

Fig. 1. Conceptual representation of the proposed on-line global power/energy management approach.

the power dissipation of the optimizer unit could be a significant component of the overall system dissipation, especially if an on-line optimization policy is being implemented for multiple components, using a conventional digital processor to solve the optimization problem continuously under varying conditions. To address this issue, we propose using analog circuit principles instead of a digital processor, and thus, saving energy. It is also a known fact that approximate, stable solutions to such multi-variable optimization problems (such as the gradient descent algorithm) can be obtained by using very compact analog circuits [10]-[14]. This approach has the potential advantage of generating sufficiently accurate solutions, while dissipating a small fraction of the power that would be needed by a digital processor to solve the same problem. Furthermore, the energy manager design shall demonstrate the continuous, real-time energy consumption optimization (being independent of the application) to be more response-time efficient than currently proposed energy management policies utilizing discrete power levels [3, 15].

This chapter will explain in detail the basis of the proposed idea of continuously (in time domain) adjusting the control knobs of the overall system in order to minimize the global energy consumption of the embedded system, subject to timing constraints, and it will present the design of the proposed central (global) optimizer unit based on simple analog circuit topologies and design aspects. To do this, the analogy between the problem of minimizing the energy dissipation on a given task graph and the fundamental electrical behavior of resistive networks will be exploited first. It will be also shown that the energy requirement of solving a multi-variable optimization problem in real-time can be minimized based on our approach.

2   Modelling the Multi-Unit Global Energy Optimization Problem

In large scale systems design, it is essential to perform pre-design analysis using extensive modelling tools in order to gain further insight in a complex system, improve understanding of the problem under consideration, find unexpected emergent properties and quantify system parameters before starting physical design. It is also important to abstract the real design problem in order to provide its complete definition and form the basis of its formulation to experiment and to represent any possible solutions. This section explores the hierarchical abstraction and formalization of the energy optimization problem in multi-core systems to consolidate the conceptual variables and constraints of the objective function.

2.1   Cost Function and Non-Linear Constraint Formulation

We start this section with the definitions as well as the key elements of the problem of discourse. It is assumed that the system is composed of real-time dependent tasks with deadlines to be executed on multiple variable-voltage PEs. Task scheduling being known a priori, the functionality of the data-flow of such systems realized as heterogeneous distributed architectures are captured as Directed Acyclic Graphs (DAG) $G_S(t, C)$, an example structure of which is shown in Fig. 2. Furthermore, the formalism describing the relationship between the application software and the time-domain scheduling of various tasks is typically supplied by the concept of task graphs (TG).

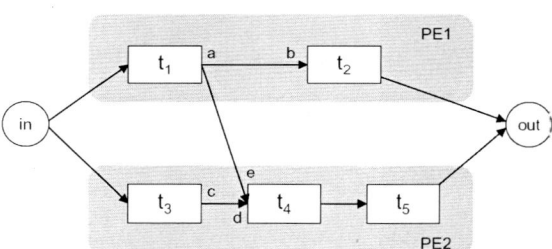

Fig. 2. Task graph of five tasks mapped on two processing element.

In the DAG in Fig. 2, each node represents computational tasks t_u, while edges indicate the data dependencies between these tasks, indicating that task v can only start after task u finishes. Tasks require a finite number of clock cycles N_u to be executed, depending on the PE on which they are mapped. Further, tasks are annotated with deadlines D_{lu} that have to be met during application run-time and T_{graph} on a task set restricts all tasks' finish time. During voltage scaling N_u remains constant, while CT cycle time and d_u task execution time of uth task change with supply voltage V_{DDu}. CT, d_u and E_{dyn} can be computed as

$$d_u = N_u\, CT \quad \text{where} \quad CT = \frac{k\, L_d\, V_{DDu}}{(V_{DDu} - V_{th})^{\alpha}} \tag{1}$$

$$E_{dyn} = N_u\, C_u\, V_{DDu}^2 \tag{2}$$

where k is a technology dependent constant, V_{th} is the threshold voltage of the devices, is a technology dependent constant, ranging from 1.2 to 2 for recent technologies, derived in *Alpha-Power Law MOSFET Model* [16, 17] and, C_u is the effective switching capacitance per cycle. Although the dynamic power dissipation is still dominating, the trend to reduce the overall circuit supply voltage and the threshold voltage is increasing concerns about the leakage currents; for advanced technologies (< 90 nm) it is expected that the leakage will account for more than 50% of the total power [18]. The leakage energy is given by Eq. 3, where V_{BSu} is the voltage applied between the body and the source of the transistor (body bias), I_{ju} is the junction leakage current, L_d, N_g are logic depth and average number of gates respectively, k, K_3, K_4 *and* K_5 are constant fitting parameters denoting circuit technology dependency [18]-[22]. Another important issue, which often is overlooked in voltage scaling approaches, is the consideration of transition overheads, i.e., each time the PE's supply voltage is altered; this change requires a certain amount of extra energy and time. The energy overhead (E_{OH}), when switching from V_{DDu} to V_{DDv} is given by Eq. 4, where C_r denotes power rail capacitance.

$$E_{leak} = \frac{L_d\, N_g\, N_u\, k\, V_{DDu} \left(V_{DDu}\, K_3\, e^{K_4\, V_{DDu}}\, e^{K_5\, V_{BSu}} + |V_{BSu}|\, I_{ju} \right)}{(V_{DDu} - V_{th})^\alpha} \tag{3}$$

$$E_{OH} = C_r\, |V_{DDu} - V_{DDv}|^2 \tag{4}$$

Hence, the total task energy and the energy of the whole system can be given as Eq. 5 and Eq. 6 respectively. Here, Eq. 6 represents the objective (cost) function to be minimized.

$$E_{task} = E_{dyn} + E_{leak} + E_{OH} \tag{5}$$

$$E_{total} = \sum_{taskcount} E_{task} \tag{6}$$

We have to guarantee that due to voltage scaling technique applied, tasks with deadlines still finish before their deadlines and the last task on each PE finishes no later than T_{graph}. This means that the sum of task durations in each path (from input to output) in the DAG should not be greater than T_{graph}. Note that the number of paths in a DAG can grow exponentially with respect to the number of edges, which implies that path-based optimization methods are not easily applicable and that voltage scaling approaches used on single PE systems cannot be readily extended to solve energy optimization problem on multiple PEs.

The execution time of task u at the highest supply voltage (V_{DDMAX}) is a constant T_u. Since *IN* and *OUT* nodes are conceptual nodes their execution times (T_{OUT}, T_{IN}) are set to 0. Besides the task execution time d_u, each task is also associated with a starting time denoted as D_u. Hence, the timing constraints on the given task graph can be modelled as

$$D_{OUT} - D_{IN} \leq T_{graph} \qquad (7)$$

$$D_v - D_u - d_u \geq 0 \; \forall \, e(u,v) \, \varepsilon \, DAG \qquad (8)$$

$$D_u + d_u \leq dl_u \; \forall \, u \, with \, deadline \qquad (9)$$

$$dl_u \geq T_u, \, and \, D_u \geq 0, \, integer \qquad (10)$$

$$V_{DD_{MIN}} \leq V_{ddu} \leq V_{DD_{MAX}} \qquad (11)$$

For a feasible scheduling, if D_{IN} set to be 0, the above constraints guarantee that tasks with deadlines will finish before their deadlines and the finish time of all tasks is not greater than T_{graph}. In order to be compatible with the working range of the PE's the supply voltage of each task is constricted between a minimum and a maximum value, according to Eq. 11. Combining the objective and the constraints given in Equations 7 to 11, we have the IP formulation for the voltage scaling problem in multi-PE systems. Trading-in the increase of delay for energy savings, the relationship between du and E_{task} has to be established. If the supply voltage can change continuously, the task execution time d_u can also be assumed to change continuously. In this case E_{task} is a convex function of d_u [22], i.e., we have $E_u = f(d_u)$, where $f(.)$ is a convex function. Substituting E_u with $f(d_u)$, the IP formulation of the problem of minimization of energy dissipation in multi-core systems for continuous voltage scaling approach is completed.

3   Energy Minimization on an Arbitrary Task Graph Using Resistive Network Analogies

In the following analysis, we will consider generic high-performance multi-core systems that are composed of a heterogeneous network of PEs. Due to the diversity of the applications that run within the system and their different degrees of parallelism, the workloads imposed on the system components are non-uniform over time. For many applications, peak performance is required only during some time intervals in such systems. This introduces slack times during which the system can reduce its performance to save energy. The key in energy-efficient designs is the ability to tune PE performance to the non-uniform workload. Here, the first goal is to explore the problem of *energy minimization* on a given TG, and to show the analogy between this problem and the fundamental electrical behavior of a resistive network. Our ultimate goal is to exploit this analogy in the form of a compact solution to the optimization problem.

In this section, an analogy will be introduced which maps the cost function of the energy minimization problem under timing constraints in multi-core systems, into the problem of minimizing power consumption in an equivalent resistive network subject to KCL. In the following, it will be assumed that the TG is mapped and scheduled onto the target architecture (multi-unit PE system), i.e., it is known a priori where and in which order tasks are executed and the communications between tasks take place.

3.1   From Task Graph to Resistive Network

In heterogeneous multi-core systems, the timing relationships (constraints) and the relative ordering between various tasks of an application are usually represented with a task graph which is used to capture the data-flow interdependencies of the entire system, as already established in the previous section. On the other hand, a resistive network is a connected graph (possibly with multiple edges) where each edge e is assigned a positive real number R_e called its resistance (in Ohms). Based on this analogy a given task graph can be mapped onto a resistive network by replacing each task (node) in the task graph with a resistor and edges of task graph as electrical connections in the resistive counter part.

It is a well-known fact that analog processing is usually more efficient than digital processing with respect to power consumption and chip area, when high precision is not required. Resistive networks have been widely used for various applications in analog VLSI [23]. A resistive network (RN) can be described by a system of linear equations based on Kirchhoff's and Ohm's laws. In a parallel resistive network that consists of n resistors, an imposed current i splits into n components proportional to branch conductances (G_n) that act as a current divider. According to Maxwell's heat theorem [24], *any network of linear resistors driven by a constant current, at steady state intrinsically minimizes the power dissipated in the form of heat in the network*. The demonstration of this theorem can be found from the book entitled *A Treatise on Electricity and Magnetism* by James C. Maxwell, first published in 1891 [24].

Tasks in real-world applications usually have control and data dependencies. Processing element sharing can be captured in a TG with multiple PEs with additional edges representing the control relation between dependent tasks. Figure 2 is a simple yet good example for such a case, where each parallel branch represents a PE. In this example, five tasks are mapped and scheduled on two PEs. Tasks mapped on different PEs can run in parallel in time as a basic consequence of parallelism, i.e., the period of each parallel branch in a TG is still equal to T. Notice that the points a through e (labelled on the TG for the sake of easy identification) actually represent the same instant in time. Also recall that t_4 can only start after processing of t_1 and t_3 are finished (DAG). The given TG indicates that t_1 and t_3 have to be finished at the same time for sake of completing work just-in-time (corresponding to minimum energy consumption), which describes a soft deadline for tasks. Similarly, execution time of tasks t_4 and t_5 mapped on the second PE should be equal to that of task t_2 running on the first PE. Hence, from the time point of view, the system shown in Fig. 2 can be presented by a simple sequential task graph as shown in Fig. 3(a). Here, original tasks t_1 and t_3 running in parallel on two different processing elements, are represented as the first combined task $W_1 = f(t_1, t_3)$, and similarly $W_2 = f(t_2, t_4, t_5)$ represents the combined tasks t_2, t_4, t_5.

It is clear from the very simple task graph shown in Fig. 3(a) that the second task (W_2) can start only after the preceding task (W_1) is finished. To optimize the overall energy consumption, these two tasks must share the available time T with respect to their workloads, where the duration of each task is indicated as

d_{W1} and d_{W2} respectively. Clearly, the time constraint: $d_{W1}+d_{W2} = T$ describes the condition that is needed in order to finish the mapped function within its deadline.

The total dissipated energy in the system can be written as the summation of the all task energies. Since the uth task is executed during the time period d_u consuming a power of P_u, the energy dissipation of the uth task can be found as $E_u = P_u d_u$. Hence, the problem of minimizing the overall dissipated energy in a given system (Fig. 3(a)), represented with its task graph under timing constraints, is formulated in Eq. 12.

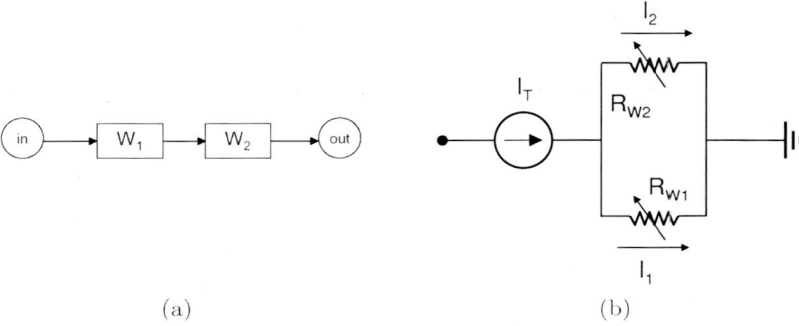

(a)   (b)

Fig. 3. (a) Simple task graph of two sequential tasks (b) and the resistive-network representation that corresponds to this task graph.

$$\min E_{total} = \min \sum_u E_u = \min \sum_u P_u d_u$$
$$subject\ to \sum_u d_u = T \tag{12}$$

At this point, we surmise that the equivalent resistive network of this specific task graph consists of two controlled resistors in parallel as shown in Fig. 3(b), where the network is supplied with a constant current I_T. This total current will be split linearly between branches proportional to branch conductances, I_1 and I_2 (Ohm's law). Due to KCL these currents must satisfy the equality $I_1+I_2 = I_T$.

Now consider the total power consumption in the equivalent RN shown in Fig. 3(b). Intrinsically, we know that the RN will consume the lowest possible power, P_{total}, at steady-state for a given driving current (I_T) according to Maxwell's Heat theorem. Due to KCL, I_T will be split into parallel branch currents (I_i) that are inversely proportional to branch resistances R_i (proportional to branch conductance G_i). Hence, it can be seen that the parallel resistive network actually realizes the solution of the following minimization problem, under the constraint that the sum of all branch currents is equal to I_T.

$$\min P_{total} \quad = \quad \min \sum_i P_i = \min \sum_i \frac{I_i^2}{G_i}$$

$$\text{subject to} \sum_i I_i \quad = \quad I_T \tag{13}$$

A comparison between Eq. 12 and Eq. 13 reveals the clear analogy between the problem of minimizing energy consumption on a complex system under timing constraints, and the problem of minimizing power dissipation in a resistive network under KCL constraint. Note that the branch currents (I_i) correspond to task durations (d_u) in the former problem. Thus, for the simple case described in Fig. 3, it is shown that the task graph can be represented by the equivalent resistive network.

Still, the simple resistive network equivalent given in Fig. 3(b) is not sufficient to model the actual behavior of the system with respect to individual tasks mapped on two processing elements. Note that, W_1 is a function of the two tasks t_1 and t_3, that must be executed in parallel on two different processing elements, i.e. these two tasks must have the same duration $(d_{W1} = d_1 = d_3)$. Similarly, in a resistive network branch consisting of two series connected resistors, each resistor must carry the same amount of branch current. Based on this analogy *all parallel tasks can be converted into series-connected branches in the equivalent resistive network*. Furthermore, W_2 is a function of two series tasks executed in parallel to a third task. In this case, execution time of task t_2 must be equal to the sum of execution times of tasks t_4 and t_5. Consequently, the amount of time necessary for execution of task W_2 will be split among t_4 and t_5 $(d_{W2} = d_2 = d_4 + d_5)$ according to the actual workload of these two tasks. Similarly, in a resistive network branch of parallel connected resistors the main branch current will be shared proportionally between the parallel branches according to KCL. Hence, *all sequential tasks can be represented by parallel-connected branches in the equivalent resistive network*.

Finally, the equivalent RN of the TG given in Fig. 4(a) can be implemented as shown in Fig. 4(b). I_T represents the overall available time, and each device current, (I_i), in the resistive network corresponds to the duration of the related task, (d_u) in the associated TG. Consequently, the problem of minimizing the sum of all task energies in a certain application is mapped onto an equivalent resistive network (Fig. 4(b)) consisting of controlled (pseudo-) conductances, where all parallel tasks are converted into series-connected branches, and all serial tasks are converted into parallel-connected branches. Note that each task sequence (or sub-sequence) in the TG corresponds to a parallel section (or sub-section) in the resistive network where KCL is valid, and is represented by a corresponding cut-set. The equivalence between the two analogous minimization problems is illustrated below.

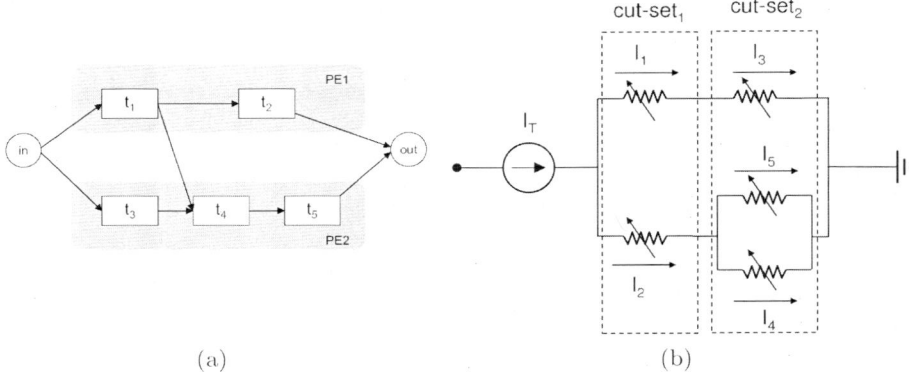

Fig. 4. (a) Task graph of five tasks mapped on two processing element (b) and its parallel-resistive network counter part.

$$\min E_{total} \quad \Leftrightarrow \quad \min P_{total}$$

$$\min \sum_u E_u = \min \sum_u \frac{P_u}{d_u} d_u^2 \quad \Leftrightarrow \quad \min \sum_i P_i = \min \sum_i \frac{I_i^2}{G_i} \qquad (14)$$

$$\text{subject to} \sum_u d_{u \in j} = T_j \quad \Leftrightarrow \quad \text{subject to} \sum_i I_{i \in k} = I_{T,k} \qquad (15)$$

$$\forall \text{ task sequence "j"} \qquad \forall \text{ cut-sets "k" that correspond}$$
$$\text{to a task sequence "j"}$$

$$\text{where } d_u \quad \Leftrightarrow \quad I_i \qquad (16)$$

$$\text{hence } \frac{d_u}{P_u} \quad \Leftrightarrow \quad G_i \qquad (17)$$

Note that the summation of branch currents, i.e. elements of the kth cut-set is always equal to the resistive network driving current (I_T). Similarly, summation of the task durations that corresponds to the jth task sequence is equal the task graph period (T) in the given TG. Besides, any subset of the imposed timing constraints, e.g. $d_4 + d_5 = d_2$, are intrinsically modelled by Eq. 15, and hence guaranteed by the resistive network implementation. Thus, the analogy between the energy minimization problem under timing constraints in a general TG and the power minimization problem under KCL constraints in an equivalent RN is demonstrated. Note that, we need to construct a structure to carry out the calculation of P_u (Eq. 17).

Assuming that V_u and I_u are the supply voltage and the current drawn from supply (including dynamic, short circuit and leakage currents) for the uth task, and using d_u as the task duration, the corresponding device conductance G_i can be mapped as follows.

$$\frac{d_u}{P_u} = \frac{d_u}{V_u I_u} \Leftrightarrow G_i \tag{18}$$

From the above explanations, the generalized steps involved for mapping the given task graph into a parallel-resistive network equivalent are as follows:

- Identify and assign the processing elements and the tasks mapped and scheduled on to the given system.
- Insert IN and OUT nodes into the given TG where edges from IN node to the first task of each PE, and edges from the last task on each PE to OUT node are added.
- If possible simplify the task graph by replacing the edges representing processing element sharing by equivalent edges capturing the data/control dependencies between PEs.
- Convert all *parallel tasks* in the simplified TG to *series-connected resistor branches* in the resistive network.
- Convert all *series tasks* in the simplified TG to *parallel-connected resistor branches* in the resistive network.
- Replace the IN and the OUT nodes of TG by a DC current source modelling the TG period and by the *ground* connection providing the necessary current path in the resistive network respectively.

However, not every task graph is in series/parallel configuration. The task graph given in Fig. 5(a) is an example of such non series/parallel configuration. Still, an equivalent resistive network can be mapped from the given TG without violating the corresponding timing constraints as shown in Fig. 5(b) where the cut-sets are highlighted by dashed lines. Here, timing constraints, e.g. $d_1 + d_4 = T$, $d_2 + d_3 + d_4 = T$ and $d_3 + d_4 = d_5$, are intrinsically satisfied due to KCL constraint in the resistive network, i.e. $I_1 + I_4 = I_T$, $I_2 + I_3 + I_4 = I_T$ and $I_3 + I_4 = I_5$ respectively. Hence, the equivalent resistive network of controlled resistors can be mapped for any arbitrary task graph, where each device current represents the available time for the corresponding task. Although the applied mapping scheme has a certain resemblance to creating the dual of a given task graph, it is important to emphasize that the mapping of a given task graph to its equivalent resistive network is based on converting the time domain relation between tasks into equivalent resistive network currents.

4   Implementation of Resistive Network for Global Energy Optimization

4.1   Closed-Loop Operation of the Resistive Network

The ultimate goal of this work is to solve the system-wide energy optimization problem continuously by means of the equivalent power minimization in

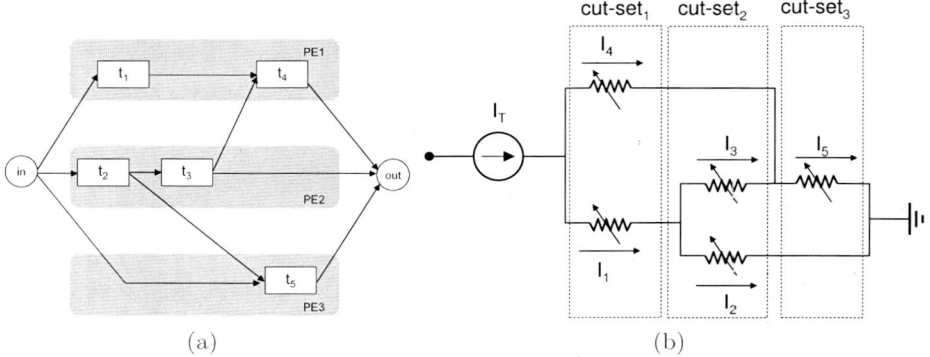

Fig. 5. (a) An example task graph of non series/parallel configuration (b) and the equivalent resistive-network of this task graph.

its resistive network model. In order to fulfil the overall goal, the variations of task-related design parameters given in Eq. 18 must be monitored. The design parameter d_u (task duration), is intrinsically embedded in the resistive-network model of the system, as indicated in Eq. 17.

The premise is that such global optimization (taking into account the complete system) will result in better solutions compared to the case where the energy dissipation of each unit is minimized separately. This concept is illustrated in a simplified manner in Fig. 6. Here, only one loop related to one of the resistive elements is shown, other feedback loops for the rest of the RN are not shown in the figure for sake of simplicity. From now on we will call the shown scheme, modelling individual tasks mapped and scheduled on to the system, as the feedback loop. The working principle of the feedback loop is as follows: due to Kirchhoff's current law (KCL) the branch currents will be divided proportionally with respect to branch conductances. Since the available duration of individual tasks, (d_u) corresponds to the device currents, I_i, the required operation frequency and the necessary minimum possible supply voltage will be calculated in the loop by means of the device currents, I_i. These values will then be used to calculate (estimate) the average supply current as a result of the dictated voltage level, modelling all relevant components such as dynamic, short circuit and leakage currents, and the device conductance will be adjusted according to Eq. 17.

Figure 6 shows the simplified block diagram implementation of the feedback loop for one device conductance, where a current-based approach is used to represent key loop variables. The simple *ghost circuit* (GC) which consists of a ring oscillator replicating the critical path of the PE, is used in each loop to continuously determine the minimum supply voltage and the supply current that correspond to a target operation frequency. The predicted workload information (N_i) is injected into each loop in the form of a 4-bit external control variable. Any

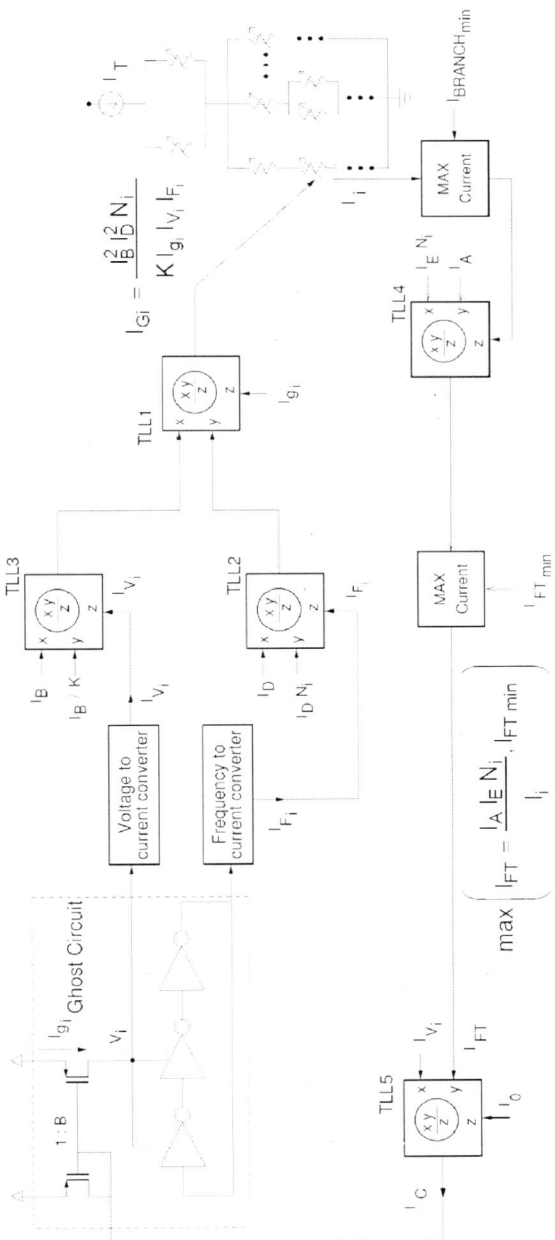

Fig. 6. Detailed representation of the constructed loop including all sub-blocks, e.g. Ghost circuit, translinear loops, voltage and frequency to current convertors and the resistive-network mapped from the given task graph.

change in the workload information (N_i) influences the current corresponding to the target operation frequency (I_{FT}) in the feedback loop. Hence, the simple GC determines the supply voltage level to be applied to the PE for achieving the target frequency as well as the resulting current consumption. These values are then converted into current representations in order to calculate the pseudo-resistor controlling currents (I_{Gi}), with several translinear loops used to carry out necessary calculations as current operators, while the device conductance value also changes according to I_{Gi}. This change in the value of device conductance forces all the branch currents in the RN to be adjusted by means of KCL. As the system settles to its new operating point, the new device currents in the pseudo-resistor network are determined by KCL, dictating the optimum task duration with the prescribed supply voltage and operating frequency for each PE to minimize system-wide energy dissipation. Implications with respect to overall stability of the loops will be discussed later in Stability Analysis Section. It can be shown that the dynamic behavior of each branch control loop is governed by a single-dominant-pole transfer function, and that the entire system always converges to a stable operating point for a given set of (N_i) values. Also, note that the GC can effectively capture the actual frequency-voltage-power relationship of the PEs, including the influence of leakage power dissipation, eliminating any analytical approximation of physical behavior that is inherently prone to inaccuracies. These circuits are capable of reflecting actual operating conditions on-chip, inherently taking into account local variations of temperature, as well as process-related fluctuations of device parameters.

In this solution, the GC is driven by its supply current (I_C) rather than the supply voltage since the instantaneous operation of the oscillator is imposed by the calculated power dissipation based on the required frequency of operation ($P = fCV^2 = IV \Longleftrightarrow I = fCV$). This is done with the assumption that the dynamic power consumption is dominant. If necessary, a static GC is added to the loop to mimic the static current consumption of the PE, proportional to the total number of gates (that may be different for different PE). Then, this current is added to the dynamic current consumption (I_{gi}).

Current-mode processing in each feedback loop is carried out by single quadrant current multiplier/dividers labelled as TLL_i. Each current operator is implemented by the simple alternating topology translinear loop (TLL) of four transistors operated in weak inversion with their bulks connected to the common substrate resulting in Eq. 19 as shown in Fig. 7(a). Thus, the entire feedback loop can be implemented with a very small number of devices, which leads to significant savings in silicon area. Here, a clockwise element (CW) is the one whose gate-to-source voltage is a voltage drop in the clockwise direction of the loop. So we shall consider a counterclockwise element (CCW) as the one whose gate-to-source voltage is a voltage increase in the clockwise direction of the loop [25].

$$I_{out} = \frac{I_x I_y}{I_z} \tag{19}$$

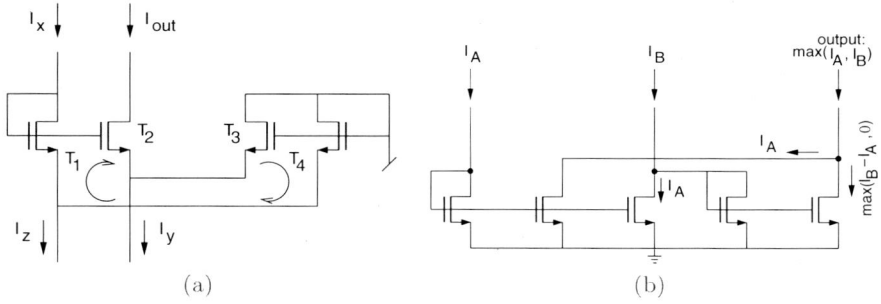

Fig. 7. (a) A subthreshold MOS translinear loop consisting of two CW transistors and two CCW transistors, constructed to operate as an inverse current multiplier, (b) A subthreshold MOS translinear loop consisting of two CW transistors and two CCW transistors, constructed to operate as an inverse current multiplier.

This is a single quadrant current multiplier/divider, thus all currents should be positive. Such multiplier/divider schemes are utilized in the feedback loop used to implement the analog optimizer for converting duration (time) to frequency (where all variables are represented as currents) and to implement the pseudo-resistor controlling current definition by means of ratio of current multiplications. The simulation results of the implemented single quadrant current multiplier/divider can be found in [26, 27].

Minimum current limiter (Maximum current selector) blocks are used to restrict the operation range to a defined value. The upper limit of the operation is intrinsically limited by the technology due to the fact that the maximum allowed core operating voltage is fixed to a constant level. In order to guarantee that the lower limit of operation, which is 1.2V or 150MHz is not violated under any circumstances two minimum current limiter blocks are used in all the feedback loops [28, 27]. For this purpose a combination of NMOS transistors is used to carry out addition/subtraction of the replicas of the two input currents as given in Fig. 7(b).

Each pseudo-resistor is realized as a single MOS transistor operating in weak inversion where the equivalent conductance value of each transistor is controlled independently by a current by means of a control transistor - thus, utilizing only a few transistors. Note that the linear pseudo-Ohm's law (Eq. 20) is valid and the network of controlled resistors remain linear with respect to currents only [23].

$$G^* \quad = \frac{1}{R^*} \quad = \frac{I_S}{V_0} \exp\left(\frac{V_G - V_{T0}}{2U_T}\right) \tag{20}$$

A resistor connected to ground potential in the RN corresponds to a saturated MOS transistor (operated in weak inversion) that provides a pseudo-ground in the equivalent pseudo-resistive network (refer to Fig. 8). Any current flowing to the pseudo-ground can be easily extracted without influencing the branch current ratio, by means of a grounded current mirror made of transistors complementary to those of the network as shown in Fig. 8 [23]. Hence, grounded current-mirrors are used to sense each device current separately, to be further used in the feedback loop.

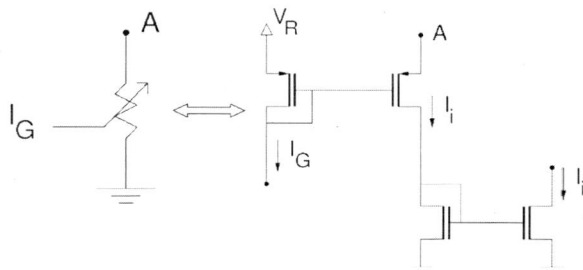

Fig. 8. Implementation of a grounded pseudo-resistors realized using CMOS process.

Figure 9 shows the simulated and the measured operation of a three-loop optimizer network which is used to model the behavior of a task graph comprising three sequential tasks. Here, the task durations (device currents) resulting in the optimum system energy dissipation are shown for various workload combinations as indicated. The workload information of three sequential tasks are shown in parenthesis for each simulation interval as (N_1, N_2, N_3) combination. The normalized workload estimations (N_i) for all tasks are updated at regular intervals of 5 μs, ranging from (2,8,4) in the first interval to (12,8,8) in the last interval. The available time is shared among the three tasks for all workload conditions; guaranteeing timing constraints and optimizing the dissipated energy in the system by means of optimally utilizing the available time. As it can be seen from the figure, the supply voltage level for the second task ($Loop_2$) varies with respect to other task workloads condition although there has not been any change in its own workload. The corresponding supply voltage and the device current (task duration) values indicate that the proposed analog optimizer is capable of responding to varying operating conditions with fast settling times and a wide dynamic range (supply voltage variation between 1.2 and 1.74 V), dictating the optimum operating voltage and duration of all three tasks mapped on the PE for minimum system energy consumption [28, 27].

The three-loop demonstrator circuit of the proposed analog optimizer architecture has been implemented using a 0.18 μm standard digital CMOS process (Fig. 10(a)). The overall circuit area of the optimizer is (250μm x 700μm) excluding decoupling capacitors, while each loop circuit occupies only (180μm x 120μm). The circuit is capable of supporting the desired frequency range of 170MHz-290MHz, as well as the voltage range of 1.2V-1.8V. The average power consumption of the entire three-loop optimizer is 6.5mW [28, 27].

Figure 10(b) shows the variation of the overall energy dissipation of the system composed of three tasks, scheduled in series and mapped on a single processor - as a function of changing workload conditions, calculated from measured

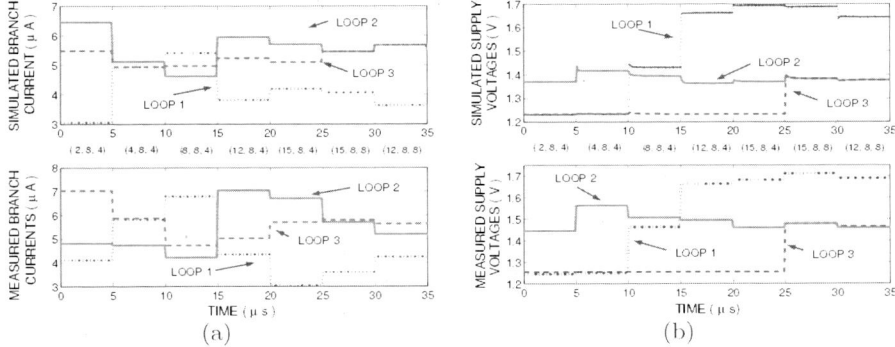

Fig. 9. Simulation and measurement results show (a) the branch currents (i.e. task duration) and (b) the corresponding supply voltages which are computed under varying workload combinations as indicated.

voltage/frequency and task duration values. To test the optimality of this solution, the device current values were slightly perturbed from their actual values (while keeping the sum constant) and the energy surface has been re-calculated. The resulting energy surface is clearly higher than the original solution for all workload combinations and for all branch current perturbations, demonstrating that the original solution indeed is the minimum energy surface.

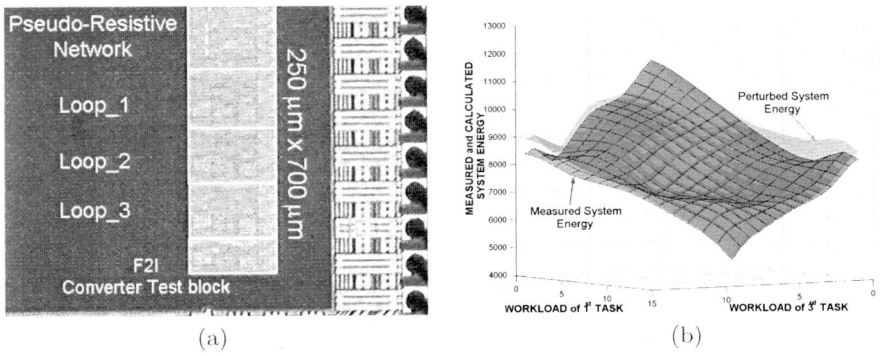

Fig. 10. (a) Chip microphotograph of the three-parallel loop optimizer. The overall circuit occupies only $250\mu m$ x $700\mu m$ and (b) Energy dissipation of the system composed of three tasks, scheduled in series and mapped on a single PE.

In Table 1 the comparison of the simulated supply voltages (V), operation frequencies (MHz) and task durations (device currents-μA) of the same system are given for the proposed global optimization approach versus local energy optimization applied to each task. Note that only the workload of first task increases throughout the table. Hence, in the local optimization scheme the core supply voltage levels and operation frequency remain constant during the second and the third tasks resulting in a higher power dissipation and energy consumption in the overall system. In contrast, when using the proposed global optimization approach, any change in workload condition of any of the tasks influences all task durations corresponding to a minimization of the total system energy dissipation by optimally using the overall available time (T). The additional energy savings varies between 11% and 20% for different cases.

Table 1. The comparison of the simulated global energy optimization approach versus local optimization.

N_i	V_{dd1} (V)	V_{dd2} (V)	V_{dd3} (V)	f_1 (MHz)	f_2 (MHz)	f_3 (MHz)	d_1 (s)	d_2 (s)	d_3 (s)	P_{total} (μW)	E_{total} (μJ)
			Global Energy Optimization (Proposed Approach)								
(2,8,4)	1.24	1.44	1.28	173.8	231.6	191.8	0.21	0.43	0.37	1087.6	385.3
(4,8,4)	1.25	1.56	1.28	175.5	252.7	192.8	0.33	0.34	0.33	1220.4	411.6
(8,8,4)	1.46	1.50	1.37	227.5	243.6	189.1	0.36	0.31	0.33	1370.1	455.3
(12,8,4)	1.66	1.35 7	1.31	273.1	242.2	191.1	0.25	0.39	0.35	1629.0	522.9
(15,8,4)	1.70	1.58	1.32	282.2	256.3	189.7	0.28	0.38	0.34	1755.9	577.7
			Local Energy Optimization								
(2,8,4)	1.23	1.49	1.25	176	255	204	0.08	0.21	0.13	1160.8	486.6
(4,8,4)	1.24	1.49	1.25	204	255	204	0.13	0.21	0.13	1228.4	478.9
(8,8,4)	1.49	1.49	1.25	255	255	204	0.21	0.21	0.13	1463.7	516.1
(12,8,4)	1.62	1.49	1.25	288	255	204	0.28	0.21	0.13	1641.8	592.4
(15,8,4)	1.71	1.49	1.25	300	255	7204	0.33	0.21	0.13	1764.6	668.4

4.2 Overall System Implementation

The proposed analog optimizer determines the supply voltage level and operation frequency of all tasks that are represented in the system task graph, simultaneously. On the other hand, tasks are to be executed in their sequential order on the PEs. This means that the individual operating voltages and frequencies will have to be assigned to the PEs according to their temporal relationships. Hence, the intended system will require an interface between the analog optimizer and the PEs. A possible candidate of such an interface is shown in Fig. 11(a). Here, a separate continuous voltage, high efficiency DC/DC converter is used for each PE individually. The supply voltage levels defined by the optimizer (per task) will be applied to the PEs through these high efficiency voltage converters during the operation of the system, sequentially. The frequency of operation on the other hand is also defined by the analog optimizer and will be used to drive the clock buffers of the PEs as indicated in the figure.

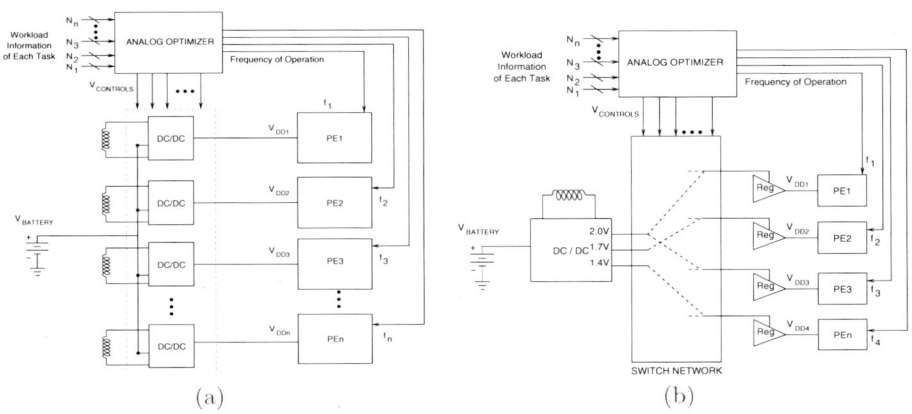

Fig. 11. (a) Block diagram representation of the system architecture in which the analog optimizer controls the individual clock frequencies and supply voltages of various PEs, and (b) Block diagram representation of the system architecture in which a single high efficiency, multiple output DC/DC converter is used to generate the supply voltage levels in a certain range.

Nevertheless, this solution could become costly due to the number of I/O pins needed for external inductors that are required to ensure the high efficiency of DC/DC converters, and silicon area (dedicated DC/DC per PE) for SoC applications employing numerous PEs. An alternative scenario for the interface between the analog optimizer and the PEs is presented in Fig. 11(b). Here, supply voltage levels defined by the analog optimizer will be applied to the PEs through voltage regulators (current efficient voltage followers) during operation. While the number of external inductors is reduced to one, it is assumed that only one DC/DC converter is utilized with three output levels (1.4V, 1.7V and 2.0V). Each output of the DC/DC converter can be used to generate the supply voltage levels in a certain range, with the help of voltage regulators, e.g. the 2.0V converter output is used to generate 1.8V - 1.51V supply voltage range. It should be noted that in this case, the energy savings obtained by utilizing the analog optimizer will be degraded due to the energy losses in the voltage regulators, by up to 33% (at the "edge" of the regulator output range). However, this drawback can be overcome by taking into account the voltage regulator supply levels in the optimization algorithm, where constant current levels can be used for I_{Vi} to represent task supply voltage levels. Hence, the final solution will still be the optimum energy dissipation for the whole system, including the regulator losses.

5　Stability Analysis

As already mentioned, the concept of system stability needs to be considered when several components adopt dynamic policies to control energy consumption and performance. Possible oscillations in energy/performance space that could be caused by applied energy management policies are undesirable, and should be avoided. In this section it will be shown that the dynamic behavior of each device control loop is governed by a single-dominant-pole transfer function, and that the entire system (the centralized optimizer unit) always converges to a stable operating point for a given set of workload (N_i) values.

In order to derive the characteristic equation of the feedback control loop, the loop is opened on the resistive network. Hence, the device current I_i is treated as the input current (variable) and the pseudo-resistor controlling current I_G is treated as the output current. Note that I_A, I_B, I_D, I_E, I_0 are constant biasing currents used in the feedback loop. From the loop dynamics, the output current I_G can be written as in Eq. 21. Note that one can show the small variations in the value of a variable as $(X + x)$, where lower case represents the variations in the value of the variable. Using this definition, the output current can be written as given in Eq. 22.

$$I_G = \frac{I_B^2 I_D^2 N}{K I_g I_V I_F} \text{ or } I_G = \frac{Constant}{I_g I_V I_F} \tag{21}$$

$$I_G \left(1 + \frac{i_G}{I_G}\right) = \frac{Constant}{I_g I_V I_F \left(1 + \frac{i_g}{I_g}\right)\left(1 + \frac{i_V}{I_V}\right)\left(1 + \frac{i_F}{I_F}\right)}$$

$$= \frac{Constant}{I_g I_V I_F}\left(1 - \frac{i_g}{I_g} - \frac{i_V}{I_V} - \frac{i_F}{I_F}\right) \tag{22}$$

Note that we can express the ratios of the current representations of the ghost circuit supply voltage and the operating frequency as well as the current consumption of the ghost circuit and their variations in terms of the ratio of the input current and its variation as given in Eq. 23.

$$\frac{i_g}{I_g} = -2\frac{i_i}{I_i} \quad \text{and} \quad \frac{i_V}{I_V} = -\frac{i_i}{I_i} \quad \text{and} \quad \frac{i_F}{I_F} = -\frac{i_i}{I_i(1 + s\tau)} \tag{23}$$

Finally, the device conductance controlling current being the output current and the related device current being the input current the small signal behavior of the feedback loop can be written as given in Eq. 24, since device conductances are linearly proportional to their controlling current I_G. Hence, it is shown that the dynamic behavior of each branch feedback loop is governed by a single-dominant-pole transfer function.

$$\frac{g_i}{G_i} = \frac{i_G}{I_G} = \frac{i_i}{I_i}\left(3 + \frac{1}{1 + s\tau}\right) \tag{24}$$

Now, consider a RN consisting of three parallel branches to illustrate the stability properties of the system. If we write the first branch current in the RN comprising three parallel branches in terms of the RN biasing current and the other branch currents, we get Eq. 25, where G_i and g_i represents the device conductance and the variations in the conductance value respectively. If we replace each $(G_i / \sum G_i)$ quantity in Eq. 25 by (I_i/I_T), and substituting Eq. 24 where ever suitable, we can finally write Eq. 26. After performing the necessary mathematical operations on the resulting equation set, the characteristic equation of the system can be expressed as Eq. 27.

$$I_1\left(1 + \frac{i_1}{I_1}\right) = I_T\left(1 + \frac{i_T}{I_T}\right) \frac{G_1\left(1 + \frac{g_1}{G_1}\right)}{(G_1 + G_2 + G_3)\left(1 + \frac{g_1 + g_2 + g_3}{G_1 + G_2 + G_3}\right)} \tag{25}$$

$$\frac{i_1}{I_1} = \frac{i_T}{I_T} + \frac{g_1}{G_1}\left(1 - \frac{I_1}{I_T}\right) - \frac{I_2}{I_T}\frac{g_2}{G_2} - \frac{I_3}{I_T}\frac{g_3}{G_3} \tag{26}$$

$$4\tau_1\tau_2\tau_3 s^3 + \left[6(\tau_1\tau_2 + \tau_2\tau_3 + \tau_1\tau_3) - 2(R_1\tau_2\tau_3 + R_2\tau_1\tau_3 + R_3\tau_1\tau_2)\right]s^2$$
$$+ \left[6(\tau_1 + \tau_2 + \tau_3) + 3(R_1\tau_1 + R_2\tau_2 + R_3\tau_3)\right]s + 9 = 0 \tag{27}$$

$$a_3 s^3 + a_2 s^2 + a_1 s + a_0 = 0 \tag{28}$$
$$\text{where} R_i = \frac{I_i}{I_T} \quad \text{thus} \quad \forall \quad \sum_i R_i = 1$$

Now, if we rewrite the characteristic equation of the system as in Eq. 28, we can check the stability of the system by applying the Routh criterion. The principal stability criterion for linear systems states that a system is stable if all poles of its transfer function lie in the left-half of the complex s-plane. Equivalently, a system is stable if the real parts of all roots of its characteristic equation are negative. Note that a root of the characteristic equation is synonymous with a system pole. To apply Routh's criterion, the Routh's Table should be created. The Routh criterion is applied by examining the sign of the coefficient in the column headed by a_3. The number of sign changes in the elements of this column, taken in order, is equal to the number of roots of the characteristic equation that have positive real parts. Hence, in order to show that the system is stable we should verify that the sign of the expression $a_1 a_2 - a_0 a_3$ is positive, since all the other components of the first column of the routh table are positive quantities. Note that all R_i and τ_i quantities are positive real values. Thus a_3, a_1 and a_0 are intrinsically positive for all R_i or τ_i values. Note that a_2 is also always positive for all R_i or τ_i values, since definition of R_i ($I_i/I_T = R_i$) guarantees that the multiplicative factor $(6 - 2R_i)$ in a_2 is always positive. After performing all the necessary multiplications it is proved that the sign of the mathematical operation $a_1a_2 - a_0a_3$ is always positive, guaranteeing that the proposed system is stable.

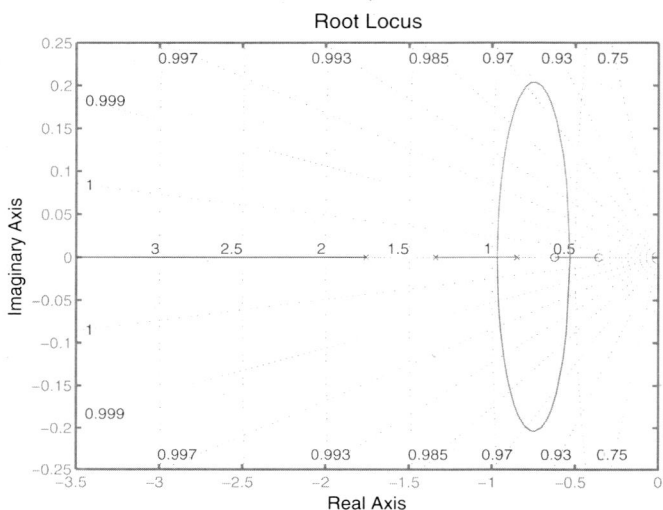

Fig. 12. The root locus plot of the system with three parallel branches for typical R_i and τ_i values.

Figure 12 shows the root locus plot for the system with the characteristic equation given in Eq. 27, displaying the closed-loop pole trajectories as a function of the feedback gain. Root loci are used to study the effects of varying feedback gains on closed-loop pole locations. In turn, these locations provide indirect information on the time and frequency responses. As can be seen from the figure all three roots of the systems lies on the left (negative) half plane, demonstrating that the system is stable. Note that the system has two negative zeros in addition to its three poles as indicated within the Fig. 12. The stability of the system is analyzed thoroughly by using MATLAB software tool, where R_i (satisfying the $\sum R_i = 1$ constraint) and τ_i values are randomly selected with 10% mismatch introduced. The graphic is obtained by sweeping the loop gain in the possible range.

6 Conclusion

In this chapter, the energy optimization problem in multi-core applications is addressed with a unique analog implementation approach. The analogy that exists between the energy minimization problem under timing constraints in a general TG and the power minimization problem under Kirchhoff's current law constraints in an equivalent RN is exploited. The principles of mapping an arbitrary task graph to an equivalent resistive network are presented. A fully analog, current-based solution to implement on-line energy minimization in complex multi-core systems under varying workload conditions is demonstrated, which achieves significant overall energy savings compared to the local energy minimization approach.

References

[1] M. Sirvastava, "Power-aware design energy consumers & sources, reduction & management." UCLA, EE202A Fall 2002 Lecture notes, 7 and 8, 2002, University of California, LA.

[2] A. Chandrakasan, R. Min, and M. Bhardwaj, "Power aware wireless microsensor systems," in *Proceedings of the 28th European Solid-State Circuits Conference*, p. In Keynote Paper, 2002.

[3] D. M. Monticelli, "Taking a system approach to energy management," in *Proceedings of the 29th European Solid-State Circuits Conference*, vol. 1, (Estoril, Portugal), pp. 15–19, Sep. 2003.

[4] A. Chandrakasan, V. Gutnik, and T. Xanthopoulos, "Data driven signal processing: an approach for energy efficient computing," in *Proceedings of the International Symposium on Low Power Electronics and Design*, (Monterey, California, United States), pp. 347–352, IEEE Press, 1996.

[5] K. Suzuki, S. Mita, T. Fujita, F. Yamane, F. Sano, A. Chiba, T. Maeda, Y. Watanabe, K. Matsuda, T. Kuroda, and T. Sakurai, "A 300 MIPS/W RISC core processor with variable supply-voltage scheme in variable threshold-voltage CMOS," in *Proceedings of IEEE Custom Integrated Circuits Conference*, pp. 587–590, May 1997.

[6] K. Usami, M. Igarashi, T. Ishikawa, M. Kanazawa, M. Takahashi, M. Hamada, H. Arakida, T. Terazawa, and T. Kuroda, "Design methodology of ultra low-power MPEG4 codec core exploiting voltage scaling techniques," in *Proceedings of the 35th Design Automation Conference*, (San Francisco, CA), pp. 483–488, June 1998.

[7] K. Flautner, D. Flynn, and M. Rives, "A combined hardware-software approach for low-power SoCs: Applying adaptive voltage scaling and intelligent energy management software," in *Proceedings of System-on-Chip and ASIC Design Conference, DesignCon*, Jan. 2003.

[8] C. Poirier, R. McGowen, C. Bostak, and S. Naffziger, "Power and temperature control on a 90nm *Itanium*®-Family Processor," in *Proceedings of IEEE International Solid-State Circuits Conference, Digest of Technical Paper*, vol. 2, pp. 304–305, Feb. 2005.

[9] X. Fan, C. S. Ellis, and A. R. Lebeck, "The synergy between power-aware memory systems and processor voltage scaling," in *Power-Aware Computer Systems, 3rd International Workshop*, pp. 164–179, Dec. 2003.

[10] D. B. Kirk, *Accurate and Precise Computation using Analog VLSI, with Applications to Computer Graphics and Neural Networks*. PhD thesis, California Institute of Technology, Pasadena, California, March 1993.

[11] G. Cauwenberghs, "A fast stochastic error-decent algorithm for supervised learning and optimization," in *Advances in Neural Information Processing Systems*, vol. 5, pp. 244–251, 1993.

[12] L. O. Chua and L. Gui-Nian, "Nonlinear programming without computation," in *IEEE Transactions on Circuits and Systems II*, vol. CAS-31, pp. 182–188, Feb. 1984.

[13] A. Dembo and T. Kailath, "Model free distributed learning," in *IEEE Transactions on Neural Networks*, vol. 1, pp. 58–70, 1990.

[14] R. Gregorian and G. C. Temes, *Analog MOS Integrated Circuits for Signal Processing*. New York: John Wiley and Sons, 1986.

[15] Technical Report, Transmeta Corporation, http://www.transmeta.com.

[16] S. Lee and T. Sakurai, "Run-time power control scheme using software feedback loop for low-power real-time application," in *Proceedings of the Asia and South Pacific Design Automation Conference*, (Yokohama, Japan), pp. 381–386, Jan. 2000.

[17] S. Lee and T. Sakurai, "Run-time voltage hopping for low-power real-time systems," in *Proceedings of the 37th Design Automation Conference*, (Los Angeles, California, United States), pp. 806–809, June 2000.

[18] A. Andrei, M. Schmitz, P. Eles, Z. Peng, and B. M. Al-Hashimi, "Overhead-conscious voltage selection for dynamic and leakage energy reduction of time-constrained systems," in *Proceedings of the Design, Automation and Test in Europe Conference and Exhibition*, (Paris, France), pp. 105–118, Feb. 2004.

[19] B. Zhai, D. Blaauw, D. Sylvester, and K. Flautner, "Theoretical and practical limits of dynamic voltage scaling," in *Proceedings of the 41st Design Automation Conference*, (San Diego, CA, USA), pp. 868–873, 2004.

[20] S. M. Martin, K. Flautner, T. Mudge, and D. Blaauw, "Combined dynamic voltage scaling and adaptive body biasing for lower power microprocessors under dynamic workloads," in *Proceedings of the International Conference on Computer Aided Design*, (San Jose, California), pp. 721–725, Nov. 2002.

[21] J. Liu, P. H. Chou, and N. Bagherzadeh, "Communication speed selection for embedded systems with networked voltage-scalable processors," in *Proceedings of the 10th International Workshop on Hardware/Software Codesign*, (Estes Park, Colorado), pp. 169–174, May 2002.

[22] Y. Zhang, X. Hu, and D. Chen, "Task scheduling and voltage selection for energy minimization," in *Proceedings of the 39th Design Automation Conference*, (New Orleans, LA, USA), pp. 183–188, June 2002.

[23] E. A. Vittoz, "Pseudo-resistive networks and their applications to analog collective computation," in *Proceedings of the 7th International Conference on Artificial Neural Networks*, (Lausanne, Switzerland), pp. 1133–1150, Oct. 1997.

[24] J. C. Maxwell, *A Treatise in Electricity and Magnetism*, vol. 1. New York, USA: Dover Publications, Inc., third ed., 1954.

[25] E. Vittoz, "Analog VLSI for collective computation," in *Proceedings of the 5th IEEE International Conference on Electronics, Circuits and Systems*, vol. 2, pp. 3–6, Sep. 1998.

[26] Z. T. Deniz, Y. Leblebici, and E. Vittoz, "Configurable on-line global energy optimization in multi-core embedded systems using principles of analog computation," in *Proceedings of 2006 IFIP International Conference on Very Large Scale Integration (VLSI-SoC)*, pp. 379–384, Oct. 2006.

[27] Z. T. Deniz, *Multi-Unit Global Energy Management and Optimization for Network-on-Chip Applications*. PhD thesis, Swiss Federal Institute of Technology (EPFL), Lausanne, Switzerland, February 2006.

[28] Z. T. Deniz, Y. Leblebici, and E. Vittoz, "On-line global energy optimization in multi-core systems using principles of analog computation," *IEEE Journal of Solid-State Circuits*, accepted for publication in July 2007 issue.

Logic Synthesis of EXOR Projected Sum of Products

Anna Bernasconi[1], Valentina Ciriani[2], and Roberto Cordone[2]

[1] Department of Computer Science, University of Pisa Italy, `annab@di.unipi.it`
[2] DTI, University of Milano Italy, {`ciriani,cordone`}`@dti.unimi.it`

Abstract. We define a new algebraic form for Boolean function representation, called *EXOR-Projected Sum of Products* (*EP-SOP*), consisting in a four level network that can be easily implemented in practice. Deriving an optimal EP-SOP from an optimal SOP form is a NP^{NP}-hard problem; nevertheless we propose a very efficient approximation algorithm, which returns, in polynomial time, an EP-SOP form whose cost is guaranteed to be near the optimum. Experimental evidence shows that for about 35% of the classical synthesis benchmarks, EP-SOP networks have a smaller area and delay with respect to the optimal SOPs (sometimes gaining even 40-50% of the area). Since the computational times required are extremely short, we recommend the use of the proposed approach as a post-processing step after SOP minimization.

1 Introduction

The classical approach to logic synthesis is the minimization of two-level SOP networks [2, 4, 13]. In this framework the resulting networks have a very low delay, thanks to the fixed number of levels, and the SOP expressions can be computed in a reasonable amount of time. To build networks with a more compact area, multi-level network synthesis has been proposed and widely studied [8, 17]. The drawbacks of this approach are the unbounded number of levels (and therefore the longer delay), as well as the much larger computational time required to synthesize the network. In an attempt to establish an effective trade-off between these two opposite approaches, recent studies have proposed the optimization of networks with a fixed number of levels (typically, three or four levels) [1, 5–7, 12, 14, 16]. Three levels of logic are enough to produce a minimal network for most of the Boolean functions; and in many cases three-level logic is a good compromise between circuit speed, circuit size, and the time needed for the minimization procedure [15]. Three and four-level logic networks are typically more compact than the corresponding SOPs, but the computational time required to compute them can be much longer.

The aim of this paper is to define a network with a bounded number of levels that can be easily implemented in practice and synthesized in a competitive time with respect to two-level synthesis. For this purpose, we propose a four-level network, *EXOR-Projected Sum of Products* (*EP-SOP*), which can be built in a very fast post-processing step from an optimal two-level SOP. We first define

Please use the following format when citing this chapter:

Bernasconi, A., Ciriani, V. and Cordone, R., 2007, in IFIP International Federation for Information Processing, Volume 249, VLSI-SoC: Research Trends in VLSI and Systems on Chip, eds. De Micheli, G., Mir, S., Reis, R., (Boston: Springer), pp. 241–257

the algebraic form of EP-SOP networks, and prove that deriving an optimal EP-SOP from an optimal SOP form is a hard problem (NP^{NP}-hard). We then describe an approximation algorithm, which returns in polynomial time an EP-SOP form whose cost is guaranteed to be near the optimum. Our experimental results show that in about 35% of the classical synthesis benchmarks the EP-SOP obtained has area and delay smaller then those of an optimal SOP form (sometimes gaining even 40-50% of the area). The computational times required are extremely short, thus recommending the use of this approach as a post-processing step after SOP minimization.

Before defining EP-SOP forms, we introduce them informally through an example. Let us consider the Boolean function f shown on the left side of Figure 1. An optimal SOP representation for f is $\phi = \overline{x}_1 x_2 \overline{x}_3 + x_1 \overline{x}_2 \overline{x}_3 + \overline{x}_1 \overline{x}_2 x_3 + x_1 x_2 x_3 + x_3 \overline{x}_4$. The right side of Figure 1 represents the projections of f onto the two subspaces where $x_1 = x_2$ and $x_1 \neq x_2$, respectively. As described in the Karnaugh maps on the right side of the figure, the projection of f onto the space $x_1 \neq x_2$ is covered by the optimal SOP form $\overline{x}_3 + x_3 \overline{x}_4$, while the projection onto the space $x_1 = x_2$ is covered by x_3. Notice that both SOP forms are much more compact than the original one, because minterms, which were not adjacent in the original Karnaugh map, now merge into new larger prime cubes. For example, the two products $\overline{x}_1 x_2 \overline{x}_3$ and $x_1 \overline{x}_2 \overline{x}_3$, which cannot be merged in the original Karnaugh map, correspond to the products $x_2 \overline{x}_3$ and $\overline{x}_2 \overline{x}_3$, which can be unified into product $\overline{x}_3$ in the lower Karnaugh map on the right side.

Since the two subspaces, $x_1 = x_2$ and $x_1 \neq x_2$, have characteristic functions equal to $(x_1 \oplus \overline{x}_2)$ and $(x_1 \oplus x_2)$, respectively, f can be expressed as $f \equiv (x_1 \oplus \overline{x}_2) x_3 + (x_1 \oplus x_2)(\overline{x}_3 + \overline{x}_4)$. Figure 2 shows how this form can be easily implemented by using a single 2-fan in EXOR gate and two PLAs.

As the previous example shows, the products of a generic SOP ϕ can be classified into two subsets: those that are entirely included into one of the two subspaces $x_1 = x_2$ and $x_1 \neq x_2$ (for example, in Figure 1 the product $\overline{x}_1 x_2 \overline{x}_3$ belongs entirely to the subspace $x_1 \neq x_2$) and those that intersect both of them, which we will call *crossing products* (for example, in Figure 1 the product $x_3 \overline{x}_4$). In general, it is not always convenient to project a crossing product, since this produces two smaller products, which reside into both subspaces. Therefore, we can choose whether projecting the crossing products or keeping them unprojected. In the second case, the resulting expression also includes a SOP form (called *remainder*) containing all the crossing products. We call the overall form *EP-SOP with remainder*.

Figure 3 reports the same example of Figure 1, in which the only crossing product $x_3 \overline{x}_4$ is not projected. In this case, the resulting EP-SOP with remainder form is $f \equiv (x_1 \oplus \overline{x}_2) x_3 + (x_1 \oplus x_2) \overline{x}_3 + x_3 \overline{x}_4$.

We can observe that EP-SOP expressions can be seen as Boolean factorized forms. Factorization of literal terms is a widely studied field in multi-level logic [3, 17]. Most of the proposed methods produce disjoint factorization (see [8] for an introduction). In contrast, the factorization of an EP-SOP form is not disjoint since a literal can stay simultaneously in the projected SOPs and in the

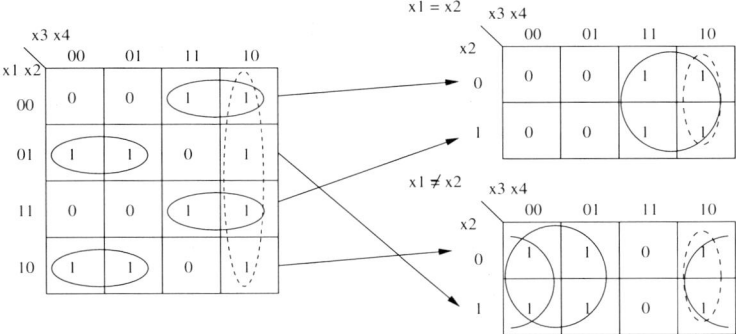

Fig. 1. *Karnaugh maps of a function f (left side) and its projections onto $x_1 \oplus \overline{x}_2$ (right side, top) and $x_1 \oplus x_2$ (right side, bottom).*

corresponding EXORs. For example, in the EP-SOP form $(x_1 \oplus \overline{x}_2)(x_2\overline{x}_4 + \overline{x}_3 x_4) + (x_1 \oplus x_2)(x_2\overline{x}_3 + \overline{x}_3\overline{x}_4)$ the literal x_2 appears both in the EXORs and in the SOPs.

Finally, EP-SOP forms share some similarities with another model of Boolean function representation: the *Linearly-Transformed BDDs* [10, 11]. LTBDDs are binary decision diagrams whose nodes are labeled with EXORs of variables, instead of just single variables. Thus, the node on the first level of a LTBDD, if labeled with an EXOR, defines the same kind of decomposition on which EP-SOPs are based.

The remainder of this paper is organized as follows. Section 2 describes the algebraic expression for EP-SOPs with and without remainder. Section 3 characterizes the computational complexity of the problem. Section 4 presents an approximation algorithm for EP-SOP synthesis, and proves that its solution is nearly optimal. In the end, Section 5 discusses the experimental results.

2   EP-SOP representation of Boolean functions

The following two sections formally describe EP-SOP expressions with and without remainder, and show how to derive them from an original optimal SOP form.

2.1   EP-SOP without remainder

Let us consider a SOP form ϕ, and a couple of variables x_i and x_j, where without loss of generality $i < j$. The space $\{0, 1\}^n$ can be partitioned into two disjoint subspaces: the space defined by the characteristic function $\chi_{\overline{\oplus}} = (x_i \oplus \overline{x}_j)$, i.e., the space where $x_i = x_j$, and its complement defined by the function $\chi_{\oplus} = (x_i \oplus x_j)$, i.e., the space where $x_i \neq x_j$.

We can write ϕ as the sum (union) of its two projections, $\phi_{\oplus}$ and $\phi_{\overline{\oplus}}$, onto these two spaces. Even if the projections allow us to eliminate a variable ad

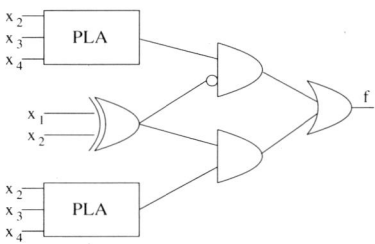

Fig. 2. *EP-SOP-network without remainder for the function in Figure 1.*

libitum between x_i and x_j, we always remove x_i (the one with lower index). In order to perform the two projections we must project one by one the products $p \in \phi$, considering four cases.

Algorithm 1 (Project. onto $(x_i \oplus \overline{x}_j)$ and $(x_i \oplus x_j)$ for EP-SOP) *Given a SOP form* $\phi = p_1 + p_2 + \cdots p_m$, *for each* p *in* $\{p_1, p_2, \ldots, p_m\}$ *project* p *in* $\phi_\oplus$ *or in* $\phi_{\overline{\oplus}}$ *using the following strategy:*

1. *If p contains both x_i and x_j (possibly complemented), i.e., $p = x_i x_j q$, where q is a product of literals, p has no projection onto the subspace where $x_i \neq x_j$. Thus, no product will be added to $\phi_\oplus$. By contrast, the projection of $p = x_i x_j q$ onto the subspace where $x_i = x_j$ gives the product $x_j q$, which will be added to $\phi_{\overline{\oplus}}$. The three other cases ($p = x_i \overline{x}_j q$, $p = \overline{x}_i x_j q$, $p = \overline{x}_i \overline{x}_j q$) can be handled in a similar way.*

2. *If p contains x_i (possibly complemented) and not x_j, i.e., $p = x_i q$, where q is a product of literals, the projection of p onto the subspace where $x_i \neq x_j$ gives the product $\overline{x}_j q$, which will be added to $\phi_\oplus$. The projection of p onto the subspace where $x_i = x_j$ gives the product $x_j q$, which will be added to $\phi_{\overline{\oplus}}$. The other case ($p = \overline{x}_i q$) can be handled in a similar way: $x_j q$ will be added to $\phi_\oplus$, and $\overline{x}_j q$ will be added to $\phi_{\overline{\oplus}}$.*

3. *If p contains x_j (possibly complemented) and not x_i, i.e., $p = x_j q$, where q is a product of literals, the projections of p onto both subspaces leave the product unchanged, thus $p = x_j q$ will be added to both $\phi_\oplus$ and $\phi_{\overline{\oplus}}$. The other case ($p = \overline{x}_j q$) can be handled in the same way, by adding p to both $\phi_\oplus$ and $\phi_{\overline{\oplus}}$.*

4. *If p does not contain x_i, $\overline{x}_i$, x_j, $\overline{x}_j$, the projections of p onto both subspaces leave the product unchanged, thus p will be added to both $\phi_\oplus$ and $\phi_{\overline{\oplus}}$.*

Observe that the last three type of products are indeed crossing products, which are projected onto the two spaces, while the products containing both x_i and x_j are projected only onto one of them.

Example 1. Let us consider the Boolean function f shown on the left side of Figure 1. An optimal SOP representation for f is $\phi = \overline{x}_1 x_2 \overline{x}_3 + x_1 \overline{x}_2 \overline{x}_3 + \overline{x}_1 \overline{x}_2 x_3 + x_1 x_2 x_3 + x_3 \overline{x}_4$. Suppose to project ϕ onto the spaces $(x_1 \oplus \overline{x}_2)$ and

$(x_1 \oplus x_2)$. The first product in ϕ contains both x_1 and x_2, thus it is not a crossing product (strategy 1 of Algorithm 1). Since x_1 is complemented and x_2 is not complemented we project this product onto the space $(x_1 \oplus x_2)$ (in fact, $x_1 \neq x_2$). The projected product is $x_2\overline{x}_3$. The unique crossing product of ϕ is $x_3\overline{x}_4$, since it does not contain x_1 and x_2. This product will be inserted in both the spaces without removing any literal. The overall projection will return the form $(x_1 \oplus \overline{x}_2)\,(\overline{x}_2x_3+x_2x_3+x_3\overline{x}_4)+(x_1 \oplus x_2)\,(x_2\overline{x}_3 + \overline{x}_2\overline{x}_3 + x_3\overline{x}_4)$. Note that the SOP forms of the projected spaces are not minimal. Minimizing them we obtain $(x_1 \oplus \overline{x}_2)\,x_3 + (x_1 \oplus x_2)\,(\overline{x}_3 + x_3\overline{x}_4)$.

We can now formally define the EP-SOP expressions. These forms can be derived starting from a SOP representation ϕ of a Boolean function f in two steps.

First we project ϕ onto the two subspaces $(x_i \oplus x_j)$ and $(x_i \oplus \overline{x}_j)$, as explained before, and we obtain the following expression.

Definition 1. *Let* $f : \{0,1\}^n \to \{0,1\}$, *and let* ϕ *be a SOP representation of* f. *Given a couple of variables* x_i *and* x_j, *the* (i, j)-*EP-SOP of* f *is the expression*

$$\xi_{ij} = (x_i \oplus x_j)\phi_\oplus + (x_i \oplus \overline{x}_j)\phi_{\overline{\oplus}},$$

where $\phi_\oplus$ *and* $\phi_{\overline{\oplus}}$ *are the projections of* ϕ *onto the spaces* $(x_i \oplus x_j)$ *and* $(x_i \oplus \overline{x}_j)$, *respectively.*

After the projection we can further minimize the two SOPs $\phi_\oplus$ and $\phi_{\overline{\oplus}}$ in order to minimize the EP-SOP ξ_{ij}.

Definition 2. *Let* $f : \{0,1\}^n \to \{0,1\}$, *and let* ϕ *be a SOP representation of* f. *Given a couple of variables* x_i *and* x_j, *the* minimal (i, j)-*EP-SOP of* f *is the expression*

$$\xi_{ij}^{(min)} = (x_i \oplus x_j)\phi_\oplus^{(min)} + (x_i \oplus \overline{x}_j)\phi_{\overline{\oplus}}^{(min)},$$

where $\phi_\oplus^{(min)}$ *and* $\phi_{\overline{\oplus}}^{(min)}$ *are two minimal SOP forms representing the projections of* ϕ *onto the spaces* $(x_i \oplus x_j)$ *and* $(x_i \oplus \overline{x}_j)$, *respectively.*

In the previous definitions we have fixed a single couple of variables, but we are interested in finding the minimal EP-SOP representation of a Boolean function, i.e., the expression containing the minimum number of products among all possible minimal EP-SOP with respect to *any* couple of variables.

Let $|\phi|$ denote the number of products in a SOP ϕ, and $|\xi| = |\phi_\oplus| + |\phi_{\overline{\oplus}}|$ the overall number of products in an EP-SOP ξ.

Definition 3. *The* minimal EP-SOP *representation of a Boolean function* f *is given by the EP-SOP expression* ξ_{MIN} *such that*

$$|\xi_{MIN}| = \min_{i,j} |\xi_{ij}^{(min)}|.$$

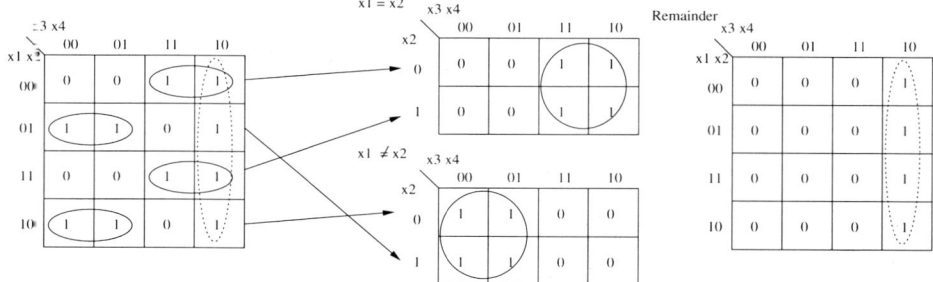

Fig. 3. *Karnaugh maps of a function f (left side), its projections onto $x_1 \oplus \overline{x}_2$ (center, top) and $x_1 \oplus x_2$ (center, bottom), and the remainder (right side).*

2.2   EP-SOP with remainder

As already noted, when we project a SOP form onto the two spaces $(x_i \oplus x_j)$ and $(x_i \oplus \overline{x}_j)$, some products will appear only once in the final expression, precisely the products containing the two literals defining the projection spaces, while the other products (crossing products) will appear twice, one in each projected SOP.

In order to keep the number of products as small as possible, we introduce the notion of *EP-SOP with remainder*.

Algorithm 2 (Proj. onto $(x_i \oplus \overline{x}_j)$ and $(x_i \oplus x_j)$ for EP-SOP with rem.)
Given a SOP form $\phi = p_1 + p_2 + \cdots p_m$, for each p in $\{p_1, p_2, \ldots, p_m\}$ project p in $\phi_\oplus$ or in $\phi_{\overline{\oplus}}$, or insert it in the remainder ρ using the following strategy:

1. *If p contains both x_i and x_j (possibly complemented), i.e., $p = x_i x_j q$, where q is a product of literals, p has no projection onto the subspace where $x_i \neq x_j$. Thus, no product will be added to $\phi_\oplus$. By contrast, the projection of $p = x_i x_j q$ onto the subspace where $x_i = x_j$ gives the product $x_j q$, which will be added to $\phi_{\overline{\oplus}}$. The three other cases ($p = x_i \overline{x}_j q$, $p = \overline{x}_i x_j q$, $p = \overline{x}_i \overline{x}_j q$) can be handled in a similar way.*
2. *Otherwise (p is a crossing product) insert p in the remainder.*

For example, let us consider the Boolean function f shown on the left side of Figure 3. The unique crossing product of ϕ is $x_3 \overline{x}_4$ since it does not contain x_1 and x_2. This product will be inserted now in the remainder. The overall projection will return the form: $(x_1 \oplus \overline{x}_2)(\overline{x}_2 x_3 + x_2 x_3) + (x_1 \oplus x_2)(x_2 \overline{x}_3 + \overline{x}_2 \overline{x}_3) + x_3 \overline{x}_4$. Minimizing the projected SOPs we obtain $(x_1 \oplus \overline{x}_2)x_3 + (x_1 \oplus x_2)\overline{x}_3 + x_3 \overline{x}_4$.

Formally we have:

Definition 4. *Let $f : \{0,1\}^n \to \{0,1\}$, and let ϕ be a SOP representation of f. Given a couple of variable x_i and x_j, the (i,j)-EP-SOP with remainder of f is the expression*

$$\psi_{ij} = (x_i \oplus x_j)\phi'_\oplus + (x_i \oplus \overline{x}_j)\phi'_{\overline{\oplus}} + \rho,$$

where $\phi'_\oplus$ and $\phi'_{\overline{\oplus}}$ are the two projections of the products of ϕ containing both x_i and x_j (possibly complemented) onto the spaces $(x_i \oplus x_j)$ and $(x_i \oplus \overline{x}_j)$, respectively, and ρ is the sum of all crossing products of ϕ.

In other words we project onto the subspaces $(x_i \oplus x_j)$ and $(x_i \oplus \overline{x}_j)$ only the products that entirely reside in them, while the crossing products are not projected, but are inserted in the remainder ρ. Again for this form, we can further minimize the projected SOPs $\phi'_\oplus$ and $\phi'_{\overline{\oplus}}$ in order to obtain a more compact expression:

Definition 5. *Let* $f : \{0,1\}^n \to \{0,1\}$, *and let* ϕ *be a SOP representation of* f. *Given a couple of variable* x_i *and* x_j, *the* minimal (i,j)-EP-SOP with remainder *of* f *is the expression*

$$\psi_{ij}^{(min)} = (x_i \oplus x_j)\phi_\oplus'^{(min)} + (x_i \oplus \overline{x}_j)\phi_{\overline{\oplus}}'^{(min)} + \rho^{(min)} ,$$

where $\phi_\oplus'^{(min)}$ *and* $\phi_{\overline{\oplus}}'^{(min)}$ *are two minimal SOP forms representing the projections of the products of* ϕ *containing both* x_i *and* x_j *(possibly complemented) onto the spaces* $(x_i \oplus x_j)$ *and* $(x_i \oplus \overline{x}_j)$, *respectively, and* $\rho^{(min)}$ *is the optimal sum of all other products of* ϕ.

The overall minimal form (with respect to any possible couple of variables) is described as follows. Let $|\psi|$ denote the number of products in an EP-SOP with remainder, i.e., $|\psi| = |\phi'_\oplus| + |\phi'_{\overline{\oplus}}| + |\rho|$.

Definition 6. *The* minimal EP-SOP with remainder *representation of a Boolean function* f *is given by the EP-SOP expression* ψ_{MIN} *such that*

$$|\psi_{MIN}| = \min_{i,j} |\psi_{ij}^{(min)}| .$$

Note that if we start from a minimal SOP, the remainder is already minimal, i.e., the number of its products cannot be further reduced: $|\rho^{(min)}| = |\rho|$.

We cannot decide in advance which one of the two EP-SOP expressions (with or without remainder) is the more compact. On one hand, if we project the crossing products in the two spaces we could further minimize them. On the other hand it could be more convenient kipping them in the remainder. For example, consider the minimal SOP form $\phi = \overline{x}_1 x_2 \overline{x}_3 + x_1 \overline{x}_2 \overline{x}_3 + x_3 \overline{x}_4$ and the couple x_1 and x_2. The minimal $(1,2)$-EP-SOP without remainder is $(x_1 \oplus \overline{x}_2) x_3 \overline{x}_4 + (x_1 \oplus x_2)(\overline{x}_3 + x_3 \overline{x}_4)$, while the minimal $(1,2)$-EP-SOP with remainder is $(x_1 \oplus x_2)\overline{x}_3 + x_3 \overline{x}_4$. In this case the form with remainder is clearly more compact.

Alternatively consider the minimal SOP form $\phi = \overline{x}_1 \overline{x}_2 x_3 \overline{x}_4 + x_1 x_2 x_3 \overline{x}_4 + \overline{x}_1 x_2 \overline{x}_3 + x_1 \overline{x}_2 \overline{x}_3 \overline{x}_4 + \overline{x}_1 \overline{x}_3 x_4 + x_2 \overline{x}_3 x_4$ and the couple x_1 and x_2. The minimal $(1,2)$-EP-SOP without remainder is $(x_1 \oplus \overline{x}_2)(x_3 \overline{x}_4 + \overline{x}_3 x_4) + (x_1 \oplus x_2)(x_2 \overline{x}_3 + \overline{x}_3 \overline{x}_4)$, while the minimal $(1,2)$-EP-SOP with remainder is $(x_1 \oplus \overline{x}_2)(x_3 \overline{x}_4) + (x_1 \oplus x_2)(x_2 \overline{x}_3 + \overline{x}_3 \overline{x}_4) + \overline{x}_1 \overline{x}_3 x_4 + x_2 \overline{x}_3 x_4$. In this case the form without remainder is more convenient.

3   Computational Complexity

In this section we analyze the computational complexity of the following problem: given a *minimal* SOP form ϕ for a Boolean function f and a couple of variables x_i and x_j, find a minimal (i,j)-EP-SOP of f.

Since projecting ϕ is easy (polynomial), as shown in Section 2, the core of the problem is the minimization of the two projected SOPs. In general $\phi_{\oplus}^{(min)}$ and $\phi_{\overline{\oplus}}^{(min)}$ are different from the projections $\phi_{\oplus}$ and $\phi_{\overline{\oplus}}$, even if ϕ is already in minimal form. Indeed, projecting the single products of ϕ, we have no guarantee that the resulting SOP forms $\phi_{\oplus}$ and $\phi_{\overline{\oplus}}$ are still prime and irredundant.

Notice that the more common projections of a minimal SOP form ϕ onto the spaces x_i and $\overline{x}_i$ (Shannon projections) are guaranteed to be minimal. For instance, the projection (Shannon decomposition) with respect to x_1 and $\overline{x}_1$ of the minimal SOP form $\phi = \overline{x}_1 x_2 \overline{x}_3 + x_1 \overline{x}_2 \overline{x}_3 + x_1 x_3 \overline{x}_4$ is $\overline{x}_1(x_2 \overline{x}_3) + x_1(\overline{x}_2 \overline{x}_3 + x_3 \overline{x}_4)$, and the two projected SOP forms are already minimal.

In [19], the decision version of the problem of finding a minimal SOP representation of a Boolean function f starting from any SOP for f (*SOP-2-MIN SOP*) has been proved to be NP^{NP}-complete. Finding $\phi_{\oplus}^{(min)}$ and $\phi_{\overline{\oplus}}^{(min)}$ from $\phi_{\oplus}$ and $\phi_{\overline{\oplus}}$ when the starting SOP ϕ is *minimal*, could nevertheless be an easy (polynomial) problem? We show here that the answer to this question is negative, since the problem under study turns out to be at least as difficult as *SOP-2-MIN SOP*. Let us first formally define the two problems.

Problem 1 (SOP-2-MIN SOP).
INSTANCE: A SOP formula ϕ and an integer k.
QUESTION: Is there a SOP ϕ' with at most k products and for which $\phi' \equiv \phi$?

Problem 2 (MIN SOP-2-MIN (i,j)-EP-SOP).
INSTANCE: Minimal SOP formula ϕ, a couple of variables x_i and x_j.
QUESTION: Find the minimal (i,j)-EP-SOP $\xi_{ij}^{(min)}$:

$$\xi_{ij}^{(min)} = (x_i \oplus x_j)\phi_{\oplus}^{(min)} + (x_i \oplus \overline{x}_j)\phi_{\overline{\oplus}}^{(min)} .$$

The proof of the hardness of *MIN SOP-2-MIN (i,j)-EP-SOP* is based on the concept of *polynomial time Turing reduction*, defined as follows.

Definition 7. *A problem Π is* Turing-reducible *to a problem Π', $\Pi \preceq_T \Pi'$, if there is an algorithm A that solves Π by using a hypothetical subroutine S for solving Π' such that, if S were a polynomial time algorithm for Π', then A would be a polynomial time algorithm for Π.*

Let us consider the complexity of the following problem:

Problem 3 (MIN SOP+PRODUCT-2-MIN SOP).
INSTANCE: A minimal SOP formula ϕ for a Boolean function f and a product p.
QUESTION: Find a minimal SOP formula for the function $f + p$

Lemma 1. *MIN SOP+PRODUCT-2-MIN SOP is a NP^{NP}-hard problem.*

Proof. We show that the problem *SOP-2-MIN SOP* is Turing-reducible to *MIN SOP+PRODUCT-2-MIN SOP*.

Consider a SOP $\phi = p_1 + p_2 + \ldots + p_m$ for a function f depending on n variables. In order to find a minimal SOP ϕ' for f, we can proceed iteratively as follows.

First we compute a minimal SOP for the function $p_1 + p_2$. Note that this corresponds to deriving a minimal SOP $\phi^{(1)}$ for the union of the minimal SOP p_1 and a product p_2.

In general, step i of this iterative process would consist in computing a minimal SOP $\phi^{(i)}$ for the function $\phi^{(i-1)} + p_i$, defined as the union of a minimal SOP, $\phi^{(i-1)}$, and the product p_i.

If we could perform this step in polynomial time, then we could solve the problem *SOP-2-MIN SOP* in polynomial time, performing $m - 1$ iterations.   ∎

Based on the previous lemma, we can now prove our main complexity result.

Theorem 1. *MIN SOP-2-MIN (i, j)-EP-SOP is NP^{NP}-hard.*

Proof. To prove the thesis, it is enough to show that the *MIN SOP+PRODUCT-2-MIN SOP* problem is Turing-reducible to *MIN SOP-2-MIN (i, j)-EP-SOP*.

Consider a minimal SOP $\phi^{(min)}$ for a function f, depending on n variables $x_1, x_2, \ldots, x_n$, and a product p. Then consider the SOP

$$\Phi = x_{n+1}\phi^{(min)} + x_{n+2}p\,,$$

where x_{n+1} and x_{n+2} are two additional variables. Suppose that $\phi^{(min)}$ contains t products.

First of all observe that Φ is a minimal SOP form. Indeed, $x_{n+1}\phi^{(min)}$ is minimal and does not cover the points of the cube described by $\overline{x}_{n+1}x_{n+2}p$. Thus we need at least a product to cover these points. This means that a minimal SOP must contain at least $t + 1$ products, and this immediately implies that Φ is minimal.

Now, let us derive an EP-SOP from Φ with respect to the couple of additional variables x_{n+1} and x_{n+2}. We get the following expression:

$$\xi_{n+1,n+2} = (x_{n+1} \oplus x_{n+2})(\overline{x}_{n+2}\phi^{(min)} + x_{n+2}p) + (x_{n+1} \oplus \overline{x}_{n+2})(x_{n+2}\phi^{(min)} + x_{n+2}p)\,.$$

If we could derive $\xi_{n+1,n+2}^{(min)}$ in polynomial time, then we would be able to minimize the two expressions $(\overline{x}_{n+2}\phi^{(min)} + x_{n+2}p)$ and $(x_{n+2}\phi^{(min)} + x_{n+2}p)$ in polynomial time. This implies that we could solve in polynomial time an instance of *MIN SOP+PRODUCT-2-MIN SOP* since, from the second expression, we have $(x_{n+2}\phi^{(min)} + x_{n+2}p)^{(min)} = x_{n+2} \cdot (\phi^{min} + p)^{(min)}$.   ∎

4   Polynomial time approximation algorithms

In the previous section we have shown that, even if we start from a minimal SOP form and we fix a couple of variables x_i and x_j, finding a *minimal (i, j)-EP-SOP* is a hard problem. In this section we will show how it is possible to find a good solution to the latter problem in polynomial time.

In a minimization framework, a p-approximation algorithm (i.e., an algorithm with approximation ratio p) guarantees that the cost C of its solution is such that $C/C^* \leq p$, where C^* is the cost of an optimal solution [9]. Both heuristics and approximation algorithms do not guarantee the minimality of their solution, but while we cannot perform any evaluation on the result of a heuristic, an approximation algorithm gives guaranteed near-optimum solutions.

We now describe a *polynomial approximation algorithm* for the problem of finding the minimal EP-SOP (minimal EP-SOP with remainder) representation of a function f starting from a *minimal SOP ϕ* for f that guarantees an approximation ratio of 4 (2). The main idea is to select the most frequent couple of variables in the minimal SOP representation, and project the expression with respect to this couple. The two projected SOPs will be further synthesized with a SOP polynomial heuristic. The overall algorithm is described below.

Algorithm 3 (Approximation Algorithm) *Given a* minimal *SOP expression ϕ:*

Step 1 *Select the couple of variables x_i and x_j simultaneously appearing (possibly complemented) with the highest frequency in the products of ϕ.*

Step 2 *Project ϕ onto the spaces $(x_i \oplus x_j)$ and $(x_i \oplus \overline{x}_j)$ as described in Algorithms 1 or 2.*

Step 3 *Minimize the two projected SOPs using a polynomial time heuristic (e.g.,* ESPRESSO NOT EXACT*).*

Notice that the two versions (with and without remainder) differ only in the projection Step 2 discussed in Section 2. The three steps can be performed in polynomial time.

We now prove that the proposed synthesis strategy is indeed an approximation algorithm for the two EP-SOP minimization problems.

Consider first the problem without remainder. In order to prove that the cost $|\xi_{ij}^{(min)}|$ of our solution is such that $|\xi_{ij}^{(min)}|/|\xi_{MIN}|$ is upper bounded by a constant, where $|\xi_{MIN}|$ is the cost of an optimal solution, we first find a lower bound for $|\xi_{MIN}|$, as shown in Lemma 2 and then an upper bound for $|\xi_{ij}^{(min)}|$, as shown in Theorem 2. We follow a similar strategy for EP-SOPs with remainder using Lemma 3 and Theorem 3.

Let us consider a minimal SOP form ϕ for a Boolean function f and a minimal EP-SOP without remainder ξ_{MIN}.

Lemma 2.
$$|\xi_{MIN}| \geq \frac{1}{2}|\phi|.$$

Proof. Let us suppose that the variables x_h and x_k are such that

$$\xi_{hk}^{(min)} = (x_h \oplus x_k)\phi_{\oplus}^{(min)} + (x_h \oplus \overline{x}_k)\phi_{\overline{\oplus}}^{(min)} = \xi_{MIN}.$$

We build a SOP ϕ_{hk} starting from $\xi_{hk}^{(min)}$. Let $\phi_{\oplus}^{(min)} = \sum_{i=1}^{|\phi_{\oplus}^{(min)}|} p_i$ and $\phi_{\overline{\oplus}}^{(min)} = \sum_{i=1}^{|\phi_{\overline{\oplus}}^{(min)}|} q_i$. Thus

$$\phi_{hk} = (x_h \oplus x_k) \sum_{i=1}^{|\phi_{\oplus}^{(min)}|} p_i + (x_h \oplus \overline{x}_k) \sum_{i=1}^{|\phi_{\overline{\oplus}}^{(min)}|} q_i$$

$$= x_h \overline{x}_k \sum_{i=1}^{|\phi_{\oplus}^{(min)}|} p_i + \overline{x}_h x_k \sum_{i=1}^{|\phi_{\oplus}^{(min)}|} p_i + x_h x_k \sum_{i=1}^{|\phi_{\overline{\oplus}}^{(min)}|} q_i + \overline{x}_h \overline{x}_k \sum_{i=1}^{|\phi_{\overline{\oplus}}^{(min)}|} q_i.$$

Since ϕ is minimal, we have that $|\phi_{hk}| \geq |\phi|$. Moreover, since

$$|\phi_{hk}| = 2|\xi_{hk}^{(min)}| = 2(|\phi_{\oplus}^{(min)}| + |\phi_{\overline{\oplus}}^{(min)}|) \geq |\phi|,$$

the thesis immediately follows:

$$|\xi_{MIN}| = |\xi_{hk}^{(min)}| = |\phi_{\oplus}^{(min)}| + |\phi_{\overline{\oplus}}^{(min)}| \geq \frac{1}{2}|\phi|.$$

■

A similar result holds for the EP-SOPs with remainder.

Lemma 3.

$$|\psi_{MIN}| \geq \frac{1}{2}(|\phi| + |\rho|),$$

where ρ is the remainder of ψ_{MIN}.

Proof. Let us suppose that the EP-SOP form

$$\psi_{hk}^{(min)} = (x_h \oplus x_k)\phi_{\oplus}'^{(min)} + (x_h \oplus \overline{x}_k)\phi_{\overline{\oplus}}'^{(min)} + \rho$$

is minimal, with respect to the overall number of products, among all other EP-SOPs with remainder, i.e., $\psi_{hk}^{(min)} = \psi_{MIN}$.

As in the proof of Lemma 2, we derive a SOP representation ϕ_{hk} for f from $\psi_{hk}^{(min)}$, and we get

$$|\phi_{hk}| = 2(|\phi_{\oplus}'^{(min)}| + |\phi_{\overline{\oplus}}'^{(min)}|) + |\rho| \geq |\phi|.$$

Thus

$$|\phi_{\oplus}'^{(min)}| + |\phi_{\overline{\oplus}}'^{(min)}| + \frac{1}{2}|\rho| \geq \frac{1}{2}|\phi|,$$

and we immediately derive

$$|\psi_{MIN}| = |\psi_{hk}^{(min)}| = |\phi_{\oplus}'^{(min)}| + |\phi_{\overline{\oplus}}'^{(min)}| + |\rho| \geq \frac{1}{2}(|\phi| + |\rho|).$$

■

We now prove that if we project the starting minimal SOP ϕ with respect to the couple of variables x_i and x_j simultaneously appearing (possibly complemented) with the highest frequency in its products, we get a solution whose approximation ratio in the worst case is bounded by 4 for the EP-SOP without remainder, and by 2 for the EP-SOP with remainder.

For a couple of variables x_i and x_j, let us denote with ν_{ij} the number of products in ϕ containing both x_i and x_j, possibly complemented.

Theorem 2. *Let ξ_{MIN} be a minimal EP-SOP of a Boolean function f, and ϕ a minimal SOP form for f. Let $\xi_{ij}^{(min)}$ be the minimal (i, j)-EP-SOP derived with respect to the couple of variables (x_i and x_j) appearing with the highest frequency in the products of ϕ. Then*

$$\frac{|\xi_{ij}^{(min)}|}{|\xi_{MIN}|} \leq \frac{|\xi_{ij}|}{|\xi_{MIN}|} \leq 4 - \frac{2\nu_{ij}}{|\phi|} \, .$$

Proof. Observe that

$$|\xi_{ij}^{(min)}| \leq |\xi_{ij}| \leq 2|\phi| - \nu_{ij} \, ,$$

since the ν_{ij} products containing the two variables x_i and x_j appear only once in ξ_{ij}, while all other products appear twice. The thesis follows since Lemma 2 implies that $|\xi_{MIN}| \geq \frac{|\phi|}{2}$. ∎

Observe that in the best case $\nu_{ij} = |\phi|$, thus the bound becomes

$$\frac{|\xi_{ij}^{(min)}|}{|\xi_{MIN}|} \leq \frac{|\xi_{ij}|}{|\xi_{MIN}|} \leq 2 \, ,$$

while in the worst case $\nu_{ij} = 1$ and we have

$$\frac{|\xi_{ij}^{(min)}|}{|\xi_{MIN}|} \leq \frac{|\xi_{ij}|}{|\xi_{MIN}|} \leq 4 - \frac{2}{|\phi|} \leq 4 \, .$$

Theorem 3 shows a similar result for the EP-SOPs with remainder.

Theorem 3. *Let ψ_{MIN} be a minimal EP-SOP with remainder of a Boolean function f, and ϕ be a minimal SOP form for f.*

Let $\psi_{ij}^{(min)}$ be the minimal (i, j)-EP-SOP with remainder derived with respect to the couple of variables x_i and x_j appearing with the highest frequency in the products of ϕ. Then

$$\frac{|\psi_{ij}^{(min)}|}{|\psi_{MIN}|} \leq \frac{|\psi_{ij}|}{|\psi_{MIN}|} \leq 2 \, .$$

Proof. First observe that

$$|\psi_{ij}^{(min)}| \leq |\psi_{ij}| = |\phi| \, ,$$

since each product of ϕ appears only once in ψ_{ij}, in one of the two factors $\phi'_\oplus$ and $\phi'_{\overline{\oplus}}$, or in the remainder ρ, and the two expressions $\phi'_\oplus$ and $\phi'_{\overline{\oplus}}$ are further minimized. Moreover, Lemma 3 implies that

$$|\psi_{MIN}| \geq \frac{1}{2}(|\phi| + |\rho|).$$

Now suppose that the projections in ψ_{MIN} are performed with respect to the variables x_h and x_k. Thus, since $|\rho| = |\phi| - \nu_{hk}$, with $\nu_{hk} \leq \nu_{ij}$ and $\nu_{ij} \leq |\phi|$, we get

$$\frac{|\psi_{ij}^{(min)}|}{|\psi_{MIN}|} \leq \frac{|\psi_{ij}|}{|\psi_{MIN}|} \leq \frac{|\phi|}{|\phi| - \nu_{ij}/2} \leq \frac{|\phi|}{|\phi| - |\phi|/2} \leq 2.$$

∎

Note that Theorem 2 and Theorem 3 show that the approximation ratios hold even if the factors $\phi_\oplus$, $\phi_{\overline{\oplus}}$, $\phi'_\oplus$ and $\phi'_{\overline{\oplus}}$ are not minimized. Therefore, the algorithms proposed are indeed polynomial approximation algorithms for the given problems. The resulting EP-SOP without remainder ξ_{ij} has a size that can be upper bounded by $(4 - 2\nu_{ij}/|\phi|)|\xi_{MIN}|$, i.e., in the worst case by $4|\xi_{MIN}|$, while the EP-SOP with remainder ψ_{ij} has a size that can be upper bounded by $2|\psi_{MIN}|$.

As a final observation, we would like to point out that the couple of variables, say x_i and x_j, with the highest frequency in general does not guarantee that $\xi_{MIN} = \xi_{ij}^{(min)}$ and $\psi_{MIN} = \psi_{ij}^{(min)}$, as the following counterexample shows.

Example 2. Let us consider the minimal SOP $\phi = x_1 x_2 x_3 x_4 + x_1 x_2 \overline{x}_3 \overline{x}_4 + \overline{x}_1 x_2$. We want to find the two minimal EP-SOP forms. The couple of variables with the highest frequency is x_1 and x_2.

The approximation algorithm computes the following form without remainder:

$$\xi_{12}^{(min)} = (x_1 \oplus x_2)x_2 + (x_1 \oplus \overline{x}_2)(x_2 x_3 x_4 + x_2 \overline{x}_3 \overline{x}_4)$$

and the following form with remainder:

$$\psi_{12}^{(min)} = (x_1 \oplus x_2)x_2 + (x_1 \oplus \overline{x}_2)(x_2 x_3 x_4 + x_2 \overline{x}_3 \overline{x}_4),$$

while the minimal solutions are $\xi_{MIN} = \xi_{34}^{(min)} = (x_3 \oplus \overline{x}_4)x_2 + (x_3 \oplus x_4)(\overline{x}_1 x_2)$ and $\psi_{MIN} = \psi_{34}^{(min)} = (x_3 \oplus \overline{x}_4)(x_1 x_2) + \overline{x}_1 x_2$, respectively.

5   Experimental results

In this section we discuss the computational results obtained by applying the polynomial approximation algorithm presented above to the standard ESPRESSO benchmark suite [20]. We consider four different variants of our algorithm. In fact, we address the minimization of EP-SOP forms both with and without remainder, in order to estimate the practical utility of either form. Moreover, as most benchmarks have multiple outputs, the definition of the most frequent couple of variables can be referred either to the whole set of outputs (*global frequency*) or to each single output (*local frequency*). In the former case, we will determine a single EP-SOP form, projecting the original minimal SOP form ϕ with respect to the couple of variables appearing in the largest number of products of ϕ. In the latter case, we will find the most frequent couple of variables for each different output and perform independent projections, obtaining separate EP-SOP forms for the outputs which have been projected onto different couples of subspaces. In both cases all the SOP forms are synthesized together with multi-output synthesis. Combining the two approaches related to the use of the remainder and the two approaches related to the global and local frequency, we obtain four different algorithms, respectively denoted as NG (no remainder and global frequency), NL (no remainder and local frequency), RG (remainder and global frequency), RL (remainder and local frequency).

All computational experiments were performed on a Pentium 1.6 GHz processor with 1 GB RAM. We report in the following a significant subset of the experiments.

Table 1 reports a cost-oriented comparison among the original optimal SOP form determined by ESPRESSO EXACT and the EP-SOP forms yielded by the four algorithms: the first column reports the name of the instance, the following five triples of columns report the computational time in seconds, together with the area and the delay of physical implementations for the five expressions. These were evaluated using a technology mapping (mcnc.genlib) provided by the SIS [18] tool.

The computational time for the EP-SOP forms does not include the time required to compute the optimal SOP form (which is shown in the second column), but only the time to factorize it and to heuristically minimize its projections. As the results show, the overhead added by the last two steps is quite limited.

Of course, the physical implementation of the EP-SOP forms also includes one or more EXOR gates, whose cost cannot be neglected, as our results clearly show. First of all, the EXOR part of the network can be expensive, depending on the technology adopted. Second, some functions benefit from the multi-output minimization: common products can be shared, thus reducing the overall area. Comparing the performances of the four algorithms one to another, we can note how this fact particularly affects the performance of the algorithms NL and RL referring to the local definition of frequency, while the algorithm performing better seems to be the RG algorithm.

It should be noticed, however, that the gain obtained by the EP-SOP form is on about the 35% of instances, and can be quite striking: the gain on instance *adr4* exceeds 50% and for many other instances (e.g., *root*, *z4*) it exceeds 40%.

Apart from algorithm NL, which only equals some best result, never hitting one alone, even the less effective of the other three algorithms, that is RL, improves by 45% the cost of instance *f51m*.

Given that the time required to obtain such improvements is rather limited, evaluating the EP-SOP forms as a possible alternative to the optimal SOP form appears to be an advisable *post-processing strategy*. We have further investigated

	min SOP			min EP-SOP											
				NG			NL			RG			RL		
Bench.	CPU	area	delay	CPU	area	delay	CPU	area	delay	CPU	area	delay	CPU	area	delay
addm4	0.14	1172	47.9	0.06	1291	52.5	0.06	975	40.4	0.04	1101	48.5	0.07	906	38.3
adr4	0.04	224	19.2	0.03	174	15.2	0.03	155	16.0	0.03	105	11.1	0.04	141	13.5
amd	0.06	1171	46.7	0.03	1082	43.5	0.05	1040	39.1	0.03	1046	42.4	0.06	1022	38.0
b2	0.23	3876	79.8	0.06	4113	81.3	0.06	4180	81.3	0.04	4169	82.6	0.04	4242	82.6
b4	3.45	645	30.5	0.01	802	33.3	0.01	841	33.1	0.01	717	34.4	0.01	779	32.8
br1	0.01	446	32.5	0.02	353	24.5	0.02	381	25.7	0.02	353	24.5	0.02	381	25.7
br2	0.01	352	26.6	0.01	292	25.5	0.01	314	30.0	0.01	292	25.5	0.01	314	30.0
chkn	0.48	717	43.6	0.04	832	42.2	0.06	777	39.2	0.01	758	36.1	0.01	764	46.7
dc2	0.04	253	23.1	0.01	286	22.4	0.01	236	19.7	0.01	263	21.7	0.01	236	19.7
exps	0.50	3932	114.5	0.06	3778	114.8	0.06	3900	104.6	0.08	3760	112.6	0.09	3877	106.4
f51m	0.09	501	31.5	0.04	413	26.2	0.04	339	26.4	0.04	311	20.5	0.04	273	19.1
in0	0.10	1214	48.3	0.03	1056	48.1	0.05	1015	42.5	0.05	1019	48.0	0.06	989	44.9
in1	0.23	3876	79.8	0.06	4113	81.3	0.06	4180	81.3	0.06	4169	82.6	0.06	4242	82.6
in2	0.09	1112	41.4	0.03	1000	36.7	0.01	1041	37.3	0.03	1002	37.3	0.03	1039	37.9
in5	0.14	905	38.5	0.01	976	39.2	0.01	1040	37.2	0.01	923	40.9	0.01	993	39.7
intb	2.96	2170	57.3	0.44	3392	75.5	0.83	2693	63.2	0.34	2466	57.6	0.67	2526	61.6
luc	0.01	806	41.0	0.01	779	52.8	0.01	883	51.8	0.01	758	52.4	0.01	862	50.6
m1	0.01	208	19.6	0.03	304	21.0	0.03	352	21.2	0.03	308	22.8	0.03	356	22.8
m2	0.01	710	37.8	0.01	833	40.9	0.01	893	40.5	0.01	861	42.5	0.01	921	41.9
m3	0.04	839	38.3	0.01	1286	48.4	0.01	1283	52.2	0.01	1172	51.7	0.01	1235	54.4
m181	0.60	166	18.4	0.01	327	22.4	0.03	311	24.9	0.01	240	22.5	0.01	267	19.8
max128	0.09	1292	58.0	0.09	2055	71.6	0.09	2194	77.6	0.07	2098	71.5	0.07	1975	72.4
mlp4	0.31	734	36.4	0.03	983	43.0	0.04	891	40.1	0.03	839	40.5	0.03	857	40.1
mp2d	0.25	362	26.0	0.01	428	25.3	0.01	420	28.9	0.01	333	23.7	0.01	360	25.5
newcond	0.01	114	17.4	0.01	132	18.6	0.01	124	18.6	0.01	119	18.2	0.01	124	18.6
p82	0.01	239	18.4	0.01	239	25.8	0.01	302	23.9	0.01	241	25.0	0.01	309	24.7
radd	0.39	183	15.7	0.01	196	18.9	0.01	181	19.5	0.01	120	15.1	0.01	158	16.8
rckl	0.04	341	49.7	0.01	495	72.3	0.01	519	72.3	0.01	495	72.3	0.01	519	72.3
rd73	0.03	220	25.6	0.03	389	27.6	0.03	308	28.4	0.03	339	26.9	0.03	264	24.1
risc	0.01	228	18.7	0.02	312	29.0	0.02	435	32.7	0.03	310	29.0	0.02	434	32.5
root	0.35	592	35.5	0.02	367	27.7	0.02	380	25.3	0.03	349	26.5	0.03	350	25.7
sqr6	0.06	278	25.5	0.01	397	27.0	0.01	462	26.2	0.01	330	24.9	0.01	405	26.2
t3	0.40	186	21.5	0.02	193	16.2	0.03	213	15.8	0.02	180	19.8	0.02	206	19.7
tms	0.03	587	35.4	0.01	675	35.2	0.01	754	35.5	0.01	675	35.2	0.01	754	35.5
vg2	0.53	341	18.6	0.04	628	25.7	0.06	581	26.0	0.03	468	22.5	0.04	500	21.4
vtx1	0.17	324	21.3	0.01	441	25.5	0.01	497	21.1	0.01	365	23.4	0.01	465	20.7
x6dn	0.18	1054	36.8	0.01	854	34.9	0.01	870	34.9	0.01	817	34.8	0.01	834	34.8
x9dn	0.20	384	23.0	0.04	496	25.4	0.06	560	24.2	0.04	424	24.7	0.03	528	22.6
z4	0.01	171	18.3	0.01	159	18.6	0.01	165	20.6	0.01	99	14.2	0.01	132	17.9

Table 1. Synthesis time, area and delay of EP-SOP and SOP forms (computed in SIS after the technology mapping).

whether the Boolean factorization proposed in the present paper actually differs from similar techniques already known in the literature and applied in synthesis tools. We have applied the multilevel synthesis routines (script.rugged) of SIS to the optimal SOP forms and to the four EP-SOP forms, in order to find out whether they end up with a similar final structure or not. The first remark that can be done is that in some cases (e.g., *b2*, *exps* and *in1*), SIS was unable to process the optimal SOP form (in a limit time of 12 hours). Starting from the EP-SOP forms, this happened only for instance *in1*, and only for the two EP-SOP forms with remainder. Only few times the final results were identical (10%), and half of the times the final result obtained starting from an EP-SOP form was better than the one obtained from the optimal SOP form, ranging from 30% better to 30% worse.

6   Conclusion

Although deriving an optimal EP-SOP form from an optimal SOP form is an NP^{NP}-hard problem, in this paper we have described a polynomial time approximation algorithm which guarantees a near-optimal solution. We propose this algorithm as a post-processing step after the SOP synthesis, in order to possibly reduce the area of the resulting networks. Our experiments show that in about 35% of the considered benchmarks the area obtained is smaller, sometimes even by 40-50%.

It could be an interesting development to study different kinds of projection, such as dividing the Boolean space into subspaces whose characteristic functions are represented by EXORs with more than two literals. Given the similar nature of the problem, it could also be interesting to study the relationship between Linear Transformed BDDs [10] and EP-SOP forms.

References

1. A. Bernasconi, V. Ciriani, F. Luccio, and L. Pagli. Three-Level Logic Minimization Based on Function Regularities. *IEEE Transactions on TCAD*, 22(8):1005–1016, 2003.
2. R. Brayton, G. Hachtel, C. McMullen, and A. Sangiovanni-Vincentelli. *Logic Minimization Algorithms for VLSI Synthesis*. Kluwer Ac. Pub., 1984.
3. G. Caruso. Near Optimal Factorization of Boolean Functions. *IEEE Transactions on CAD*, 10(8):1072–1078, 1991.
4. O. Coudert. Two-Level Logic Minimization: an Overview. *INTEGRATION*, 17:97–140, 1994.
5. D. Debnath and T. Sasao. A Heuristic Algorithm to Design AND-OR-EXOR Three-Level Networks. In *Asia and South Pacific Design Automation Conference*, pages 69–74, 1998.
6. D. Debnath and Z. Vranesic. A Fast Algorithm for OR-AND-OR Synthesis. *IEEE Transactions on CAD*, 22(9):1166–1176, 2003.
7. E. Dubrova, D. Miller, and J. Muzio. AOXMIN-MV: A Heuristic Algorithm for AND-OR-XOR Minimization. In *4th Int. Workshop on the Applications of the Reed Muller Expansion in circuit Design*, pages 37–54, 1999.
8. M. Fujita, Y. Matsunaga, and M. Ciesielski. Multi-Level Logic Optimization. In S. Hassoun and T. Sasao, editors, *Logic Synthesis and Verification*, pages 29–63. Kluwer Academic Publishers, 2002.
9. M. Garey and D. Johnson. *Computer and Intractability: A Guide to the Theory of NP-completeness*. W.H. Freeman and Company, 1979.
10. W. Günther and R. Drechsler. On the Computational Power of Linearly-Transformed BDDs. *Information Processing Letters*, 75(3):119–125, 2000.
11. W. Günther and R. Drechsler. Efficient Minimization and Manipulation of Linearly-Transformed Binary Decision Diagrams. *IEEE Transaction on Computers*, 52(9):1196–1209, 2003.

12. F. Luccio and L. Pagli. On a New Boolean Function with Applications. *IEEE Transactions on Computers*, 48(3):296–310, 1999.
13. P. McGeer, J. Sanghavi, R. Brayton, and A. Sangiovanni-Vincentelli. ESPRESSO-SIGNATURE: A New Exact Minimizer for Logic Functions. *IEEE Transactions on VLSI*, 1(4):432–440, 1993.
14. M. Perkowski. A New Representation of Strongly Unspecified Switching Functions and its Application to Multi-Level AND/OR/EXOR Synthesis. In *IFIP WG 10.5 Workshop on Applications of the Reed-Muller Expansion*, pages 143–151, 1995.
15. T. Sasao. On the Complexity of Three-Level Logic Circuits. In *Int. Workshop on Logic Synthesis*, 1989.
16. T. Sasao. A Design Method for AND-OR-EXOR Three Level Networks. In *Int. Workshop on Logic Synthesis*, pages 8:11–8:20, 1995.
17. T. Sasao. *Switching Theory for Logic Synthesis*. Kluwer Academic Publishers, 1999.
18. E. Sentovich, K. Singh, L. Lavagno, C. Moon, A. S. R. Murgai, H. Savoj, P. Stephan, R. Brayton, and A. Sangiovanni-Vincentelli. SIS: A system for sequential circuit synthesis. Technical report, 1992.
19. C. Umans, T. Villa, and A. Sangiovanni-Vincentelli. Complexity of Two-Level Logic Minimization. *IEEE Transactions on Computer-Aided Design*, to appear.
20. S. Yang. Logic synthesis and optimization benchmarks user guide version 3.0. User guide, Microelectronic Center, 1991.

A Method for I/O Pins Partitioning Targeting 3D VLSI Circuits

Renato Hentschke, Sandro Sawicki *, Marcelo Johann, and Ricardo Reis

UFRGS - Universidade Federal do Rio Grande do Sul - Instituto de Informatica
Av. Bento Goncalves, 9500. CEP 91501-970. Porto Alegre, Brazil
{renato,sawicki,johann,reis}@inf.ufrgs.br

Abstract. This paper presents an algorithm for I/O pins partitioning and placement targeting 3D circuits. The method starts from a standard 2D placement of the pins around a flat rectangle and outputs a 3D representation of the circuit composed of a set of tiers and pins placed at the four sides of the resulting cube. The proposed algorithm targets a balanced distribution of the I/Os that is required both for accommodating the pins evenly as well as to serve as an starting point for cell placement algorithms that are initially guided by I/O's locations, such as analytical placers. Moreover, the I/O partitioning tries to set pins in such a way the it allows the cell placer to reach a reduced number of 3D-Vias. The method works in two phases: first the I/O partitioning considering the logic distances as weights; second, fix the I/Os and perform partitioning of the cells. The experimental results show the effectiveness of the approach on balance and number of 3D-Vias compared to simplistic methods for I/O partitioning, including traditional min-cut algorithms. Since our method contains the information of the whole circuit compressed in a small graph, it could actually improve the partitioning algorithm at the expense of more CPU time. Additional experiments demonstrated that the method could be adapted to further reduce the number of 3D-Vias if the I/O pin balance constraint can be relaxed.

1 Introduction

Many of existing design issues rely on wiring problems. Issues like delay, variability and manufacturability are highly valuable research subjects in the present days. Timing is importantly affected by wires that contribute to more than 50% of the critical path delay. The power consumption produced by the switching activity of wires, specially clock signals, also contribute to a very large chunk of total power dissipated. Reliability and manufacturability are also related to chip wires.

3D circuits appear as a change of design paradigm, providing higher integration and reducing wire lengths [10]. By either analytical methods [5] [8] [19] [20] and practical experimentation [10] [9] [15] [2], it is well known that 3D circuit technology has the potential of providing many improvements to VLSI circuits,

---

*On leave from UNIJUI - Universidade Reg. do Noroeste do Estado do RS - Brazil

Please use the following format when citing this chapter:

Hentschke, R., Sawicki, S., Johann, M. and Reis, R., 2007, in IFIP International Federation for Information Processing, Volume 249, VLSI-SoC: Research Trends in VLSI and Systems on Chip, eds. De Micheli, G., Mir, S., Reis, R., (Boston: Springer), pp. 259–279

including: reduction of the size of the longest wires [9]; average wire length reduction (from 15% to 50%) [9]; dynamic Power reduction of up to 22% [20] [10]; chip area reduction [19].

Among the new issues introduced by 3D circuits, the communication elements (known as 3D-Vias) between adjacent tiers impose several constraints to the physical design of those circuits. Firstly, their electrical characteristics are differentiated from regular wires. From the routing perspective, in order to connect to a 3D-Via, a wire is required to cross all metal layers. More importantly, 3D-Vias require significant sizes for design rules such as minimum pitch. As more detailed in section 2.2, face-to-back communication imposes more restrictions, since it digs a hole through the Bulk of a tier occupying active area and compromising reliability. All those factors make 3D-Via planning a complex issue that must be addressed by CAD tools.

The new 3D issues must be addressed with proper CAD tools able to synthesize in a new design paradigm to take full advantage of 3D integration. Among possible design methodologies, the integration granularity will impact possible benefits and the type of problems to be solved. Initially consider a *tier level integration*, which stacks separated tiers of different nature. It is the most coarse level granularity and do not severely affect existing design methodologies, since each tier can be designed separately with a simple glue logic to integrate them. Secondly, consider an *ip core level integration* that partitions big circuit blocks (ip cores) into different tiers, providing a tighter integration (more communication between tiers). Finally, *random logic level* partitioning breaks random logic into 3D. Figure 1 illustrates a random logic block broken into 2 tiers.

Basically, the finer the integration grain is, the bigger the potential vertical communication requirement is, causing two effects: 1) more potential benefits (as listed above); 2) more complex 3D-Via related problems to solve. The higher complexity of 3D-Via planning must be addressed by physical design algorithms, encouraging research on this field. The random logic integration granularity with the usage of more 3D-Vias while optimizing wire length of a block on 3D leads to a better usage of the 3D resources and helps reducing wire length, as demonstrated by [9].

The problem of partitioning a block into 2 or more tiers starts with the definition of an I/O interface. Although all the existing 3D placement literature ignores this problem, possibly using some simplistic solution, an appropriate placement of the I/Os in the boundary of the block has a very important impact on the cells placement. I/O pins play two important roles in the placement of a block: first, I/Os limit the area boundary of the block; second, the pins are used as tips for many placement algorithms to reduce wire lengths. Consider the Quadratic Placement algorithm [4], that is used by the leading industry and most of the existing academic cell placers. It requires I/Os at the boundary in order to compute a solution. If all I/Os are assigned to stay in a unique tier, the quadratic placement method will not be able to move the cells in 3D.

This paper proposes a method for the I/O partitioning of a random logic block based on the logic distance of the I/Os as partitioning criterion. Summarizing the motivation, the goal is to find a good partitioning method for the I/Os that is able to maintain a good I/O pins balance leading to area balance between the tiers. At the same time, we indirectly address the reduction of 3D-Vias. Our insight is that a low 3D-Via starting point leaves more room for a 3D placer to insert 3D-Vias while improving wire length if there is available space. The rest of the paper is organized as follows. Section 2 presents a few details on 3D VLSI circuits that will be helpful to understand the experimental results and motivation. Section 3 defines the problem we are addressing. Section 4 presents the I/O partitioning algorithm. Sections 5 and 6 present experimental results while conclusions are discussed in section 7.

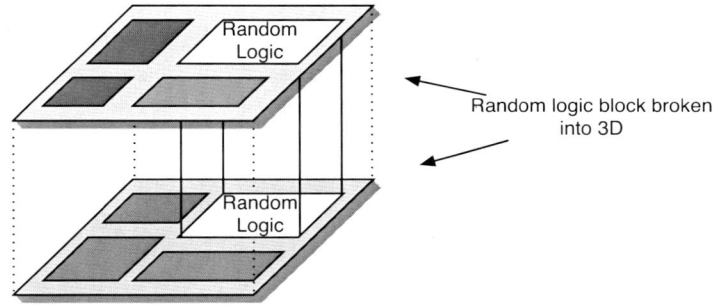

Fig. 1. Random logic blocks could be broken into 3D.

1.1 Related Work

Because of the high penalties imposed by 3D-Vias, a common approach in the placement phase is to minimize them by using min-cut partitioning. The works from [2], [10], [11] for instance, apply min-cut partitioning (usually with hMetis tool [14]) to assign cells into tiers, minimizing the number of 3D-Vias. A subsequent step performs 2D placement on each tier separately; the already placed tiers can serve as a guide to subsequent tiers in order to minimize wire length. However, [15] [16] [7] already identified that this approach leads to worse results in terms of wire length. We call True 3D placer a method that is able to both measure and optimize wire length in all the axis at the same time.

Liu et. al [16] built a two step 3D placement flow similar to the one mentioned above using hMetis for partitioning the cells into tiers. They argue that building a True 3D flow is very hard and for this reason they concentrate on improving the partitioning step. They observed that the insertion of 3D-Vias could potentially improve wire length. For this reason, their cell partitioner does not perform min-cut partitioning, but tries to maximize the 3D-Vias under an upper bound constraint. In fact, since face-to-face integration allows 3D-Vias with no cost to yield or area, they could be inserted freely in order to improve wire length. Some preliminary evaluation could be performed to analise a reasonable upper bound for those 3D-Vias. Liu's algorithm cannot achieve the exact via count provided, but tries to get a close approximation by an iterative algorithm. After the tier assignment, the algorithm uses the Capo tool [6] to place the cells in each tier.

Das et. al [7] [9] builded a true 3D partitioning based placement engine. It recursively cuts the placement cube performing min-cut partitioning. A wire length and 3D-Via trade-off can be obtained by controlling the point at which the cut is performed into the Z axis (i.e. the point at which the design is partitioned into tiers). The optimal solution for wire length is obtained when the aspect ratio drives the cut direction. The solution with fewer 3D-Vias can be obtained in the case where the first cut is made on the Z axis (method that would be equivalent to the ones based on hMetis assignment mentioned above).

Goplen and Sapatnekar [12] formulate the 3D placement problem as a True 3D placement. They provide an analytical force directed algorithm that minimizes the squared 3D wire length. Their method is iterative; at each iteration repulsive forces related to thermal issues or cell overlaps are inserted in the system. This process makes cells spread into the placeable volume. The authors do not detail how they handle I/Os into the tiers; however, on quadratic placement methods the cells will not move in the Z axis unless the I/Os are placed in different tiers. In this case, it can be understood that the repulsive forces are responsible for moving cells into other tiers. After placement is completed, the cells are sorted in the Z axis and finally assigned to a circuit tier. This method may fall into a false wire length optimization since actually cells cannot be placed into continuous coordinate; the rounding of their coordinated could potentially decrease circuit wire length.

Obenaus et. al, in [17], present an iterative force directed method for 3D placement. Different from Goplen's placer, it is not an analytical method but moves all cells (cell-by-cell) to an optimal position according to its connections. They define the 3D placement problem to minimize wire length only, which handles the problem as true 3D method. 3D-Via costs and constrains are not considered. No repulsive forces are added to the system, but a bucket re-scaling method similar to cell shifting from [21] spreads out the cells.

2   3D VLSI Integrated Circuits

A 3D circuit can be defined as a VLSI chip with stacked active layers called **tiers**. In the following sections, more details of the 3D fabrication and impacts on design methodologies will be presented. Figure 2 provides a didactic view of a 3D Chip with active layers and metal layers. Depending on the integration strategy used there may be or not metal layers above the last tier of active area. Also, depending on the integration strategy, more metal layers can be contained between a pair of adjacent tiers. More details on the technologies and how they are manufactured are provided in the following sections.

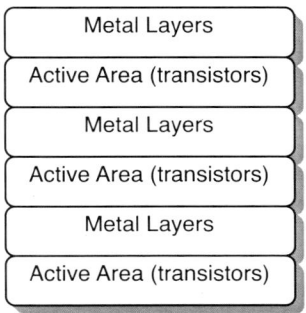

Fig. 2. A didactic picture of 3D circuit, tiers of active area and metal layers

2.1 Manufacturing technologies

According to [18], the assembly of 3D Chips is performed in different integration granularities.

Chip stacking is simply the vertical stacking of fully pre-manufactured chips. The chips have regular buffered I/O connections integrated usually by wire bonding [10]. Since all inter-chip communication must pass through the I/O buffers going outside of the chip, this methodology does not provide any advantage to circuit performance and power, reducing only the area occupied by the chip on the board. This technique is applied for cell-phones and other portable devices.

Die-on-wafer stacking is performed by stacking individual tested dies into a host wafer. Positions of the host wafer can also be pre-tested. The individual dies are placed using a pick-and-place equipment, that is a bottleneck for the cost, quality and size of inter-chip communication. Patti [18] reports that the placement misalignment today is about 10 μm.

Wafer-level stacking bonds entire wafers into a stack. Tezzaron is one company working with this kind of integration. Compared to die-on-wafer stacking, Tezzaron's technology [13] achieves better alignment (1 μm) and a more planar surface, leading to more integrated communication.

Finally, the **transistor stacking** methodology is an ideal integration of active layers fabricated in the same die, dismissing all equipment for wafer alignment. Today, those devices cannot be fabricated mainly due to high temperature process during the wafer manufacturing. Basically, the technology for fabricating high-performance transistors demands temperatures that would destroy any copper or aluminum used to manufacture metal layers bellow it. There is ongoing research in order to solve this issue and in the future this technology is very promising.

According to [10] the types of 3D-Vias can be classified into wire bounded, microbump, contactless and through vias.

In **wire bonded** technology, tiers of different sizes are stacked and I/O Pads are placed in the boundary of the tiers in such a way that they are not blocked by the upper tier. The main disadvantage of this technology is that wires are out of the chip scope, so they must be buffered and the pads consume very large areas.

Microbump technology provides micro contacts (bumps) placed in the top metal layer (sometimes the top two metal layers may be blocked for other routing). For this technology, chips can be stacked face-to-back and the package itself can provide routing space (3D package). On the other-hand, stacking the chips in a face-to-face fashion provides simpler (and consequently better) routing requiring no wiring channels in the package. The tiers are placed in such a way that their respective bumps are physically connected. Face-to-face integration is limited to two tiers.

The **contactless** technologies can be summarized as capacitive and inductive coupling. The capacity coupling technologies require the chips to be placed face-to-face because the contacts have a very tight proximity constraint. Inductive coupling is usually integrated face-to-back.

Finally, **Through Vias** consists of digging a hole though the tier for face-to-back comunication. Sometimes, such as in MITLL 3D technology [10], the first two tiers are integrated face-to-face while the rest of the tiers are stacked face-to-back. Even two chips connected face-to-face will need face-to-back comunication with the I/O pads. Due to silicon polishing issues, the traditional Bulk technologies requires a much larger pitch compared to SOI processes for 3D, such as in the MITLL. But still in the face-to-face integration, the technology for digging the hole in the oxide and depositing metal is similar. So far, this kind of technology is the one that provides the tighter integration between tiers because they are assembled in the wafer level.

Figure 3 illustrates a 3D circuit layout with Though Vias and Microbump technology for face-to-face connection. Note that there are microbumps in the top of the last metal layers that serves the purpose of connecting the tier to its neighbors.

2.2 Summary of 3D-Vias related information

In this section, some important data for the development of the paper is summarized. 3D-Vias are classified according to the following characteristics:

- The strategy used to integrate the tiers connected by the 3D-Via, that can be either face-to-face, face-to-back or back-to-back;
- The pitch of the 3D-Via;
- Whether the 3D-Via occupies active area or not;

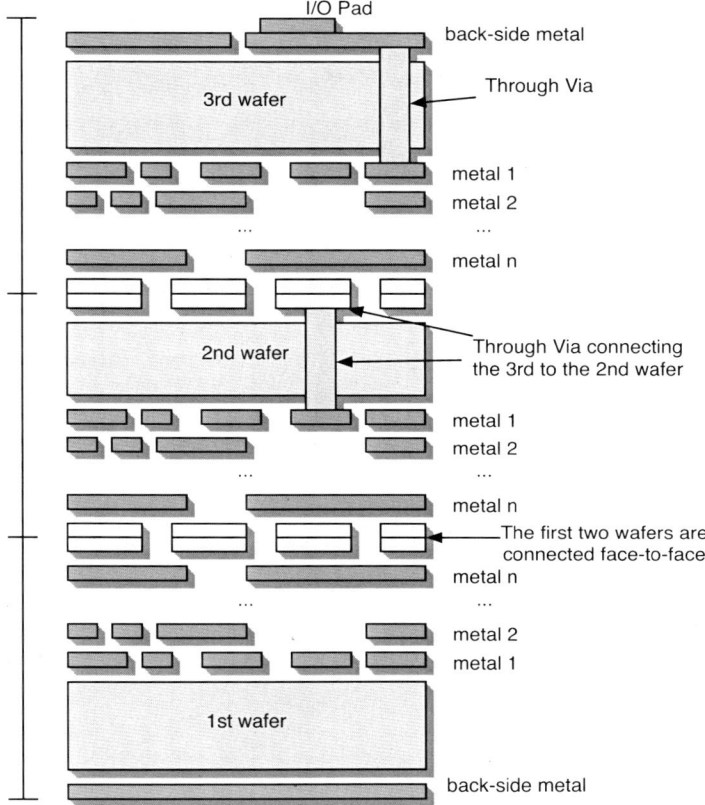

Fig. 3. The layout of a 3D circuit, containing three tiers integrated face-to-face and face-to-back respectively.

A list of some 3D-Vias and its characteristics is presented in table 1.

We can observe that there is a variety of pitches while some 3D-Vias occupy active areas. The methodology for introducing 3D-Vias during 3D placement must be subject to the 3D-Via characteristics. Consider, for instance, the face-to-face 3D-Vias, that can reach pitches in the order of 1 μm. For such technology, 3D-Vias could be used plenty. On the other hand, a face-to-back 3D-Via of, for instance, 50μm would require a huge amount of active area; for this example it would be reasonable to strongly reduce their count.

Table 1. Summary of collected data for 3D-Vias.

3D-Via	Integration Strategy	3D-Via Pitch	Occupy Active Area
Tezzaron (Copper Pads)	face-to-face	2.4 μm	no
Tezzaron (Projected)	face-to-face	1.46 μm	no
Microbump	face-to-face	10-100 μm	no
Contactless (Capacitive)	face-to-face	50-200 μm	no
MIT (Copper/Tantalum Pads)	face-to-face	5 μm	no
TSV face-to-face	face-to-face	0.5 μm	no
Tezzaron Super-Via^{TM}	face-to-back	6.08 μm	yes
Tezzaron Super-Contact^{TM}	face-to-back	¡ 4 μm	yes
Microbump 3D Package	face-to-back	25-50 μm	no
Contactless Inductive	face-to-back	50-150 μm	yes
MITLL Through Via (SOI)	face-to-back	5 μm	yes
Through Via (regular Bulk)	face-to-back	50 μm	yes
Back-to-back 3D-Via	back-to-back	15 μm	yes

3   I/O Partitioning and Placement Problem

Given a 2D placement netlist with pre-placed I/O pins at the boundary of the region available for Standard Cell placement, the migration to a 3D netlist (ready for 3D placement) has the following goals:

- Area allocation: the width and height of the tiers must be calculated according to the number of tiers.
- I/O partitioning: the I/Os must be partitioned into different tiers.
- I/O placement: the I/Os must be placed at the boundary of the block, delimiting the area for Standard Cell placement.

We understand the the I/O partitioning problem should not determine the cells partitioning as well; it is a task of the cell placement. Figure 4 illustrates the I/O pins migration. As formulated formally in the next section, the netlist migration preserves some properties of the 2D solution, such as whitespace, aspect ratio, I/O pins orientation and ordering. Our objective is to provide a migration algorithm that facilitates the 3D-Via minimization. From the perspective of the I/O pin partitioning our idea is to provide a good starting point for the cell partitioning. The algorithm should provide good I/O pins balance and respect the mentioned properties.

Once the netlist is migrated (the I/O pins are placed) we follow the methodology found in [3] that performs min-cut partitioning for the cells and tier assignment with Simulated Annealing as illustrated by figure 6. In our case, though, the min-cut have initially pre-placed fixed pins (I/Os). In this paper, we propose to study the impact of the 3D-Vias in the tier area.

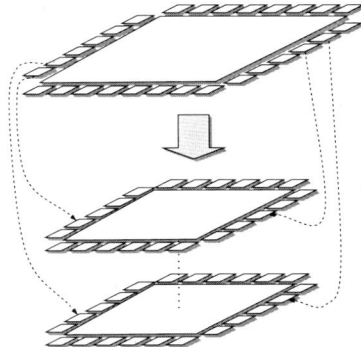

Fig. 4. Migration (from 2D to 3D) of a netlist with pre-placed I/O Pins

3.1   Formal definition

Before placement, a 2D circuit netlist Nl composed by a set of gates $G = \{g_1, g_2, g_3, , g_n\}$, a set of I/O pins $P = \{p_1, p_2, p_3, , p_m\}$ and a set of nets connecting them $N = \{n_1, n_2, n_3, , n_o\}$. A hypergraph Hg represents the netlist, where $G \bigcup P$ is the set of nodes and N is the set of hyperedges. The fixed position of each I/O pin p_i is given by $X[i]$ and $Y[i]$ ($i \leq m$) and its orientation by $Or(p_i) \epsilon \{north, south, east, west\}$. The area A (height H and width W having its bottom left corner at coordinate (x_{ini} ,y_{ini}) position) inside the I/O pins is assigned for cell placement. The whitespace ratio S on the placement area is achieved by subtracting the total gate area (Ga) from the area available inside the I/Os and dividing the result by Ga. The aspect ratio Ar is computed by W divided by H.

Let Z be the set of tier numbers $\{1, 2, ..., z\}$. The problem to be solved is defined as follows: given a 2D placement netlist Nl with fixed I/O pins, find a set of tiers $T = \{t_1, t_2, , t_z\}$ (z is the number of tiers) and their correspondent A_i, Ar_i, Ga_i, W_i, H_i, P_i, S_i, Or_i, X_i and Y_i ($i \leq z$) such that equations 1-8 hold.

$$P_1 \cup P_2 \cup ... \cup P_z = P \quad (1)$$

$$(\forall a, b \epsilon Z)(a \neq b \rightarrow P_a \bigcap P_b = \emptyset) \quad (2)$$

$$(\forall i \epsilon Z)(Wh_i \approx Wh) \quad (3)$$

$$(\forall i \epsilon Z)(Ar_i \approx Ar) \quad (4)$$

$$(\forall i \epsilon Z)(\forall j \epsilon Z)(W_i = W_j \wedge H_i = H_j) \quad (5)$$

$$(\forall i \epsilon Z)(\forall a \epsilon P_i)(Or_i(a) = Or(a)) \quad (6)$$

$$(\forall i \epsilon Z)(\forall a \epsilon P_i)(\forall b \epsilon P_i)(Or(a) = Or(b) \wedge X_i[a] < X_i[b] \rightarrow X[a] < X[b]) \quad (7)$$

$$(\forall i \epsilon Z)(\forall a \epsilon P_i)(\forall b \epsilon P_i)(Or(a) = Or(b) \wedge Y_i[a] < Y_i[b] \rightarrow Y[a] < Y[b]) \quad (8)$$

In other words, each tier will have its own set of I/O pins and no tier will share an I/O; the whitespace and aspect ratio must be evenly allocated; the orientation and ordering of the pins must be preserved.

4   Proposed algorithm

Let $Ld(p_i, p_j)$ be the length of the shortest path in Hg from p_i to p_j (e.g. the logic distance between p_i and p_j). The algorithm for I/O partitioning is described as follows.

Algorithm 1 I/O Pins Partitioning and Placement algorithm

1: Compute $Ld(i, j) \forall i, j \epsilon P$
2: Create a complete graph Pg such that P is the set of nodes and $Ld(i, j)(i, j \epsilon P)$ is the cost of the edge connecting nodes i and j.
3: Perform the partitioning of Pg into $P_1, P_2, , P_z$ configured to perform min-cut optimization at a 1% maximum unbalance ratio.
4: Compute Ga_i $(i \epsilon Z)$ by Ga divided by z
5: Compute A_i by adding Wh_i to Ga_i
6: Compute the dimensions of the tiers based on equation 9.
7: Place the I/O pins around the boundary of the block by simple stretching according to equation 10.
8: Legalize I/O Positions

$$W_i = \sqrt{A_i} \times Ar_i \qquad (9)$$
$$H_i = \frac{\sqrt{A_i}}{Ar_i}$$

$$(\forall i \epsilon z)(\forall p \epsilon P_i) X_i[p] = \frac{(X[p] - x_{ini}) \times W_i}{W} \qquad (10)$$
$$(\forall i \epsilon z)(\forall p \epsilon P_i) Y_i[p] = \frac{(Y[p] - y_{ini}) \times H_i}{H}$$

The first step of the algorithm is illustrated in 5.(a). Considering that in a real circuit net fanouts are limited, node degrees can be considered bounded or constant for the sake of complexity analysis. Thus, a single BFS search has an $O(n)$ complexity. The algorithm can be performed by m^2 BFS searches in Hg resulting in a $O(m^2 n)$ time complexity. Since the number of I/O pins do not exceed a few thousand, it is feasible to use BFS. By using a single search to compute the distance from a pin p_i to every $p \epsilon P$, the complexity can go down to $O(mn)$.

On step2, the values of Ld are used to create a Pg graph connecting all pairs of I/O pins, as shown in figure 5.(b).

For the third step, we used the hMetis tool [14]. The tool accepts cell weights. We assigned the inverse of the edge costs as their weights and imposed a very tight balance in order to keep a similar amount of I/Os in each tier. In section 6.3 the effects of unbalancing the I/O pins are discussed.

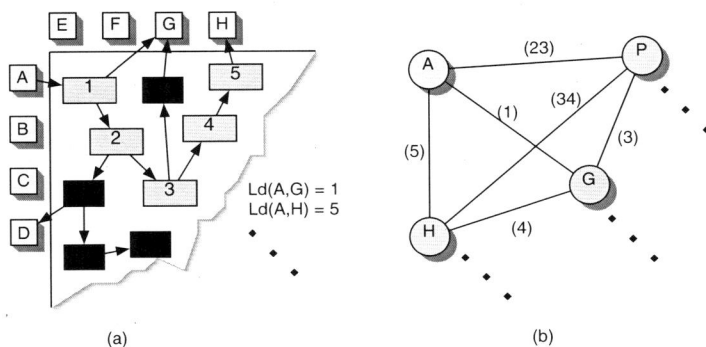

Fig. 5. An ilustration of the logic distance between I/O pins (a) and a part of the correspondent complete graph (b)

The forth step can be accomplished by a simple division of the total gate area by the number of tiers. So far, it is not possible to know whether such perfect cells partitioning will be achievable, but it is a reasonable assumption. Nevertheless, S_i could be changed to compensate the Ga_i inaccuracy.

The steps 5 and 6 compute the area of the tiers such that aspect ratio and whitespace are preserved from the original 2D circuit. At this point, new aspect ratio or whitespace could be used.

Finally, the steps 7 and 8 compute the x and y coordinates of the I/Os to their target tiers. The original orientation and ordering is preserved, since the I/O placement is a mapping from their original position into a smaller area. A legalization (step 9) is performed at the end to assure that the I/Os do not overlap.

5   Experimental Setup

The goal is to study the impact of the I/O pin partitioning in the area, number of vias and I/O pin balance. For that, we defined a simplistic 3D placement flow as follows:

1. Initially the I/O partitioning algorithm under study is performed.

2. A min-cut partitioning of Hg into z partitions is performed. The I/O pins, that have already an assigned partition, are used as fixed nodes. The hMetis tool is applied for this step. The tool is configured to keep the area as balanced as possible (maximum 1% unbalance).
3. A tier assignment (similar to the one from [3]) problem maps the sets $P_1, P_2, ..., P_z$ into tiers $t_1, t_2, ..., t_n$. A Simulated Annealing engine is used (see figure 6).
4. Cells could be placed separately in each tier. We skip this step since our goal at this point is to evaluate the number of 3D-Vias.

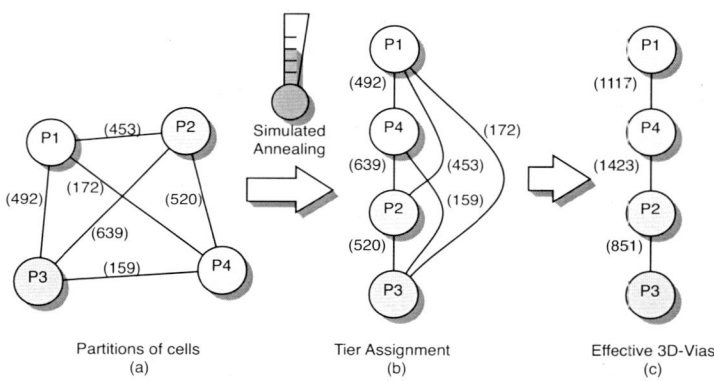

Fig. 6. A group of partitions (a) are assigned to tiers (b) using Simulated Annealing; the effective number of 3D-Vias is shown in (c)

As there is no published previous work on I/O pins handling, the proposed I/O partitioning algorithm is compared with two other simplistic algorithms that follow the same formulation described in section 3. The first algorithm is called *AlternatePins*, on figure 7.(a). This method is a pseudo-random partitioning that goes thought the boundary line of the chip picking nodes for each partition alternatively. The *AlternatePins* replaces steps 1,2 and 3 of the flow keeping steps 4,5,6,7 and 8 untouched in order to maintain the same I/O placement policy.

The idea behind the *AlternatePins* method is to provide an optimal solution in terms of balancing the I/Os. Balancing is important for the subsequent placement stage because the I/Os play a very important role in the quadratic placement engine [4]. This algorithm computes an optimal solution for the cell placement based on attraction forces between connected cells. I/O pins, placed at the boundary, are responsible for the spreading of the cells, since otherwise they would be placed at the center point.

The second method is called *UnlockedPins*, illustrated in figure 7.(c). In this method, we allow hMetis to partition the I/Os as free nodes, replacing the steps 1,2 and 3 of our algorithm. The following steps of our algorithm are done for the *UnlockedPins* as well.

The idea behind the *UnlockedPins* method is to provide a favorable solution in terms of 3D-Via minimization. Since hMetis is a leading edge hyper-graph partitioner, it will generate a netlist partitioning with close to optimal number of 3D-Vias. On the other hand, I/O pins will not be spread evenly.

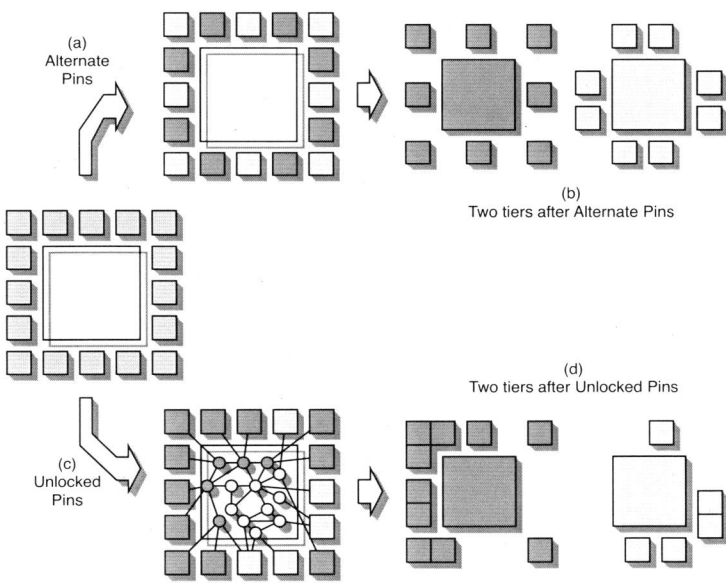

Fig. 7. An illustration of the Alternate Pins algorithm (a) resulting in a two tier circuit (b) with perfectly balance I/O pins; the Unlocked Pins algorithm (b) uses hMetis to partition the whole Netlist, which could result in unbalanced pins (d).

The method proposed here aims at a good solution both in terms of 3D-Vias and balancing. Section (6) presents experimental results comparing the algorithm under these metrics.

6   Experimental Results

6.1   Effect on 3D-Vias

Experiments measuring the amount of 3D-Vias and the balancing of the algorithm are presented in this section. Tables 2, 3 and 4 report our experimental results. ISPD 2004 benchmarks [1] are used targeting circuits with two, three, four and five tiers.

First, table 2 reports the I/O balancing measured by the standard deviation of the number of I/O pins averaged from the whole IBM benchmark suite. The average number of I/O pins from the IBM benchmarks is 264. The method *AlternatePins* delivers the optimal solution while *UnlockedPins* is very unbalanced. In some situations, the strong unbalance practically invalidates the method. The proposed algorithm has close to optimal pin balancing.

Table 2. Comparison of the I/O pins distribution in the tiers considering the three studied algorithms averaged from ibm01 to 1bm18.

# tiers	Algorithm	σ # I/Os
2	Our Algorithm	7
	UnlockedPins	233
	AlternatePins	0.4
3	Our Algorithm	6
	UnlockedPins	252
	AlternatePins	0.4
4	Our Algorithm	5
	UnlockedPins	177
	AlternatePins	0.4
5	Our Algorithm	6
	UnlockedPins	189
	AlternatePins	0.4

Tables 3 and 4 presents our experimental results for the total number of 3D-Vias for the whole IBM benchmark suite. The *AlternatePins* method has the worst results under this metric, which is expected since it is a pseudo-random partitioning. This fact enforces the conclusion that a simplistic I/O partitioning leads to a worse cut size. On the other hand, the method *UnlockedPins*, which was expected to have the best cut among the three methods was outperformed by our algorithm. This fact can be explained by our pre-processing stage that computes the logic distance between I/Os. It seems that the logic distance is a way to summarize the information of the whole graph into a single edge that connects I/O pins (step 2 of the algorithm). Since the graph into this step is very small compared to the whole netlist hyper-graph, the partitioning algorithm (hMetis in this case) could achieve a good partitioning for the pins and for the netlist as well. This computation requires intensive CPU usage. To overcome this problem, the distances are pre-computed and stored in a file so that the I/O partitioning runtimes are not harmed.

Tables 5 and 6 present experimental results for the maximum number of 3D-Vias between pairs of tiers.

Table 3. Total number of 3D vias for the proposed algorithm.

# tiers	Our Algorithm # 3D-Vias			
	2	3	4	5
ibm01	374	525	837	1162
ibm02	396	747	1156	1533
ibm03	1064	2174	2610	3974
ibm04	735	1511	2371	2852
ibm05	2258	4311	6489	9193
ibm06	1059	1642	2934	3477
ibm07	992	2050	3219	4400
ibm08	1298	2697	4018	5346
ibm09	699	1872	2495	3343
ibm10	1490	2661	4004	5216
ibm11	1190	2240	3685	4620
ibm12	2293	4094	6581	8191
ibm13	1042	1893	3099	3742
ibm14	2121	3886	5342	6667
ibm15	3002	4827	7022	9283
ibm16	2102	4316	5774	7172
ibm17	2769	5611	8526	10114
ibm18	1676	3591	4985	6581
Average	1476	2814	4175	5381

6.2 Studding the area effect of 3D-Vias

Table 7 presents an area impact study of the 3D-Vias considering the three algorithms (the numbers are averaged for all benchmarks). The column "Max # 3D-Vias" reports the maximum number of 3D-Vias connecting pairs of adjacent tiers; this data is extracted from tables 5 and 6. This number will impact the area requirements for 3D-Vias. The area study supposes 3D-Vias measuring $5\mu m$ and $50\mu m$, which represent a good 3D-Via pitch and a huge 3D-Via pitch respectively.

The following facts can be observed on table 7:

- The *big* 3D-Vias, that could be Bulk based face-to-back vias, suffer from a very high penalty for the 3D-Vias. With 2 tiers, there is a penalty of around 53% of the tier area (note that our algorithm results in less 3D-Vias and also less tier area than the others). For the cases with 4 and 5 tiers, the 3D-Via area is larger than the tier area. The important conclusion here is that when targeting a big via technology it is mandatory to minimize the number of 3D-Vias in order to obtain a feasible solution. As seen in previous tables (5 and 6) the proposed algorithm can save up to 34% which translates to area savings in the order of an entire tier.
- Technologies with small vias suffers from around 2% of the area penalty for the 3D-Vias, leaving room for more 3D-Vias if they are helpful.

Table 4. Comparison of the total number of 3D vias for the three studied algorithms for I/O pin partitioning over the others.

# tiers	UnlockedPins # 3D-Vias				AlternatePins # 3D-Vias			
	2	3	4	5	2	3	4	5
ibm01	441	857	838	1439	428	881	977	1372
ibm02	547	882	1214	1600	503	829	1340	1691
ibm03	1146	2282	2693	4020	1099	2530	3602	4366
ibm04	628	1583	2516	3202	750	1619	2461	4275
ibm05	2417	5372	6653	9651	2576	5428	7037	12400
ibm06	1057	1827	3128	3566	1075	1729	3429	3507
ibm07	880	3242	3302	4605	1049	3423	3482	6523
ibm08	1324	2814	4184	5698	1307	3431	4183	6327
ibm09	806	2828	2763	3518	780	2186	3757	3556
ibm10	1771	3565	4675	7116	1821	4062	4358	8492
ibm11	1490	3477	3958	5697	1494	3629	4923	7437
ibm12	2594	5350	7259	9158	2556	5569	8996	12515
ibm13	1193	3037	3264	4557	1170	2912	4618	4874
ibm14	2171	4561	6584	8085	2310	5090	7564	10113
ibm15	2890	7863	9082	11707	3126	7970	11144	13857
ibm16	2237	5816	6235	9300	2280	6216	9525	10903
ibm17	2539	7695	8733	10845	2847	8402	11420	14080
ibm18	1835	4686	5229	9072	1704	3899	5268	8193
Average	1554	3763	4573	6269	1604	3879	5449	7466
Our Improv.	5.29%	33.74%	9.53%	16.49%	8.72%	37.84%	30.52%	38.73%

6.3   Unbalancing the I/O pins

In the previous section we could observe that there is a trade-off between the I/O pins balance and the resulting number of 3D-Vias. The proposed algorithm for pin partitioning aims at good balance. However, it is well known that a tight balance requirement over-constraints the partitioning process [14]. In the proposed algorithm, the I/O balance can be controlled in step 3 that is performed by HMetis.

HMetis allows the user to configure the balance constraint for each bisection based on equation 11 where u is the unbalance parameter and n is the number of vertices on the hyper-graph.

$$[\frac{(50-u) \times n}{100}; \frac{(50+u) \times n}{100}]$$

(11)

For example, let $u = 10$, then the bisection balance will range from 40%-60% to 60%-40%. Now suppose that we have four partitions, then an unbalancing factor 10 will result in partitions that can contain between $0.402 \times n = 0.15 \times n$ and $0.602 \times n = 0.35 \times n$ vertices.

Table 5. Maximum number of 3D-Vias for proposed algorithm.

# tiers	2	3	4	5
ibm01	374	330	370	400
ibm02	396	413	403	594
ibm03	1064	1112	1088	1260
ibm04	735	887	992	887
ibm05	2258	2203	2469	2729
ibm06	1059	849	1135	948
ibm07	992	1332	1433	1524
ibm08	1298	1448	1397	1610
ibm09	699	1057	1008	1075
ibm10	1490	1450	1590	1750
ibm11	1190	1485	1605	1719
ibm12	2293	2278	2422	3173
ibm13	1042	1269	1548	1781
ibm14	2121	2272	2248	2459
ibm15	3002	2857	3199	3395
ibm16	2102	2164	2212	2625
ibm17	2769	3150	3601	3105
ibm18	1676	1871	1754	1782
Average	1476	1579	1693	1823

Our experimental results (averaged from all benchmark circuits) are reported on table 8 and figure 8. Table 8 presents the I/O pin unbalance measured by Standard Deviation. Figure 8 presents the benefits of unbalancing the I/Os to the 3D-Via count.

7   Conclusions

A method for the partitioning and placement of the I/O pins of a 2D block to a 3D circuit was proposed. An interesting analysis in our method lies in the fact that it actually improved the hypergraph partitioning algorithm cut by performing only shortest path analysis. Note that the method works in two phases: first the I/O partitioning considering the logic distances as weights; second, fix the I/Os and perform partitioning of the cells. In the first phase, the I/Os are arranged in a small graph (containing only the I/Os) weighted by the logic distance on the original graph. The edge weights actually contain information of the whole netlist, compressed in the small I/O graph. In the second phase, the whole netlist is partitioned, however some nodes (the I/Os) are fixed reducing the problem complexity and more importantly providing tips to the partitioning algorithm. We conclude that the reduced problem sizes with compressed information of the whole netlist actually improved the partitioning algorithm at the expense of more CPU time.

Empirically, we showed that doing the partitioning of I/O together with the cells (UnlockedPins method) leads to strongly unbalanced number of pins,

Table 6. Comparison of the maximum number of 3D vias for the three studied algorithms for I/O pin partitioning over the others.

# tiers	UnlockedPins Max # 3D-Vias				AlternatePins Max # 3D-Vias			
	2	3	4	5	2	3	4	5
ibm01	441	467	377	573	428	483	406	480
ibm02	547	496	485	552	503	469	498	553
ibm03	1146	1143	1021	1334	1099	1320	1485	1210
ibm04	628	862	1067	1039	750	913	1033	1454
ibm05	2417	2765	2478	2712	2576	2814	2526	3974
ibm06	1057	924	1134	935	1075	915	1193	937
ibm07	880	1980	1510	1525	1049	2050	1590	2402
ibm08	1324	1436	1445	1788	1307	1919	1448	1833
ibm09	806	1598	1092	1249	780	1356	1684	1137
ibm10	1771	1883	1741	1986	1821	2247	1724	2898
ibm11	1490	1909	1810	2230	1494	1856	1802	2610
ibm12	2594	2820	2747	2962	2556	3160	3113	4205
ibm13	1193	1606	1500	1755	1170	1611	1905	1954
ibm14	2171	2375	2307	2881	2310	2619	3274	3283
ibm15	2890	4188	3377	4099	3126	4207	4385	4163
ibm16	2237	3185	2266	3355	2280	3704	3794	3443
ibm17	2539	4165	3526	2990	2847	4539	5245	5053
ibm18	1835	2652	1852	2810	1704	2127	1856	2552
Average	1554	2025	1763	2043	1604	2128	2165	2452
Our Improv.	5.29%	28.24%	4.14%	12.06%	8.72%	34.76%	27.85%	34.51%

which invalidates the method. We also demonstrated the pseudo-random I/O partitioning approaches (such as AlternatePins) leads to a higher number of 3D-Vias. The proposed method demonstrated good effectiveness both in terms of I/O balance and resultant number of 3D-Vias (5% to 33% improvement on 3D-Via count compared to hMetis), outperforming both algorithms in both metrics.

After that, the area impact was studied under our simplified placement flow that minimizes the number of 3D-Vias. It was verified that the area overhead caused by 3D-Vias is prohibitively high for big (50μm pitch) 3D-Vias (in the order of 50% of the active area and up), requiring more research on via minimization methods. On the other hand, for small (5μm pitch) 3D-Vias, the impact was small (around 2% of the active area), leaving room for additional 3D-Vias if it can improve circuit performance. Any intermediary case would be able to trade 3D-Vias for performance limited by the area occupied by the 3D-Vias.

Finally, we investigated ways to further minimize the cut by working with the I/O pin balancing. We relaxed the I/O pin balance constraint keeping the area evenly distributed since the second partitioning process is still highly constrained. Adding up the advantage reported in previous works with the improvements achieved on this paper, we can outperform hMetis partitioning from 5.5% to 34% in average.

Table 7. Comparison of the 3D-Vias Area Impact Considering the Three Algorithms.

# tiers	Algorithm	Area Tier	Max # 3D-Vias	Area 3D-Vias (big - 50μm)		Area 3D-Vias (small - 5μm)	
2		6,934,347	1,476	3,690,000	53%	36,900	1%
3	OurAlgorithm	4,660,116	1,579	3,947,500	85%	39,475	1%
4		3,490,471	1693	4,232,500	121%	42,325	1%
5		2,821,087	1823	4,557,500	162%	45,575	2%
2		6,936,553	1,554	3,885,000	56%	38,850	1%
3	UnlockedPins	4,658,909	2,025	5,062,500	109%	50,625	1%
4		3,481,276	1,763	4,407,500	127%	44,075	1%
5		2,817,413	2,043	5,107,500	181%	51,075	2%
2		6,926,117	1,604	4,010,000	58%	40,100	1%
3	AlternatePins	4,640,572	2,128	5,320,000	115%	53,200	1%
4		3,489,458	2,165	5,412,500	155%	54,125	2%
5		2,816,187	2,452	6,130,00	218%	61,300	2%

Table 8. The Unbalance of the I/O pins measured by the Standard Deviation

	2 tiers	3 tiers	4 tiers	5 tiers
u=1	7	6	5	6
u=10	64	54	41	48
u=25	158	141	100	103

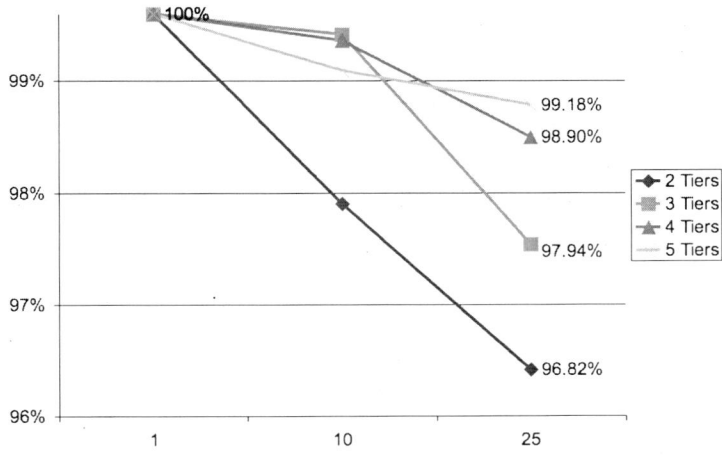

Fig. 8. The percentage improvement on 3D-Via count of unbalancing the I/O pins.

8   Acknowledgments

The authors would like to thank Robert Patti for providing valuable information about Tezzaron technology.

Bibliography

[1] Ispd04 - ibm standard cell benckmarks with pads., 2006.

[2] C. Ababei, Y. Feng, B. Goplen, H. Mogal, T. Zhang, K. Bazargan, and S. Sapatnekar. Placement and routing in 3d integrated circuits. *Design and Test of Computers*, pages 520–531, Nov-Dec 2005.

[3] C. Ababei, H. Mogal, and K. Bazargan. Three-dimensional place and route for fpgas. In *ASP-DAC '05: Proceedings of the 2005 conference on Asia South Pacific design automation.* IEEE Press, 2005.

[4] C. J. Alpert, T. Chan, D. J.-H. Huang, I. Markov, and K. Yan. Quadratic placement revisited. In *DAC '97: Proceedings of the 34th annual conference on Design automation*, pages 752–757, New York, NY, USA, 1997. ACM Press.

[5] K. Banerjee, S. Souri, P. Kapur, and K. Saraswat. 3d-ics: A novel chip design for improving deep submicrometer interconnect performance and systems on-chip integration. *Proceedings of IEEE*, 89:602–633, 2001.

[6] A. E. Caldwell, A. B. Kahng, and I. L. Markov. Can recursive bisection alone produce routable placements? In *DAC '00: Proceedings of the 37th conference on Design automation*, pages 477–482, New York, NY, USA, 2000. ACM Press.

[7] S. Das, A. Chandrakasan, and R. Reif. Design tools for 3-d integrated circuits. In *ASPDAC: Proceedings of the 2003 conference on Asia South Pacific design automation*, pages 53–56, New York, NY, USA, 2003. ACM Press.

[8] S. Das, A. Chandrakasan, and R. Reif. Calibration of rent's rule models for three-dimensional integrated circuits. *IEEE Transactions on Very Large Scale Integration (VLSI) Systems*, 12:359–366, 2004.

[9] S. Das, A. Fan, K.-N. Chen, C. S. Tan, N. Checka, and R. Reif. Technology, performance, and computer-aided design of three-dimensional integrated circuits. In *ISPD '04: Proceedings of the 2004 international symposium on Physical design*, pages 108–115, New York, NY, USA, 2004. ACM Press.

[10] W. Davis, J. Wilson, S. Mick, J. Xu, H. Hua, C. Mineo, A. Sule, M. Steer, and P. Franzon. Demystifying 3d ics: The pros and cons of going vertical. *Design and Test of Computers*, pages 498–510, Nov-Dec 2005.

[11] Y. Deng and W. P. Maly. Interconnect characteristics of 2.5-d system integration scheme. In *ISPD '01: Proceedings of the 2001 international symposium on Physical design*, pages 171–175, New York, NY, USA, 2001. ACM Press.

[12] B. Goplen and S. Sapatnekar. Efficient thermal placement of standard cells in 3d ics using a force directed approach. In *ICCAD '03: Proceedings of the 2003 IEEE/ACM international conference on Computer-aided design*, page 86, Washington, DC, USA, 2003. IEEE Computer Society.

[13] S. Gupta, M. Hilbert, S. Hong, and R. Patti. Techniques for producing 3d ics with high-density interconnect, 2005. Available at: ¡http://www.tezzaron.com/¿. Access on: Aug. 2005.

[14] G. Karypis, R. Aggarwal, V. Kumar, and S. Shekhar. Multilevel hypergraph partitioning: Applications in vlsi domain. *IEEE Transactions on Very Large Integration (VLSI) Systems*, 7:69–79, March 1999.

[15] I. Kaya, S. Salewski, M. Olbrich, and E. Barke. Wirelength reduction using 3-d physical design. In *Integrated Circuit and System Design - Power and Timing Modeling, Optimization and Simulation; Proceedings of 14th International Workshop, PATMOS 2004*, 2004.

[16] G. Liu, Z. Li, Q. Zhou, X. Hong, and H. H. Yang. 3d placement algorithm considering vertical channels and guided by 2d placement solution. In *ASICON 2005: 6th International Conference On ASIC*, pages 24–27, 2005.

[17] S. Obenaus and T. Szymanski. Gravity: Fast placement for 3-d vlsi. *ACM Transacions on Design Automation of Electronic Systems*, 8:69–79, March 1999.

[18] R. Patti. Three-dimensional integrated circuits and the future of system-on-chip designs. *Proceedings of IEEE*, 94:1214–1224, 2006.

[19] A. Rahman and R. Reif. System-level performance evaluation of three-dimensional integrated circuits. *IEEE Transactions on Very Large Scale Integration (VLSI) Systems*, 8, December 2000.

[20] A. Rahman and R. Reif. Thermal analysis of three-dimensional (3-d) integrated circuits (ics). In *Proceedings of the IEEE 2001 International Interconnect Technology Conference*, pages 157–159, 2001.

[21] N. Viswanathan, M. Pan, and C. C.-N. Chu. Fastplace: an analytical placer for mixed-mode designs. In *ISPD '05: Proceedings of the 2005 international symposium on physical design*, pages 221–223, New York, NY, USA, 2005. ACM Press.

CAT Platform for Analogue and Mixed-Signal Test Evaluation and Optimization*

Ahcène Bounceur[1], Salvador Mir[1], Luis Rolíndez[1,2] and Emmanuel Simeu[1]

[1] TIMA Laboratory,
46, av. Félix Viallet,
38031 Grenoble Cedex, France
[2] ST Microelectronics,
850, rue Jean Monnet,
38926 Crolles, France
{Ahcene.Bounceur,Salvador.Mir,Emmanuel.Simeu,Luis.Rolindez}@imag.fr

Abstract. This paper introduces a Computer-Aided-Test platform that has been developed for the evaluation of test techniques for analogue and mixed-signal circuits. The CAT platform, integrated in the Cadence Design Framework Environment, includes tools for fault simulation, test generation and test optimization for these types of circuits. Fault modeling and fault injection are simulator independent, which makes this approach flexible with respect to past approaches. In this paper, the use of this platform is illustrated for test optimization for the case of a fully differential amplifier. Test limits are set using a statistical circuit performance analysis that accounts for process deviations, as a trade-off between estimated test metrics at the design stage. Specification-based tests are next optimized in terms of their capability of detecting catastrophic and parametric faults.

1 Introduction

The test of integrated analogue and mixed-signal circuits differs importantly from the test of digital circuits. The major difference stems from the need to consider continuous signals and circuit parametric deviations, in addition to just catastrophic faults (opens and shorts). For digital circuits, structural testing has provided cost efficient solutions that target the test of catastrophic faults rather than the test of the circuit functionality. Thus, fault coverage is the major test metric in this domain and is somehow independent from the specifications. For analogue circuits, the need to consider parametric deviations has lead to the definition of analogue test metrics that take into account also the circuit functionality. In other words, even when a parametric fault-based test approach is considered for analogue circuits, test metrics such as fault coverage cannot be calculated without knowing the performance specifications [1].

---

*This research work was supported by European MEDEA+ program under the project NanoTEST-2A702.

Please use the following format when citing this chapter:

Bounceur, A., Mir, S., Rolíndez, L. and Simeu, E., 2007, in IFIP International Federation for Information Processing, Volume 249, VLSI-SoC: Research Trends in VLSI and Systems on Chip, eds. De Micheli, G., Mir, S., Reis, R., (Boston: Springer), pp. 281–300

The domain of integrated analogue and mixed-signal testing has always tried to cope with a controversy between functional and structural testing. Functional testing is practically always considered but research on structural testing continue to make progress. In fact, it has appeared clear for some test users that to find manufacturing faults such as shorts, opens and misloaded components in mixed-signal circuits is essential, and this comforted the proposal of the IEEE 1149.4 Analogue Boundary-scan mixed-signal test architecture [2]. Also, it has been shown that the study of catastrophic faults helps in identifying reliability problems in mixed-signal circuits, in particular redundant components [3]. In general, since it is possible to define a fault list for catastrophic faults, the study of catastrophic faults helps also for the generation and optimization of test patterns, even under the presence of process deviations [4].

The case of parametric faults has been considered by many authors by simply modifying the nominal values of a design parameter, and considering Monte Carlo simulations. In this way, parametric fault lists have been built in a rather arbitrary way. Recently, [1] introduced a different way of defining parametric faults. A parametric fault is considered as the minimum deviation of a design parameter that results in a circuit specification being violated. In this approach, parametric faults are obtained by transient simulations, without recurring to time consuming Monte Carlo simulations. This approach is quite acceptable when faults are considered the result of a single parameter deviation, while the other parameters remain at their nominal values. However, it cannot deal properly with the case of device misbehaviour resulting from the combination of multiple small deviations.

An early approach to avoid Monte Carlo simulation was based on the use of sensitivity analysis to deterministically identify the bounds on circuit parameters [5]. Process information and the sensitivity of the circuit principal components have been recently considered in [6] for generating the statistical models of the fault-free and faulty circuits, which is then used for test vector generation. These models are obtained using a statistical approach and a linear estimation, rather than Monte Carlo simulations. Another statistical approach is considered in [7]. Here, however, parametric faults are injected by swapping transistors, one at a time, by a transistor whose process parameters are shifted by 3σ and a sensitivity analysis is performed only in the DC domain. The problem with these approaches is again that the misbehaviour resulting from the combination of multiple small deviations cannot be evaluated properly.

In this work, we will introduce a Computer-Aided-Test platform for analogue and mixed-signal circuits. The CAT platform, integrated in the Cadence Design Framework Environment, includes tools for fault simulation, test generation and test optimization. Aspects on fault simulation and test optimization will be illustrated in this paper for the case of a fully differential amplifier. We will consider a statistical analysis that is based on Monte Carlo simulations. This analysis will allow the calculation of analogue test metrics under process deviations, and this will be used for setting test limits. These test limits will then be used for the evaluation of test metrics under catastrophic and parametric faults.

Specification-base tests will then be optimized according to fault coverage for a fully differential amplifier.

2   The CAT platform

2.1   Architecture of the platform

Figure 1 shows a simplified architecture of the proposed CAT platform. It is composed of three separate tool sets. Fault modelling, fault injection and fault simulation are carried out using the tool set FIDESIM. The results are saved in a database that can be read by the other tool sets, in particular the OPTEVAL tool set for test evaluation and the OPTEGEN tool set for test generation.

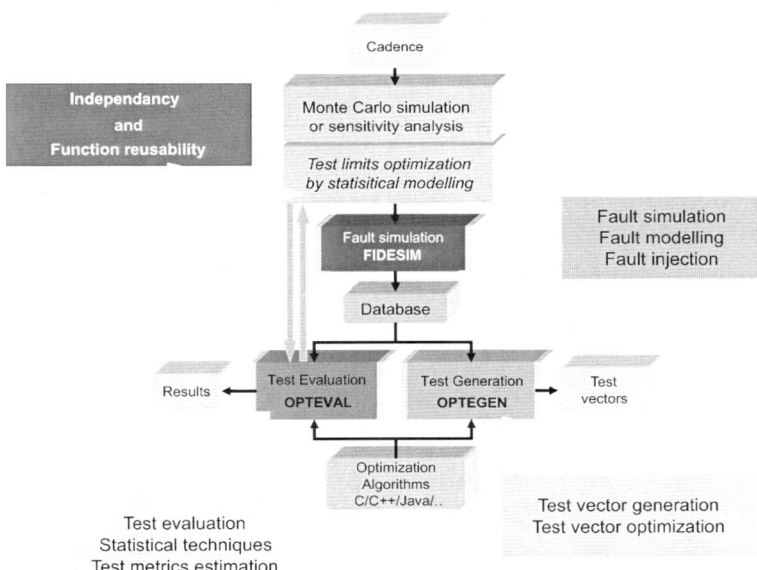

Fig. 1. Simplified architecture of the CAT platform.

In this paper, we will illustrate the use of the FIDESIM and OPTEVAL tool sets. The tool set OPTEGEN is the subject of further work. It currently includes three tools. The first tool is used for compaction of analogue functional tests. A second tool is used for the generation of multi-frequency test sets using the fault-based test approach described in [8]. This technique is valid for linear time-invariant circuits and allows the generation of a minimal set of test vectors for maximum fault coverage and, if required, maximal diagnosis. A third tool is available for the coding of analogue test patterns as optimized bit streams, as described in [9], following an approach first presented by [10].

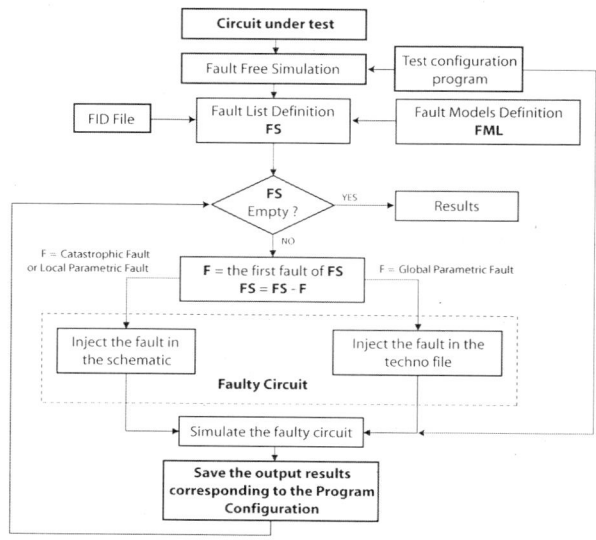

Fig. 2. Fault simulation procedure used by FIDESIM.

2.2   Fault modeling and fault injection tools

Several tools for analogue fault modelling, fault injection and fault simulation have appeared in the literature. Most of these tools are in-house developments. For example, in [7] a fault simulator called DOTTS is used for both catastrophic and parametric faults under process variations. It is also being considered for RF circuits [12, 13]. Catastrophic faults under process variations are considered by the ANTICS fault simulation environment [4]. Another in-house development called SWITTEST has been presented for fault simulation of parametric and catastrophic faults in switched capacitor systems [14]. A commercial tool for parametric fault simulation and test vector generation exploiting sensitivity analysis and statistical modelling has been commercialised [6]. Several other tools have been developed, especially for academic research, and it is not our aim to describe all of them. The common point of all these tools is that they modify the netlist of the circuit to perform the fault injection in a way that is dependent on the simulator netlist under use.

However, [15] presents a tool where the fault models are added, before simulation, in the schematic of the design (in Cadence®), and the faults are injected by changing the parameters of each fault model. The injection of fault models into the circuit schematics is also considered in the tool described in [16]. The netlist for fault simulation is then generated after the schematics, and thus can be independent of the simulator under use. The fault simulation tool set FIDESIM is based on this earlier development. A detailed description of fault model building and fault injection is given in [16]. The Fault simulation procedure is shown in Figure 2.

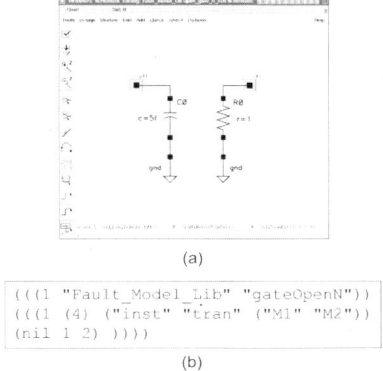

(a)

```
((1 "Fault_Model_Lib" "gateOpenN"))
(((1 (4) ("inst" "tran" ("M1" "M2"))
(nil 1 2) ))))
```

(b)

Fig. 3. Description of a fault model: (a) fault model, and (b) fault injection description file.

```
SPECIFICATIONS
pr1:(70,85)
TC_TOLERANCES
tc1:(200M,260M)
OUTPUTS
tm1 = dB20(VF("/out1")-VF("/out2"))
tm2 = phase(VF("/out1")-VF("/out2"))
PERFORMANCES
pr1 = value(tm1 100)
TEST_CRITERIA
tc1 = root(tm2 0 1)
```

Fig. 4. Test Program example.

The test engineer designs a set of fault models under the Cadence® DFII (Design Framework II) environment. A fault model is saved in a library just as a Cadence cellview. These fault models must observe some rules to allow the automatic fault injection, in particular relating to the pinout. Thus, for each fault model, the injection procedure is described using a pseudo code called FID (Fault Injection Description) stored in a file. Local, and global parametric faults can be considered as well. For example, Figure 3(a) describes the circuit that corresponds to an open in an NMOS transistor gate. The FID file for describing an injection of this fault is shown in Figure 3(b). This fault model is a cellview stored in the library "Fault_Model_Lib".

The different test benches for the circuit under test (CUT), the calculation of the circuit performances and the calculation of the proposed test measurements or test criteria are described as a pseudo code called *Test Configuration*. Figure 4 shows an example of this where one performance, two test measures and one test criteria are defined. The test measures are used to define the performances and the test criteria on which a tolerance test threshold is given.

The CUT may require several test benches and different types of analysis to measure its performances and test criteria. Thus, it generally needs to be simulated in multiple test benches under the same fault injection. FIDESIM

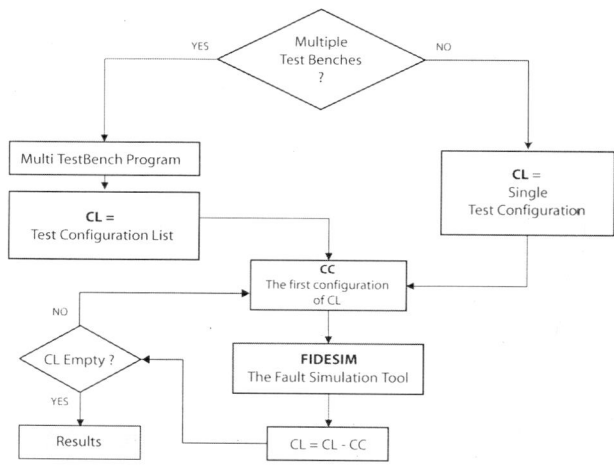

Fig. 5. Architecture of the multiple testbench procedure.

is able to perform this by describing the different test benches in the form of a pseudo code called *Multi TestBench Program*. This is illustrated by the procedure shown in Figure 5. This feature will be specially important during Monte Carlo simulations, since an instance generated during Monte Carlo will be simulated for all different test benches, before proceeding with the next instance.

3   Test metrics estimation

The test evaluation and optimization tool set OPTEVAL is developed to evaluate test techniques by estimating analogue test metrics. The estimation of test metrics such as defect level, test yield or yield loss is important in order to quantify the quality and cost of a test approach. For design-for-test purposes (DFT), this is important in order to select the best test measurements but this must be done at the design stage, before production test data is made available. In the analogue domain, previous works have considered the estimation of these metrics for the case of single faults, either catastrophic or parametric. The consideration of single parametric faults is sensible for a production test technique if the design is robust. However, in the case that production test limits are tight, test escapes resulting from multiple parametric deviations may become important. In addition, aging mechanisms result in field failures that are often caused by multiple parametric deviations. In the CAT platform presented here, we will consider the estimation of analogue test metrics under the presence of multiple parametric deviations (or process deviations) and under the presence of faults. A statistical model of a circuit is used for setting test limits under process deviations as a trade-off between test metrics calculated at the design stage. This model is obtained from a Monte Carlo circuit simulation, assuming Gaussian Probability Density Functions (PDFs) for the parameter and performance deviations.

After setting the test limits considering process deviations, the test metrics are calculated under the presence of catastrophic and parametric single faults for different potential test measurements. We will illustrate the technique for the case of a fully differential operational amplifier, proving the validity in the case of this circuit of the Gaussian PDF.

3.1 Definition of test metrics

The test metrics considered for analogue circuits are [1]: Yield Y, Test Yield Y_T, Yield Coverage Y_C, Yield Loss Y_L, Defect Level D and Fault Coverage F where:

$$Y = \text{Proportion of the functional (or good) circuits}$$
$$= P(\text{circuit is functional})$$
$$Y_T = \text{Proportion of the circuits that pass the test}$$
$$= P(\text{circuit passes the test})$$
$$Y_C = \text{Proportion of the pass circuits that are functional}$$
$$= P(\text{circuit passes the test/is functional})$$
$$Y_L = \text{Proportion of the fail circuits that are functional}$$
$$= 1 - Y_C$$
$$D = \text{Proportion of the faulty circuits that pass the test}$$
$$= 1 - P(\text{circuit is functional/passes the test})$$

where a functional (or good) circuit is the one for which all its performances are inside their specifications and a faulty circuit is the one for which at least one of its specifications is violated.

The definition of parametric fault coverage will be detailed later. For catastrophic faults, as mentioned earlier, device functionality is not considered and fault coverage is just defined as the ratio of detected faults with respect to the total number of injected faults.

3.2 Test metrics theoretical computation

Assume that we have n performances and m test criteria. Let $A = (A_1, \cdots, A_n)$ be the set of the specifications of the performances and $B = (B_1, \cdots, B_m)$ the test limits (intervals of the accepted values of the test criteria). The test metrics are then calculated theoretically as follows:

$$Y = \int_A f_S(s) \, ds \tag{1}$$

$$Y_T = \int_B f_T(t) \, dt \tag{2}$$

$$Y_C = \frac{\int_A \int_B f_{ST}(s,t) \, ds \, dt}{Y} \tag{3}$$

$$D = 1 - \frac{\int_A \int_B f_{ST}(s,t) \, ds \, dt}{Y_T} \tag{4}$$

where, $f_S(s) = f_S(s_1, s_2, \cdots, s_n)$ is the joint probability density of the performances, $f_T(t) = f_T(t_1, t_2, \cdots, t_m)$ is the joint probability density of the test criteria and $f_{ST}(s,t) = f_{ST}(s_1, s_2, \cdots, s_n, t_1, t_2, \cdots, t_m)$ is the joint probability density of the performances and the test criteria.

For the case of catastrophic faults, fault coverage is the major metric and this can be readily computed. For the case of single parametric faults, for which a fault list is available, test metrics can be computed following, for example, the technique described in [1]. However, the analysis of faulty behaviour resulting from process deviations (multiple small parametric deviations) has not been properly studied in the past, since it is impossible to produce an actual fault list. We will next describe the statistical analysis performed in the tool set for evaluating test metrics and setting test limits under process deviations. The use of these tools will be illustrated later for the case of a test vehicle.

3.3   Test metrics computation under process deviations

Given a vector $X = (X_1, X_2, ..., X_p)^T$ composed of random variables, where X_j for $j = 1, 2, ..., p$, is a one-dimensional random variable, the covariance of X_i and X_j is a measure of dependency between these random variables and is defined by:

$$\nu_{X_i X_j} = Cov(X_i, X_j) = E(X_i X_j) - E(X_i)E(X_j) \tag{5}$$

where $E(.)$ denotes the expected value. If X_i and X_j are independent of each other, the covariance $\nu_{X_i X_j}$ is necessarily equal to zero. The converse is not true. The covariance of a random variable X_i with itself is the variance:

$$\nu_{X_i X_i} = Cov(X_i, X_i) = \nu_{X_i} \tag{6}$$

The correlation between two variables X_i and X_j is defined from the covariance as follows:

$$\rho_{X_i X_j} = \frac{\nu_{X_i X_j}}{\sigma_{X_i} \sigma_{X_j}} \tag{7}$$

where the standard deviation is defined by $\sigma_{X_i} = \sqrt{\nu_{X_i}}$

The advantage of the correlation is that it is independent of the scale, i.e., changing the scale of measurement of the variables does not change the value of the correlation. Therefore, the correlation is more useful as a measure of association between two random variables than the covariance. The correlation is in absolute value always less than 1, close to zero if the random variables X_i and X_j are independent of each other.

An empirical estimation of these quantities require a number of observations. Suppose that $\{x_i\}_{i=1}^n$ is a set of n observations of a variable vector X in $\Re^p$. Each observation x_i has p dimensions: $x_i = (x_{i_1}, x_{i_2}, ..., x_{i_p})$, and it corresponds to an observed value of a variable vector $X \in \mathbb{R}^p$. The covariance of two random variables is then estimated as:

$$V_{X_i X_j} = \frac{1}{n-1} \left(\sum_{k=1}^n x_{i_k} x_{j_k} - n \overline{x}_i \cdot \overline{x}_j \right) \tag{8}$$

and the variance of a random variable is estimated as:

$$V_{X_i} = \frac{1}{n-1} \left(\sum_{k=1}^{n} x_{i_k}^2 - n\bar{x}_i^2 \right) \tag{9}$$

The correlation of two random variables is then given by:

$$r_{X_i X_j} = \frac{V_{X_i X_j}}{s_{X_i} s_{X_j}} \tag{10}$$

with $s_{X_i} = \sqrt{V_{X_i}}$.

The theoretical covariances among all the random variables can be put into matrix form, i.e. the covariance matrix:

$$\Sigma = \begin{pmatrix} \nu_{X_1} & \cdots & \nu_{X_1 X_p} \\ \vdots & \ddots & \vdots \\ \nu_{X_1 X_p} & \cdots & \nu_{X_p} \end{pmatrix} \tag{11}$$

The estimated (empirical) version of the covariance matrix is then given by:

$$S = \begin{pmatrix} V_{X_1} & \cdots & V_{X_1 X_p} \\ \vdots & \ddots & \vdots \\ V_{X_1 X_p} & \cdots & V_{X_p} \end{pmatrix} \tag{12}$$

Let X be a p-dimension random variable of expected value $\mu = (\mu_{i_1}, \mu_{i_2}, ..., \mu_{i_p})^T$ and covariance matrix Σ. If X has a multinormal distribution, then X has a probability density function (PDF) $f(x)$ defined by:

$$f(x) = \frac{1}{\sqrt{det(2\pi\Sigma)}} \cdot exp\left[-\frac{(x-\mu)^T \Sigma^{-1}(x-\mu)}{2} \right] \tag{13}$$

The probability of any subset $A \in \mathbb{R}^p$ is given by the following multiple integration formula:

$$P(A) = \frac{1}{\sqrt{det(2\pi\Sigma)}} \int_{A_1} \cdots \int_{A_p} exp\left[-\frac{(x-\mu)^T \Sigma^{-1}(x-\mu)}{2} \right] dx_1 \cdots dx_p \tag{14}$$

Thus, using the multinormal hypothesis, it is possible to estimate the actual probability density functions which must be integrated considering the actual boundaries of the random variables in order to compute the test metrics. The multinormal assumption can be validated by computing the correlation coefficients for different standard deviations in Monte Carlo simulation. We can use a Monte Carlo circuit simulation, under process deviations, to calculate the statistical parameters of the multinormal law (mean and covariance matrix) which are required.

When the number of the specifications and test criteria is important (for example, larger than 3), the number of sub-integrals in Equation (14) for the exact computation of test metrics is too large. The computation is then impossible. To overcome this problem, as the joint PDF of performances and test criteria is assumed multinormal, a simple program implemented in Matlab (or R) is used to generate about one million instances from the multinormal distribution using a Monte Carlo technique. Next, the test metrics can be directly estimated using the following estimators:

$$\hat{Y}^D = \frac{\text{Number of functional circuits}}{N} \qquad (15)$$

$$\hat{Y}_T^D = \frac{\text{Number of pass circuits}}{N} \qquad (16)$$

$$\hat{Y}_L^D = \frac{\text{Number of fail functional circuits}}{\text{Number of functional circuits}} \qquad (17)$$

$$\hat{D}^D = \frac{\text{Number of pass faulty circuits}}{\text{Number of pass circuits}} \qquad (18)$$

where N is the number of generated circuits.

We use the index D to indicate that the metrics are estimated at the design stage using process deviations.

3.4   Test metrics computation under the presence of faults

For single parametric faults in physical parameters, test metrics can be calculated using the methodology presented by [1] where partial detectability of the parametric faults is considered. A fault is defined as the minimum value of a physical parameter i that causes any performance specification to fail. This will help to calculate the probability of the occurrence p_i^{spec} of this fault, which represents the probability that the value of this physical parameter is greater than v_i^{spec} (Figure 6).

We calculate also the probability p_i^{test} to detect a fault in this parameter, which represents the probability of this parameter to be greater than v_i^{test}, where v_i^{test} is the minimum value of the process parameter that causes the test criteria to fail (Figure 6).

Considering theses definitions, we can write the analogue test metrics, based on the probabilities p_i^{spec} and p_i^{test} as follows [1]:

$$Y^F = \prod_{i=1}^{n}(1 - p_i^{spec}) \qquad (19)$$

$$Y_T^F = \prod_{j=1}^{m}(1 - p_j^{test}) \qquad (20)$$

$$Y_C^F = \frac{G_P^F}{Y^F} = 1 - Y_L^F \qquad (21)$$

$$D^F = 1 - \frac{G_P^F}{Y_T^F} \qquad (22)$$

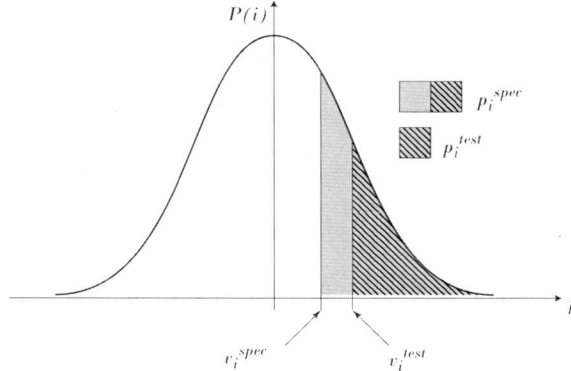

Fig. 6. Distribution of a parameter i with associated probabilities.

where, $G_P^F = \prod_{i=1}^{n}(1 - \max(p_i^{spec}, p_i^{test}))$, which represents the probability that a circuit is functional and passes the test.

The fault coverage F for the parametric faults is calculated using the following equation [1]:

$$F^F = \frac{\sum_{i=1}^{n} \ln(1 - \min(p_i^{spec}, p_i^{test}))}{\sum_{i=1}^{n} \ln(1 - p_i^{spec})} \tag{23}$$

We use the index F to indicate that the metrics are calculated under the presence of faults.

4   Test vehicle

4.1   Description of the circuit

The test vehicle is a fully-differential operational amplifier. This amplifier has been designed in a $0.18\mu m$ CMOS technology from ST Microelectronics. The amplifier is formed of four main blocks: a bias circuit, a start-up circuit, a common-mode control circuit and a differential amplifier circuit. Figure 7 illustrates this circuit.

First, we use the statistical analysis to calculate the test limits for the test criteria under process deviations by calculating the analogue test metrics, in particular the defect level and the yield loss at the design stage. Then, taking into account these test limits, we calculate the fault coverage in order to test the capability of the test technique for fault detection considering both catastrophic and parametric faults. Finally, we will find the minimal set of specifications and test criteria which give the best fault coverage.

In order to find the different fitted Gaussian distributions of each circuit performance and test criteria we performed a Monte Carlo circuit simulation (1000 iterations). The comprehensive set of performances and test criteria considered is given in Tables 1 and 2, where a_1 and a_2 represent the specifications, that

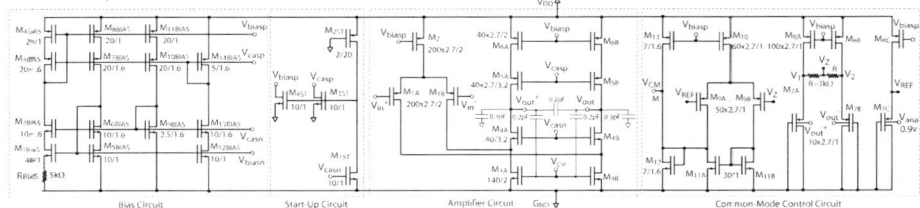

Fig. 7. Folded cascode fully differential amplifier. The dimensions of each transistor $W_r L$ are expressed in multiples of the unity size transistor ($W_{UNIT} = 0.28\mu m$ and $L_{UNIT} = 0.18\mu m$).

Performance	Test bench	μ	σ	Specification	
				a_1	a_2
A_D		76.60dB	0.493dB	74.49dB	78.71dB
GBW_D	1	330MHz	18.14MHz	252.36MHz	407.64MHz
Phase Margin ϕ_D		63.33°	0.45°	61.40°	65.26°
CMRR	2	−42.76dB	1.02dB	−47.13dB	−38.39dB
PSRR (G_{ND})	3	−29.99dB	3.65dB	−45.61dB	−14.37dB
PSRR (V_{DD})	4	−28.21dB	3.75dB	−44.26dB	−12.16dB
THD	5	66.19dB	2.38dB	56.00dB	76.38dB
Current (I_{DD})		2.48mA	0.21mA	1.58mA	3.38mA
Intermodulation	6	67.57dB	1.09dB	62.90dB	72.24dB
SR + ($C_L = 1pF$)	7	73.14V/μs	5.55V/μs	49.38V/μs	96.88V/μs
SR − ($C_L = 1pF$)		73.14V/μs	5.55V/μs	49.38V/μs	96.88V/μs
In Ref. Noise (BW = 20kHz)	8	39.22μV	0.5μV	37.08μV	41.36μV

Table 1. The performances of the amplifier with their Gaussian parameters and the specifications set at 4.3σ.

is, the bounds of each performance. The specifications of the amplifier are not known a priori, since the actual system application of the device is not considered in this work. Thus, we have set ourselves the specification bounds in order to have a high yield at the design stage of $Y^D = 99.99\%$ when all performances are considered. This requires a tolerance interval of $\mu \pm 4.3\sigma$ for each performance. The test limits b_1 and b_2 will be calculated by the technique which will be presented below.

Different test benches have been used to calculate the different performances. For example, Figures 8(a) and (b) show the test benches n° 1 and n° 7, respectively. Each test bench allows the calculation of one or more performances and test criteria. Table 1 shows the specifications with the actual test bench used for the calculation. Eight different test benches are required for the performances.

For the actual test of the fully differential amplifier, the measurement of the $SNDR$ of the amplifier is considered using a sine-wave fitting technique

| Test criteria | Test bench | μ | σ | Test limits | |
				b_1	b_2
SNDR	9	68.85 dB	2.19 dB	To determine	To determine
Offset		0 μV	7.69 μV	To determine	To determine

Table 2. The test criteria of the amplifier with their Gaussian parameters.

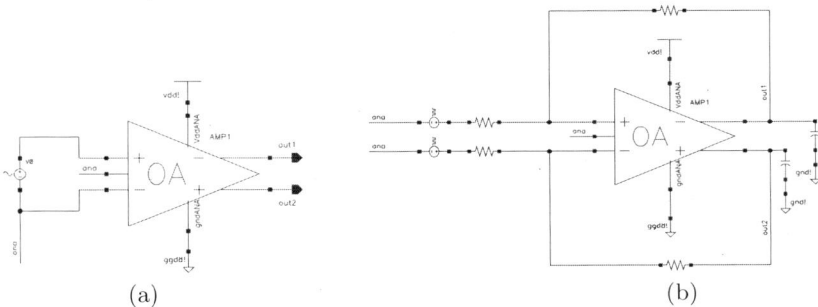

(a)   (b)

Fig. 8. Example of the test bench n°1(a) and n°7(b) of the amplifier.

described in [11, 17]. The DC Offset of the amplifier is also considered as a possible test criteria that can be measured using the sine-wave fitting technique. To simulate these test measurements, an additional test bench is required.

The total Monte Carlo circuit simulation time of 1000 instances, considering the 8 test benches for the performances and the 9th test bench for the test criteria, is 3 hours. The overall process is fully automated using the CAT platform.

As we can see in Figures 9(a) and 9(b), the distributions of the *Gain* and the *THD* are very close to the multinormal one. The same results are obtained for the other performances and test criteria.

4.2   Precision on test metrics estimation

We need to find the test limits to separate the faulty circuits from the fault-free ones, as a function of the required test metrics. A trade-off between Defect level and Yield loss must be considered under process deviations, and this will set the actual test limit.

Using the equations (1) to (4) presented in Section 3.2 these test metrics at the design stage can be theoretically calculated. However, in our case, we have 12 specifications, and it is not feasible to perform the integration with such a large number of integrals. Thus, we will use the Monte Carlo method of estimation proposed in Section 3.3.

In order to see the accuracy of this method, we will first illustrate a simpler case when only two performances, *Phase Margin* and *THD*, are considered together with the test criterion *SNDR*. In this case, the metrics presented by

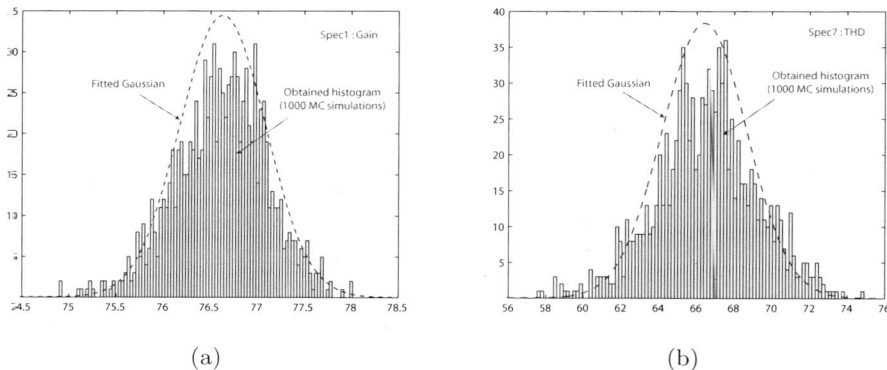

(a)   (b)

Fig. 9. The distribution of the *Gain* A_D (a) and the *THD* (b) of the amplifier.

(1) to (4) are calculated as follows:

$$Y^D = \int_{A_3} \int_{A_7} f(s_1, s_2) \, ds_1 \, ds_2 \tag{24}$$

$$Y_T^D = \int_{B_1} f(t_1) \, dt_1 \tag{25}$$

$$Y_L^D = 1 - \frac{G_P}{Y} \tag{26}$$

$$D^D = 1 - \frac{G_P}{Y_T} \tag{27}$$

where

$$G_P^D = \int_{A_3} \int_{A_7} \int_{B_1} f(s_1, s_2, t_1) \, ds_1 \, ds_2 \, dt_1$$

is the probability that the circuit is functional and passes the test, s_1 is the *Phase Margin* value, s_2 is the *THD* value, t_1 is the *SNDR* value, A_i is the ith specification, B_1 is the test limit of the *SNDR* and $f(.)$ is calculated by (13). The covariance matrix Σ is estimated by S given by (12).

For a given test limit, these metrics can be calculated exactly in this case, because the number of integrals is small.

Figure 10(a,b) shows that the estimated (Monte Carlo algorithm) and the theoretical values of the metrics are very close for the case of Defect level and Yield loss.

For comparison, Figure 11(a) shows the distributions of the *Phase Margin* and the *SNDR* for the case of 1000 instances obtained via Monte Carlo circuit simulation and for the case of 1000 instances generated from the multinormal distribution with a Monte Carlo technique. From this Figure, it is clear that both distributions are the same. The same results have been obtained for the other performances and test criteria. In addition, Figure 11(b) shows the generation of

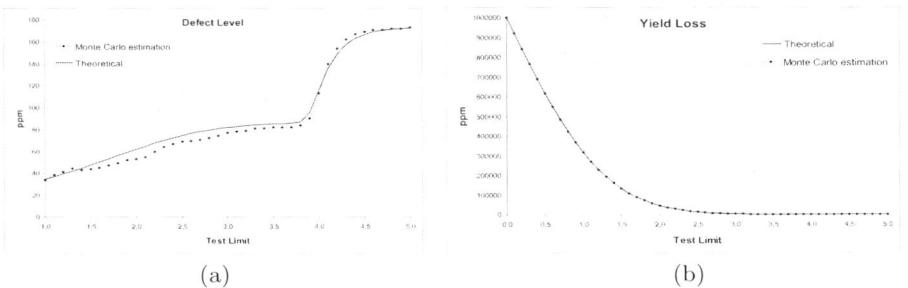

Fig. 10. Comparison of the estimated and the theoretical test metrics for the case of two specifications and one test criterion: (a) Defect level and (b) Yield loss.

1000 and 1 million circuit instances generated from the multinormal distribution. It is clear that with 1 million instances we will reach the required ppm precision.

4.3   Test limit setting under process deviations

The setting of test limits is always a trade-off between test cost and test quality. Here, just as an example for our case-study, we will consider setting the test limits that simultaneously try to minimise both Defect Level and Yield Loss. Figure 12(a) shows Defect Level and Yield Loss together as a function of the test limits of $SNDR$ and I_{DD}, where the intersection between them is of interest to us.

We have introduced here as test criterion the current consumption I_{DD}, since we will se it is important fault detection. The test limits for I_{DD} is given by $(\mu_{IDD} \pm k_{IDD}\ \sigma_{IDD})$.

This intersection (points where the Defect Level is equal to the Yield Loss) is redrawn in Figure 12(b). We have chosen as trade-off of the test limits of the $SNDR$ and the I_{DD} the minimum of these points which is equal to $55ppm$. This results in a test limit of 4.0σ for the $SNDR$ and 4.1σ for the I_{DD}.

4.4   Fault coverage under the presence of catastrophic faults

We have considered catastrophic faults that result in shorts and opens in all the transistors, resistances and capacitances of the amplifier. This results in 160 catastrophic faults. Figure 13 shows the fault coverage given by the measurement of each performance and several possible test criteria with the test limits fixed as explained before. The performances allow to detect 98.12% of faults where the undetected faults occur in the Bias block. On the other hand, the test criterion $SNDR$ allows the detection of 89.38% of the faults. The undetected faults occur also in the Bias block. Maximum fault coverage can be achieved if power consumption (I_{DD}) is considered in addition to the $SNDR$ measurement.

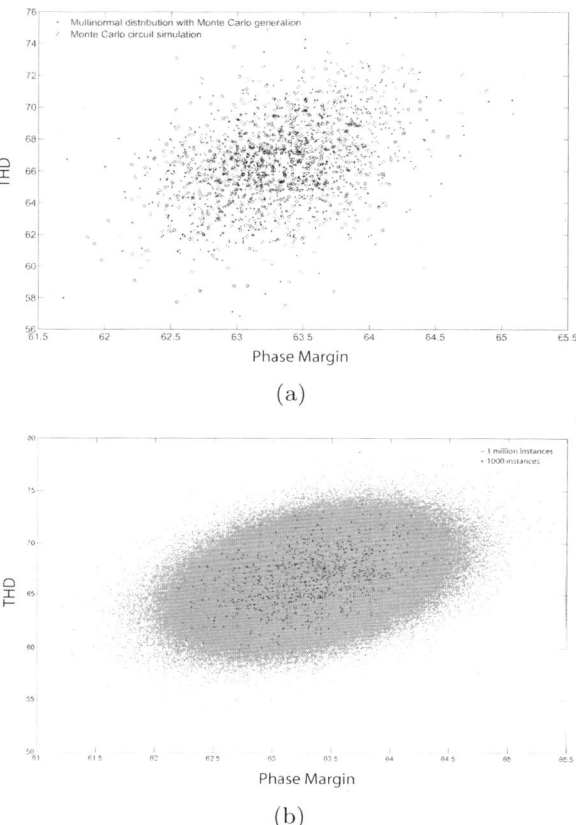

Fig. 11. (a) Distribution of 1000 circuits generated by Monte Carlo circuit simulation and from the multinormal law and (b) generation of 1000 and 1 million circuits from the multinormal distribution.

4.5 Test metrics under the presence of parametric faults

We consider parametric faults as a result of a physical parameter deviating beyond an acceptable value. The physical parameters considered include the L and W of the PMOS and NMOS transistors and the resistance and capacitance values. We note that L and W of the transistors are not process parameters, and thus their deviations are not included under process deviations. On the other hand, resistance and capacitance values deviate under process deviations.

In order to calculate the probabilities p_i^{spec} and p_i^{test} for each potentially faulty physical parameter, we have to obtain by simulation its limit deviation values. The distribution of each physical parameter is considered as Gaussian with mean equal to the typical value of this parameter. A standard deviation of $10nm$ is taken for L and W of the transistors and 5% for the resistances and the

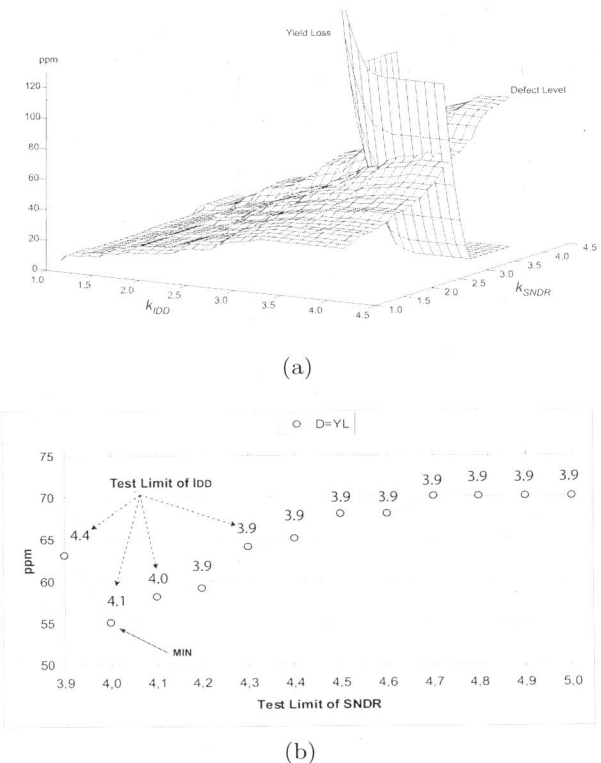

(a)

(b)

Fig. 12. (a) The Defect Level and the Yield Loss as a function of the $SNDR$ and I_{DD} test limits, (b) The test limits of the $SNDR$ and the I_{DD} where the Defect Level and the Yield Loss are equal.

capacitances. Thus, for each varying physical parameter, we have injected in the amplifier deviations from -20% to 100%. For each value of a physical parameter, all test benches must be simulated. A dichotomic search is applied to find the limit deviation for each physical parameter.

This process has resulted in the consideration of 180 potentially faulty physical parameters, where only 13 of them result in a specification violation. These faults are listed in Table 3 together with their probabilities. The other faults have a negligible probability of occurrence ($p_i^{spec} = 0$). We note here that deviations in transistors that are matched cannot be considered individually. They are considered by injecting the same deviation in all matched transistors. We have seen that faults in all matched transistors have a negligible probability. In order to consider the faulty behaviour resulting from mismatch, it is necessary to use the mismatch option in the Monte Carlo circuit simulation. In this work, we have used the process option for this, but a similar analysis could be performed for mismatch deviations.

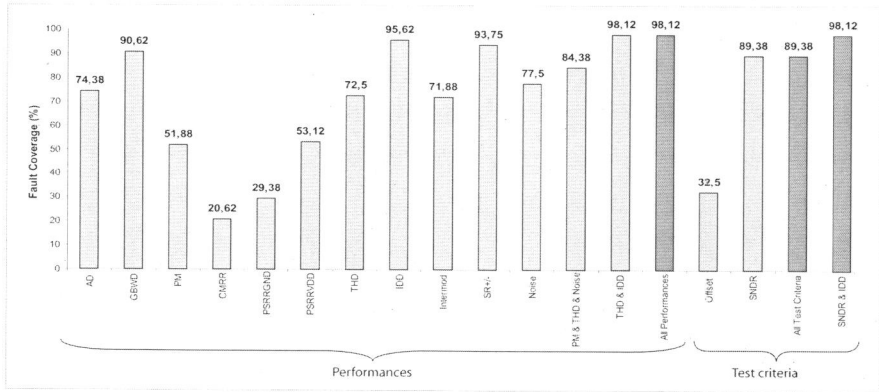

Fig. 13. Catastrophic fault coverage of the different specifications and several possible test criteria.

Using these parameters and Equations (19) to (23), the values of the different test metrics are given in Table 4. We consider two cases: in the first case, all 13 parametric faults are considered. Only 2 of these faults are not detected by the test criteria (SNDR, Offset and Current Consumption), this is why the fault coverage does not reach 100%. The fault coverage F^F is high because the undetected faults have lower probability. Notice that the yield Y^F is lower than 60%, much lower than the design yield Y^D that has a value above 99%. This is because deviations of physical parameters such as L and W are not considered under process deviations. The second case of Table 4 does not consider deviations in the physical parameters L and W, but only in the resistance values (faults due to capacitance deviations have negligible probabilities). In this case, fault coverage F^F reaches 100% with a yield Y^F above 99%. Since these deviations are also considered as process deviations, we obtain similar results between Y^F and Y^D.

5   Conclusions and Future Work

This paper has introduced a CAT platform for analogue and mixed-signal test evaluation. In particular, fault simulation and test optimization are considered. Catastrophic faults can be tackled in a similar way as for digital circuits, regardless of the circuit specifications, although process deviations may need to be considered. Single parametric faults have been considered (based on the approach of [1]). A statistical approach for the estimation of test metrics and the setting of test limits under process deviations is the major contribution of this paper. The data for this approach is obtained from a Monte Carlo simulation. Test optimization for a fully differential amplifier has been illustrated as a case-study.

No	Component (Parameter)	Fault	p^{spec}	p^{test}
1	MP1 (l)	+12.21%	0.013996	0.013996
2	MP3 (l)	+7.32%	0.093690	0.093690
3	MP5 (l)	+22.95%	0.000009	0.000009
4	MP1 (l)	−3.18%	0.283305	0.283305
5	MP3 (l)	−11.70%	0.017604	0
6	MP18 (l)	−15.76%	0.002270	0
7	MN2 (l)	−16.23%	0.001736	0.001736
8	MN4 (l)	−6.46%	0.122278	0.122278
9	R1 (r)	+15.14%	0.001227	0.001227
10	R4 (r)	+17.09%	0.000308	0.000308
11	R7 (r)	+16.11%	0.000628	0.000628
12	R4 (r)	−12.79%	0.005249	0.005249
13	R7 (r)	−16.15%	0.000611	0.000611

Table 3. Parameters used to calculate the metrics for the case of parametric faults.

Metric	All parametric faults	Resistor faults only
F^F	96.69%	100%
Y^F	54.56%	99.22%
Y_T^F	55.56%	99.22%
Y_C^F	100%	100%
D^F	1.98%	0%

Table 4. Test metrics values for single parametric faults.

Acknowledgements

The authors would like to acknowledge the fruitful discussions on analogue parametric testing with Jean Louis Carbonéro, ST Microelectronics, France.

References

1. S. Sunter and N. Nagi. Test metrics for analog parametric faults. *In Proc. VTS*, 1999, pp. 226-234.
2. K.P. Parker. *The Boundary Scan Handbook*. Third Edition. Kluwer Academic Publishers, 2003.
3. S. Mir, A. Rueda, D. Vázquez and J.-L. Huertas. Switch-level fault coverage analysis of switched-capacitor systems. *In IEEE Design Automation and Test Conference in Europe DATE-98*, pages 810-814, Paris, France, February 1998.
4. S.J. Spinks, C.D. Chalk, I.M. Bell ans M. Zwolinski. Generation and Verification of Tests for Analog Circuits Subject to Process Parameter Deviations, *Journal of Electronic Testing: Theory and Applications*, v.20 n.1, February 2004, pp. 11-23.

5. N. Ben-Hamida and B. Kaminska. Analog circuit testing based on sensitivity computation. *IEEE International Test Conferece*, Baltimore, October 1993, pp. 652-661.

6. K. Saab, N. Ben Hamida and B. Kaminska. Parametric Fault Simulation and Test Vector Generation. *In IEEE Design and Test Conference in Europe*, 2000, pp. 650-656.

7. A. Zjajo, J. Pineda de Gyvez, G Gronthoud. A quasi-static approach detection and simulation of parametric faults in analog and mixed-signal circuits. *In 11th IEEE International Mixed-Signals Testing Workshop*, Cannes, France, June, 2005, pp. 155-164.

8. S. Mir, M. Lubaszewski and B. Courtois. Fault-based ATPG for linear analogue circuits with minimal size multifrequency test sets. *Journal of Electronic Testing: Theory and Applications*, August/October 1996, pp. 43-57.

9. A. Bounceur, S. Mir and E. Simeu. Optimization of digitally coded test vectors for mixed-signal components. *In 19th Conference on Design of Circuits and Integrated Systems*, Bordeaux, France, November, 2004, pp. 895-900.

10. B. Dufort and G.W. Roberts. On-chip analog signal generation for mixed-signal Built-In Self-Test, *IEEE Journal of Solid-State Circuits*, Vol. 34, No. 3, pp. 318-330, March 1999.

11. L. Rolíndez, S. Mir, A. Bounceur and J.-L. Carbonéro. A SNDR BIST for $\Sigma\Delta$ Analogue-to-Digital Converters. In *24th VLSI Test Symposium*, Berkeley, California, USA, April-May 2006, pp. 314-319.

12. A. Zjajo and J. Pineda de Gyvez. Evaluation of Signature-Based Testing of RF/Analog Circuits. *In European Test Symposium*, Tallin, Estonia, May 2005, pp. 62-67.

13. J. Pineda de Gyvez, G. Gronthoud and R. Amine. VDD Ramp Testing for RF Circuits. *In Proc. IEEE Int. Test Conf. (ITC)*, 2003, pp. 651-658.

14. S. Mir, A. Rueda, T. Olbrich, E. Peralías and J.-L. Huertas. SWITTEST: automatic switch-level fault simulation and test evaluation of switched-capacitor systems. *Proceedings of the 34th annual conference on Design automation conference*, June 09-13, 1997, Anaheim, California, United States, pp. 281-286.

15. Y. Eben Aimine, A. Richardson, C. Descleves and K. Sommacal. GDS FaultSim, a Mixed-Signal IC Computer-Aided-Test (CAT) Tool, *In IEEE Design and Test Conference in Europe*, Munich, Germany, March, 1999, pp. 232-238.

16. C. Roman, S. Mir and B. Charlot. Building an analogue fault simulation tool and its application to MEMS. *Microelectronics Journal*, 34(10), 2003, pp. 897-906.

17. IEEE Std. 1057-1994. *IEEE Standard for Digitizing Waveform Recorders*, IEEE Press, Dec. 1994.

Broadside Transition Test Generation for Partial Scan Circuits through Stuck-at Test Generation*

Tsuyoshi Iwagaki[1], Satoshi Ohtake[2], and Hideo Fujiwara[2]

[1] School of Information Science, Japan Advanced Institute of Science and Technology (JAIST), 1-1 Asahidai, Nomi, Ishikawa 923-1292, Japan
iwagaki@jaist.ac.jp
[2] Graduate School of Information Science, Nara Institute of Science and Technology (NAIST), Kansai Science City 630-0192, Japan
{ohtake, fujiwara}@is.naist.jp

Abstract. This paper presents a method of broadside transition test generation for partial scan circuits. The proposed method first transforms the kernel circuit of a given partial scan circuit into some combinational circuits. Then, by performing stuck-at test generation on the transformed circuits, broadside transition tests for the original circuit are obtained. This method allows us to use existing stuck-at test generation tools in order to generate broadside transition tests. It is shown that the proposed scheme is effective in area overhead and test generation time by experiments. In this paper, some variations of broadside transition testing of partial scan circuits are also discussed in terms of different test application strategies and fault sizes.

1 Introduction

Scan design is widely accepted by industry as an effective design for testability (DFT) method for delay faults as well as stuck-at faults. There is an essential difference between scan testing for stuck-at faults and that for delay faults. Unlike stuck-at testing, an additional consideration must be taken into account for delay testing using scan methodology. That is, to detect a delay fault, two consecutive vectors (two-pattern test) are needed to be applied to the faulty site in a scan environment. This can be done by using enhanced scan technique [7] or standard scan technique such as skewed-load technique [13] and broadside technique [14].

In [7], all the flip-flops (FFs) in a given circuit are replaced with enhanced scan FFs (ESFFs). Since each ESFF can store any two consecutive vectors, any two-pattern tests can be applied to the circuit. Although this method can drastically reduce the test generation complexity of a given circuit, its use is limited

---

* This work was supported in part by Research Promoting Expenses for Associates of JAIST and in part by Japan Society for the Promotion of Science under Grants-in-Aid for Young Scientists (B) (No. 17700062) and for Scientific Research B(2) (No. 15300018).

Please use the following format when citing this chapter:

Iwagaki, T., Ohtake, S. and Fujiwara, H., 2007, in IFIP International Federation for Information Processing, Volume 249, VLSI-SoC: Research Trends in VLSI and Systems on Chip, eds. De Micheli, G., Mir, S., Reis, R., (Boston: Springer), pp. 301–316

because of the considerable area and delay penalties incurred by ESFFs. For delay faults as well as stuck-at faults, full scan design is widely used as a DFT method. In delay testing, as mentioned before, two-pattern tests are required to detect delay faults, and they have to be applied by using scan FFs (SFFs), which can store any one vector. The skewed-load technique and broadside technique have been proposed as techniques to apply two-pattern tests to full scan circuits. In both of the techniques, the first vectors of two-pattern tests can freely be set to the SFFs through the scan chain. The second vectors are derived by shift operation in the skewed-load technique. In contrast, the broadside technique creates the second vectors by normal operation. In terms of feasibility, the broadside technique is more desirable than the skewed-load technique. This is because, in skewed-load testing, the scan signal is operated at the rated speed and it forces the scan chain to be designed judiciously. So far, there have been proposed several broadside test generation methods for full scan circuits [6, 17, 18 20, 15, 2, 21].

Partial scan methodology is a viable solution to reduce the test generation effort of sequential circuits with reasonable area and delay overheads. For stuck-at faults, many researchers have considered partial scan design from various aspects. However, there are few works for delay faults in partial scan circuits. A transition test generation method, which is based on skewed-load testing, for partial scan circuits has been proposed in [5]. As mentioned previously, since skewed-load testing has some undesirable properties, a test generation method based on broadside testing is also needed for partial scan circuits. However, there have so far been no systematic approaches to generate broadside transition tests for partial scan circuits. In this paper, we tackle this problem. It is notable that broadside transition testing of partial scan circuits has a possibility of alleviating over-testing, which is one of the main concerns during testing [12, 1], in addition to reducing the penalties of area and delay.

In this paper, we propose a method to generate broadside transition tests for partial scan circuits. This method targets partial scan circuits whose kernel circuits are acyclic. To generate broadside transition tests for a partial scan circuit, we transform its kernel circuit into some combinational circuits. This transformed circuits are constructed by using a time-expansion model [8] of the kernel circuit. All the broadside transition tests are generated by performing constrained stuck-at test generation on the transformed circuits. Our method is effective in terms of ease of use because commercial stuck-at test generation tools, which are usually capable of handling combinational stuck-at test generation efficiently, can be used to generate broadside transition tests. By experiments, we show that our method can reduce area overhead and can generate broadside transition tests for partial scan circuits efficiently. In this paper, we also discuss some variations of broadside transition testing for partial scan circuits in terms of different test application strategies and fault sizes.

The rest of this paper is organized as follows. In Sect. 2, our target circuits and faults are explained, and previous work related to this paper is described. Section 3 presents a new test generation model to generate broadside transition

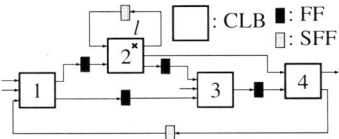

Fig. 1. Partial scan circuit: S.

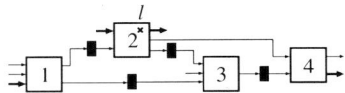

Fig. 2. Kernel circuit of Fig. 1: S_K

tests for partial scan circuits. Then, we give a test generation procedure using the new model, and the correctness of the procedure is proven. Experimental results are also presented in Sect. 3. We discuss some variations of broadside transition testing for partial scan circuits in Sect. 4. Section 5 concludes the paper and describes our future work.

2   Preliminaries

2.1   Target Circuits and Faults

In this paper, we handle partial scan circuits whose kernel circuits are acyclic. A sequential circuit can be represented as combinational logic blocks (CLBs) connected with each other directly or through FFs. A CLB is a region of connected combinational logic gates. An example of a partial scan circuit S and its kernel circuit S_K are shown in Figs. 1 and 2, respectively. The input (resp. output) of an SFF in Fig. 1 is treated as a primary output (PO) (resp. primary input (PI)) in Fig. 2, which is represented as a bold arrow and called a pseudo PO (PPO) (resp. pseudo PI (PPI)).

This paper tackles a broadside test generation problem for transition faults in a partial scan circuit. There are two transition faults associated with each line in a circuit: a slow-to-rise fault and a slow-to-fall fault. It is assumed that, under the transition fault model, the extra delay caused by a transition fault is large enough to prevent the transition through the faulty site from reaching any FF or any PO within a specified period. Note that, in a sequential circuit, different faulty behaviors can happen depending on the size of a transition fault [5, 21]. The size of a transition fault is defined as the amount of extra delay caused by the defect, and it is quantized by the number of clock cycles [5]. In this paper, although we concentrate on a transition fault whose size is one, the case where the size of a transition fault is more than one will be discussed in Sect. 4. This paper

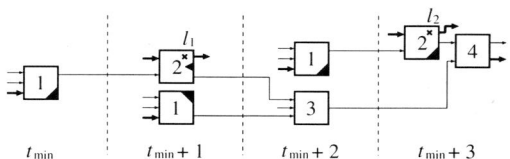

Fig. 3. Time-expansion model of Fig. 2: $C^T(S_K)$

assumes that transition faults in a partial scan circuit are tested in the slow-fast-slow testing manner [10] where a slow clock is used in the both of the fault initialization and fault effect propagation phases except in the fault activation phase. Under this assumption, we can consider a sequential circuit to be delay fault-free in both of the fault initialization and fault effect propagation phases. There are two possible strategies to apply a broadside transition test to a partial scan circuit. One strategy is called scan-per-vector [11] where scan operation is always allowed except in the fault activation phase. The other strategy is called scan-per-test [11] where scan operation is allowed only at the beginning of the fault initialization phase and at the end of the fault effect propagation phase. The former strategy is used in this paper. The discussion about the latter strategy will appear in Sect. 4.

2.2   Related Work

In this paper, we borrow an idea of a double time-expansion model, which is used to generate transition tests for an acyclic sequential circuit, from [9]. In [9], given an acyclic sequential circuit, a double time-expansion model of the circuit is constructed from a time-expansion model (TEM) [8] of the circuit. In the following paragraphs, we briefly explain those two models.

A TEM of an acyclic sequential circuit is a combinational circuit where the behavior of the original circuit within a specific time span is simulated. Figure 3 is a TEM $C^T(S_K)$ of the kernel circuit S_K shown in Fig. 2. TEM $C^T(S_K)$ is a combinational circuit derived by connecting CLBs according to their sequential depths. A sequential depth between two CLBs is defined as the number of FFs on a path between the CLBs. If a CLB has paths to another CLB in S_K whose sequential depths are different, the CLB is duplicated in $C^T(S_K)$. In Fig. 2, for example, since CLB 2 has two paths to CLB 4 whose sequential depths are different, CLB 2 is duplicated in $C^T(S_K)$. A shaded part of a CLB in Fig. 3 represents a portion of the lines and gates being removed. There is no path from the portion to any input of CLBs or any PO and PPO of $C^T(S_K)$. The character placed at the bottom of each frame in Fig. 3 is the label of CLBs in the frame, where $t_{\min}$ denotes an arbitrary integer. The label of a CLB v is denoted as $t(v)$ which corresponds to a specific time.

For an acyclic sequential circuit, its double time-expansion model is defined as follows [9].

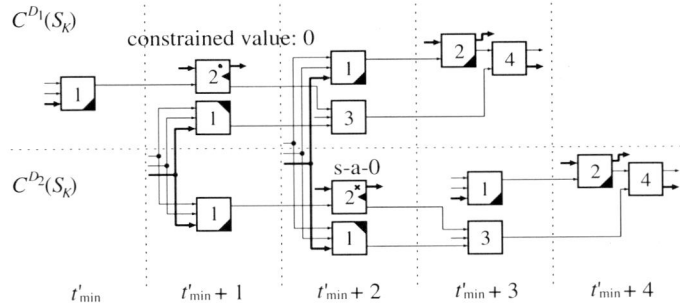

Fig. 4. Double time-expansion model of Fig. 2: $C^D(S_K)$

Definition 1. *Let S be an acyclic sequential circuit, and $C^T(S)$ be a TEM of S. Then, a combinational circuit obtained by the following procedure is said to be a double time-expansion model (DTEM) $C^D(S)$ of S.*

Step 1: *Make two copies of $C^T(S)$: $C^{D_1}(S)$, $C^{D_2}(S)$.*
Step 2: *Connect each pair of PIs u in $C^{D_1}(S)$ and v in $C^{D_2}(S)$ such that $t(u) - t(v) = 1$ and $l(u) = l(v)$, and feed a new primary input w into them, where $l(u) = l(v)$ means that u and v are identical in S.* □

According to the above definition, a DTEM $C^D(S_K)$ of S_K (Fig. 2) is constructed as Fig. 4. Note that, although two copies of CLB 1 in $t'_{\min} + 1$ (also in $t'_{\min} + 2$) can be merged into one CLB, $C^D(S_K)$ is expressed as Fig. 4 to differentiate $C^{D_1}(S_K)$ and $C^{D_2}(S_K)$ from each other. If one wants to test the slow-to-rise fault on line l in S_K, test generation for one of the corresponding stuck-at 0 fault is performed on $C^D(S_K)$ under the constrained value of 0 that must be satisfied during test generation. In this way, transition tests for an acyclic sequential circuit can be generated by using a DTEM.

In [9], an acyclic sequential circuit is assumed to be obtained as a kernel circuit of a given circuit by using enhanced scan technique. Thereby, two consecutive vectors V_1 and V_2 to be applied to PPIs at the times corresponding to $t'_{\min} + 1$ and $t'_{\min} + 2$ in Fig. 4 can be stored in ESFFs. Here, suppose a given circuit is designed by using standard scan technique. In this case, V_1 and V_2 for PPIs cannot be stored in SFFs but only V_1 can be stored. Consequently, V_2 must be justified by using some technique. In the next section, we discuss this problem.

3   Proposed Method

3.1   Broadside Test Generation Model

As explained in Sect. 2.2, in a DTEM, vectors for PPIs in a frame where a stuck-at fault exists must be justified by using some technique. Note that this

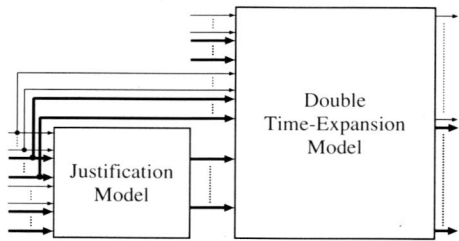

Fig. 5. Sketch of a broadside test generation model

frame is called a test frame. To achieve this requirement, we propose a new test generation model called a broadside test generation model. The sketch of a broadside test generation model is shown in Fig. 5. A broadside test generation model is composed of a DTEM and a justification model which is used for the above requirement. We first define a justification model as follows.

Definition 2. *Let S and S_K be a partial scan circuit and its acyclic kernel circuit, respectively. Let $C^T(S_K)$ and $C^D(S_K)$ be a TEM of S_K and a DTEM of S_K, respectively. Let t be the label value of a test frame in $C^D(S_K)$. Then, a combinational circuit obtained by performing the following procedure is said to be the justification model (JM) $C_t^J(S_K)$ with respect to t.*

Step 1: *For each PPI which belongs to only $C^{D_2}(S_K)$ in t, extract the logic cone of the corresponding PPO in $C^T(S_K)$. Also, for each PPI shared by $C^{D_1}(S_K)$ and $C^{D_2}(S_K)$ in t, extract the logic cone of the corresponding PPO in $C^T(S_K)$.*

Step 2: *For each pair of the logic cones, connect each pair of PIs (resp. PPIs) u in one cone and v in the other cone such that $t(u) = t(v)$ and $l(u) = l(v)$, and feed a new PI (resp. PPI) w into them.* □

By using a JM and a DTEM, a broadside test generation model is defined as follows.

Definition 3. *Let S and S_K be a partial scan circuit and its acyclic kernel circuit, respectively. Let $C^D(S_K)$ and $C_t^J(S_K)$ be a DTEM of S_K and the JM with respect to the label value t of a test frame in $C^D(S_K)$. Then, a combinational circuit obtained by performing the following procedure is said to be the broadside test generation model (BTGM) $C_t^B(S_K)$ with respect to t.*

Step 1: *For each PPI which belongs to only $C^{D_2}(S_K)$ in t, connect the corresponding PPO of $C_t^J(S_K)$ to the PPI. Also, for each PPI shared by $C^{D_1}(S_K)$ and $C^{D_2}(S_K)$ in t, connect the corresponding PPO of $C_t^J(S_K)$ to the PPI.*

Step 2: *Connect each pair of PIs (resp. PPIs) u in $C_t^J(S_K)$ and v in $C^D(S_K)$ that $t(u) = t(v)$ and $l(u) = l(v)$, and feed a new PI (resp. PPI) w into them.* □

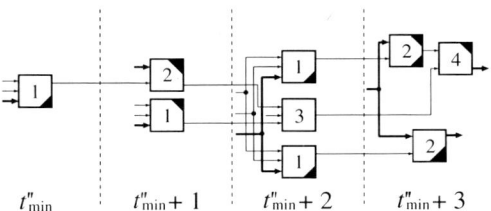

$t''_{\min}$ $t''_{\min}+1$ $t''_{\min}+2$ $t''_{\min}+3$

Fig. 6. Justification model with respect to $t'_{\min}+2$ in Fig. 4: $C^J_{t'_{\min}+2}(S_K)$

Notice that, for a given circuit, $d+1$ JMs are created, where d denotes the sequential depth of its kernel circuit. Hence, $d+1$ BTGMs are also created.

Figure 6 shows the JM $C^J_{t'_{\min}+2}(S_K)$ of Fig. 4. This JM is composed of the logic cone of the PPO of CLB 4 in $t_{\min}+3$ (Fig. 3) and that of the PPO of CLB 2 in $t_{\min}+3$. Note that although those two logic cones can share CLBs 1 and 2, we explicitly express the two logic cones for simplicity. Figure 7 shows the BTGM $C^B_4(S_K)$ of Fig. 4. In creating this BTGM, the value of 2 is assigned to $t'_{\min}$ of Fig. 4 and the value of 0 is assigned to $t''_{\min}$ of Fig. 6. As shown in Fig. 7, CLBs in a frame are not shared to differentiate the DTEM and the JM. Patterns that are needed to activate stuck-at faults in a test frame and propagate those effects to a PO or a PPO can be justified by using its JM.

3.2 Test Generation Procedure

Given a partial scan circuit S whose kernel circuit S_K is acyclic, broadside transition tests for S are generated as follows.

Step 1: Create a transition fault list F^T of S.

Step 2: Construct $d+1$ BTGMs $C^B_{t_1}(S_K),\ldots,C^B_{t_{d+1}}(S_K)$ of S_K, where d is the sequential depth of S_K.

Step 3: Create stuck-at fault lists F^S_1 for $C^B_{t_1}(S_K),\ldots,F^S_{d+1}$ for $C^B_{t_{d+1}}(S_K)$ corresponding to F^T, and constrained value lists C_1 for $F^S_1,\ldots,C_{d+1}$ for F^S_{d+1}.

Step 4: For each stuck-at fault $f^S \in F^S_i$ $(i=1,\ldots,d+1)$,
  (a): generate a test pattern t^S under the corresponding constraint $c \in C_i$, and
  (b): transform t^S into a broadside test t^T for the corresponding transition fault $f^T \in F^T$ according to the label information of $C^B_{t_i}(S_K)$.

Note that, in Step 3, even if a transition fault in a given circuit corresponds to some stuck-at faults in its BTGMs, we can handle the respective stuck-at faults one by one. This is because generated broadside transition tests are applied in the slow-fast-slow testing manner. In Step 4, if all the stuck-at faults corresponding to a transition fault are identified as untestable, the transition

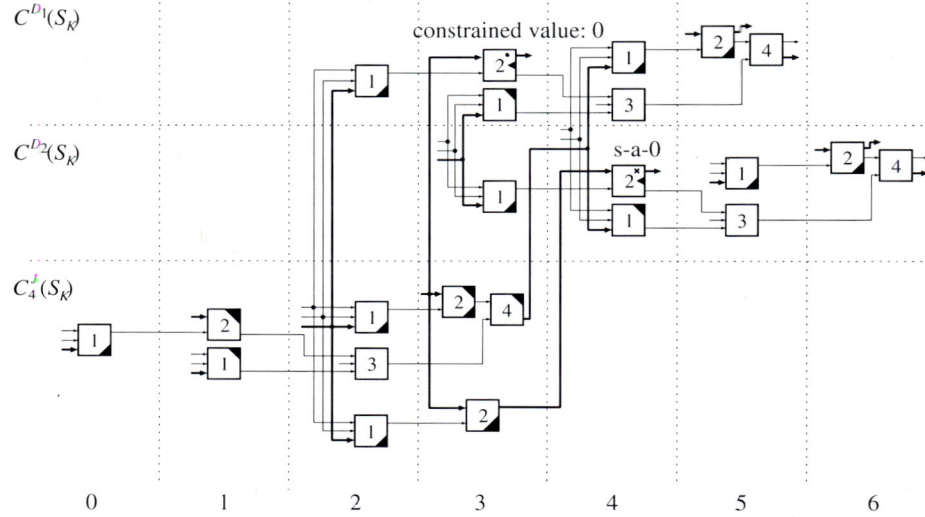

Fig. 7. Broadside test generation model with respect to $t'_{\min} + 2$ ($t'_{\min} = 2$) in Fig. 4: $C_4^B(S_K)$

fault is also untestable. Furthermore, it is sufficient to generate a test pattern for one of the stuck-at faults corresponding to a transition fault. In Step 4 (b), t^S is transformed into t^T as follows. For example, in Fig. 7, a pattern for each of the PIs and the PPI of CLB 1 in frame 0 is transformed into a pattern for each of the PIs of CLB 1 and the corresponding SFF at time 0 in Fig. 2. Notice that, the pattern for the SFF is set by scan-in operation before time 0. Other patterns in frames from 1 to 6 are transformed in the same way.

The following theorem guarantees the correctness of our test generation method.

Theorem 1. *Let S and S_K be a partial scan circuit and its kernel circuit which is acyclic, respectively. Let $f_\uparrow^T$ (resp. $f_\downarrow^T$) be a slow-to-rise (resp. slow-to-fall) transition fault in S. Let $F_{s\text{-}a\text{-}0}^S$ (resp. $F_{s\text{-}a\text{-}1}^S$) be the set of stuck-at 0 (resp. 1) faults corresponding to $f_\uparrow^T$ (resp. $f_\downarrow^T$). Then, $f_\uparrow^T$ (resp. $f_\downarrow^T$) is testable under the broadside testing manner if and only if at least one $f_{s\text{-}a\text{-}0}^S \in F_{s\text{-}a\text{-}0}^S$ (resp. $f_{s\text{-}a\text{-}1}^S \in F_{s\text{-}a\text{-}1}^S$) in the corresponding BTGM $C_t^B(S_K)$ is testable under the constrained value of 0 (resp. 1).*

Proof. Broadside test generation for $f_\uparrow^T$ (resp. $f_\downarrow^T$) in S can be viewed as test generation for the stuck-at 0 (resp. 1) fault in S corresponding to $f_\uparrow^T$ (resp. $f_\downarrow^T$) in a situation where (a) the constrained value of 0 (resp. 1) has to be set to the faulty site at time t_{1st}, and (b) no scan operation has to be performed between t_{1st} and t_{2nd}. Here, t_{2nd} denotes a time at which the stuck-at 0 (resp. 1) fault in S is activated, and $t_{1st} = t_{2nd} - 1$. In [8], it has been shown that the stuck-at test generation problem for an acyclic sequential circuit can be reduced to that for its TEM. The properties of a TEM still hold in a BTGM because the BTGM is constructed by using the TEM. Hence, to demonstrate this theorem, we need to show that (a) and (b) are satisfied in test generation for the BTGM.

First, under the slow-fast-slow testing manner, it is sufficient to consider whether at least one $f_{\text{s-a-0}}^S \in F_{\text{s-a-0}}^S$ (resp. $f_{\text{s-a-1}}^S \in F_{\text{s-a-1}}^S$) is testable. Since, in $C_t^B(S_K)$, stuck-at test generation for $f_{\text{s-a-0}}^S$ (resp. $f_{\text{s-a-1}}^S$) is performed under the constrained value of 0 (resp. 1), (a) is satisfied. Furthermore, since patterns for $f_{\text{s-a-0}}^S$ (resp. $f_{\text{s-a-1}}^S$) in the test frame of $C_t^B(S_K)$ are justified by its JM, (b) is also satisfied. Thus, the theorem is demonstrated.   □

3.3   Test Application

This subsection describes how to apply broadside transition tests to a partial scan circuit.

Broadside transition tests generated by the method of Sect. 3.2 are applied to a partial scan circuit S whose kernel circuit S_K is acyclic as follows. Let $C^D(S_K)$ be a DTEM of S_K, and t be the label value of a test frame. In test application, the circuit is operated at a slow clock speed except when its rated clock is applied at the time corresponding to t. If there exists a PPI in a frame before the test frame, scan-in operation is performed before the corresponding time. Also, if there exists a PPI which belongs to only $C^{D_2}(S_K)$ in a frame after the test frame t, scan-in operation is performed before the corresponding time. Scan-out operation is performed after the corresponding time if there exists a PPO which belongs to only $C^{D_2}(S_K)$ in a frame between the test frame t and the last frame. Note that, in order to keep the values of normal FFs during scan operation, the system clock must be separated from the scan clock or all the normal FFs have to be redesigned such that the values can be held during scan operation. For example, a broadside transition test generated by performing test generation on the BTGM $C_4^B(S_K)$ shown in Fig. 7 is applied to the partial scan circuit shown in Fig. 1 as follows. Scan-in operation is performed before each time from 0 to 3, then the circuit is operated at a slow clock speed. The transition to activate a fault is created between times 3 and 4, then between times 4 and 5, its fault effect is captured at the rated clock speed. Before each time of 5 and 6, scan-in and scan-out operations are performed simultaneously, then the circuit is operated at the slow clock speed. After time 6, scan-out operation is performed. Let d be the sequential depth of S_K. The length of a broadside transition test can range from $d + 2$ to $2d + 2$. In the case of Fig. 2, it ranges from 5 to 8.

3.4   Experimental Results

Here, we evaluate the proposed method in terms of area overhead, fault coverage, fault efficiency and test generation time.

The following experiment was performed on a Sun Fire V890 workstation (CPU: UltraSPARC IV 1.35GHz × 8, Memory: 64GB). TetraMAX from Synopsys was used as a stuck-at test generation tool, and its backtrack limit was set to

Table 1. Circuit characteristics

Circuit	#PIs	#POs	#FFs	Area
EWF	57	32	352	9,276
IIR	48	32	224	16,519
JWF	44	32	224	6,947
LWF	35	32	96	2,614
Paulin	41	64	192	19,174
Tseng	104	32	160	12,150

Table 2. Area overheads

Circuit	Area OH [%]		
	ES	SS	Ours
EWF	64.5	26.6	16.9
IIR	23.1	9.5	5.4
JWF	54.8	22.6	16.1
LWF	62.4	25.7	8.6
Paulin	17.0	7.0	4.7
Tseng	22.4	9.2	3.7

100. We applied our method to six 32bit datapath circuits [16]. The characteristics of the circuits are shown in Table 1. Columns "#PIs," "#POs" and "#FFs" list the number of PIs, POs and FFs, respectively. Column "Area" gives the area of a circuit which is estimated by Design Compiler from Synopsys, where the area of a 2-input NAND gate is considered to be 2. In this experiment, we compared the proposed method to fully enhanced scan testing and broadside testing based on the full scan method.

We first show area overheads needed for the three methods considered. In our method, acyclic kernel circuits for all the circuits were obtained by using the exact algorithm in [4]. Table 2 lists area overheads. In the table, fully enhanced scan testing, broadside testing based on the full scan method and the proposed one are denoted by "ES," "SS" and "Ours," respectively. In estimating area overhead, the areas of an ESFF and an SFF were 27 and 17, respectively. For all the circuits, we achieved the lowest area overheads. Since the proposed method is based on partial scan design, we can achieve low area overhead compared with the other methods.

Next, we show test generation results. In this experiment, we compared fault coverage, fault efficiency and test generation time of our method with those of the other two methods, and fault simulation was not invoked. In "ES," to generate transition tests, constrained stuck-at test generation were performed on a combinational circuit that consists of two independent copies of the combinational part of a given circuit. For example, to generate a two-pattern test for a slow-to-rise transition fault, we performed stuck-at test generation for the stuck-at

Table 3. Test generation results

Circuit	Method	#flts	FC [%]	FE [%]	TGT [s]	Model Size
EWF	ES		99.86	100.00	27.69	11,512
	SS	17,646	99.86	100.00	23.34	11,512
	Ours		99.86	100.00	32.62	26,268
IIR	ES		99.85	100.00	106.31	28,558
	SS	38,444	99.85	100.00	104.27	28,558
	Ours		99.85	100.00	229.43	83,574
JWF	ES		99.88	100.00	15.76	9,414
	SS	13,692	99.88	100.00	14.65	9,414
	Ours		99.88	100.00	14.35	16,788
LWF	ES		99.83	100.00	3.51	3,308
	SS	4,804	99.81	100.00	3.64	3,308
	Ours		99.81	100.00	2.77	6,171
Paulin	ES		100.00	100.00	165.33	34,508
	SS	46,248	100.00	100.00	164.12	34,508
	Ours		100.00	100.00	252.15	51,762
Tseng	ES		100.00	100.00	83.07	21,100
	SS	28,592	99.68	100.00	101.81	21,100
	Ours		99.68	100.00	154.58	33,051

0 fault in the second copy under the following constraint: the value of 0 must be set to the corresponding site in the first copy. Similarly, in "SS," we performed constrained stuck-at test generation on a combinational circuit corresponding to the two time frames of a given circuit. For example, to generate a two-pattern test for a slow-to-rise transition fault, we performed stuck-at test generation for the stuck-at 0 fault in the second time frame under the following constraint: the value of 0 must be set to the corresponding site in the first time frame. Table 3 lists the test generation results. Column "#flts" represents the number of targeted transition faults. Columns "FC [%]," "FE [%]" and "TGT [s]" denote fault coverage, fault efficiency and test generation time, respectively. The last column "Model Size" represents the average area of broadside test generation models in "Ours," and the area of the test generation model used in each case of "ES" and "SS," which are estimated by Design Compiler. In Table 3, all the methods achieved complete fault efficiency. However, in the case of "LWF" and "Tseng," some untestable faults in "SS" and "Ours" were unintentionally detected in "ES." Thus, in terms of over-testing, "ES" is not desirable. Since our broadside test generation model is larger (about 2.0 times larger on average) than the test generation models used in the other two methods, the test generation time of our method increased in some circuits. However, we consider our method to be comparable to the other two methods in test generation time. The reason is as follows. In [19], the time complexity for practical instances of the test generation problem for combinational circuits was claimed to be $O(n^3)$, where n is the size of a combinational circuit. Nevertheless, it was not observed

in our method. For example, in "IIR," the test generation time of our method was only about 2.2 times longer than that of the other two methods, although the size of our broadside test generation model was about 2.9 times larger than that of the test generation models used in the other two methods.

From the above results, we can see that our method can provide a good trade-off between area overhead and test generation effort. It is conceivable that the proposed method can also work efficiently for more complex circuits because combinational stuck-at test generation is performed.

4 Variations of Broadside Transition Testing of Partial Scan Circuits

4.1 Two Test Application Strategies

As mentioned in Sect. 2.1, the scan-per-vector strategy or the scan-per-test strategy can be used during test application. By using the iterative array model [3] of a partial scan circuit, the two test application strategies can be represented as Fig. 8. In Fig. 8, each box represents the combinational part of the partial scan circuit. In the case of the scan-per-vector strategy, inputs (resp. outputs) corresponding to SFFs shown in Fig. 8(a) are considered to be primary inputs (resp. primary outputs) except in the fault activation phase. For the stuck-at $c \in \{0, 1\}$ fault in the iterative array model of Fig. 8(a), a test pattern which detects the stuck-at fault with satisfying the constrained value of c is equivalent to a broadside test for the corresponding transition fault. Since we consider a partial scan circuit whose kernel circuit is acyclic, the fault initialization and fault effect propagation phases are bounded. Therefore, the length of a broadside transition test is at most $2d + 2$ where d is the sequential depth of the kernel circuit. Indeed, our test generation model proposed in Sect. 3.1 can be interpreted as a compact and sophisticated model of the iterative array model.

In the case of the scan-per-test strategy, inputs (resp. outputs) corresponding to SFFs shown in Fig. 8(b) are considered to be primary inputs (resp. primary outputs) only in the first time frame (resp. last time frame). This test application strategy has some advantages against the scan-per-vector strategy. Since few scan operations are required compared to the scan-per-vector strategy, the scan-per-test strategy is effective in test application time. Furthermore, over-testing can be alleviated compared to the scan-per-vector strategy, because the circuit behavior under the scan-per-test strategy is more similar to the original circuit behavior than that under the scan-per-vector strategy. Clearly, the set of untestable transition faults under the scan-per-test strategy is a superset of the set of untestable transition faults under the scan-per-vector strategy. It is notable that, unlike the scan-per-vector strategy, there are no restrictions on the scan clock and the normal FFs in the scan-per-test strategy. Thus, since the scan-per-test strategy has some desirable properties, an efficient method to generate broadside transition tests under the scan-per-test strategy should also be investigated in the future.

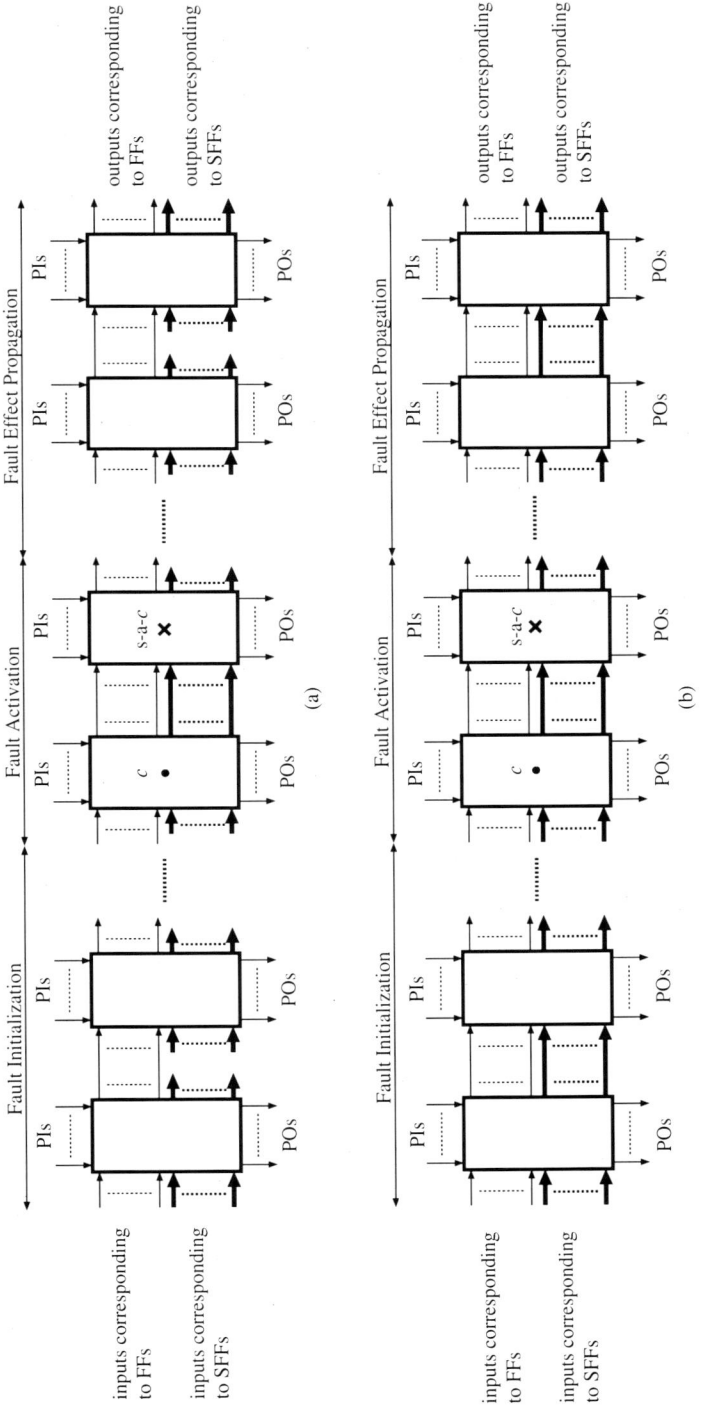

Fig. 8. Two test application strategies: (a) scan-per-vector and (b) scan-per-test

4.2   Fault Sizes

In this work, we only target transition faults whose sizes are one. However, it is important to consider fault sizes during test generation to detect large delay defects. Here we mention how to detect transition faults whose sizes are more than one in a partial scan environment.

In [21], transition faults whose sizes are more than one in a full scan circuit were handled. The basic idea used in [21] can easily be adapted to partial scan circuits. Again, we use the iterative array model of a partial scan circuit to explain how transition faults whose sizes are more than one are handled. Figure 9 shows the iterative array models under the scan-per-vector and scan-per-test strategies to generate a broadside test for a transition fault whose size is more than one. For the stuck-at $c \in \{0, 1\}$ fault in Fig. 9(a) (also (b)), a test pattern which detects the stuck-at fault with satisfying the constrained values of $c, \bar{c}, \ldots, \bar{c}$ corresponds to a broadside test for the corresponding transition fault whose size is more than one. For example, in the case of a slow-to-rise fault whose size is three, four clock cycles are required to activate the transition fault. In the iterative array model, a test pattern for the corresponding stack-at 0 fault sets the constrained values of $0, 1, 1$ to the time frames corresponding to the fault activation phase. In this way, transition faults whose sizes are more than one can be handled for partial scan circuits. However, a more precise analysis will be needed in future work.

5   Conclusions and Future Work

In this paper, we investigated broadside transition testing of partial scan circuits. The proposed scheme can utilize existing combinational stuck-at test generation tools to generate broadside transition tests. From a practical point of view, this feature is very useful because existing techniques for combinational stuck-at test generation reach a mature level. Through experiments, we showed that our method can reduce area overhead and can generate broadside transition tests in reasonable test generation time.

As mentioned in Sect. 4, broadside transition testing of partial scan circuits under the scan-per-test strategy should be investigated in the future. Moreover, fault sizes should be taken into account in future work. We also plan to extend the proposed method so that the path delay fault model can be handled.

References

1. *International technology roadmap for semiconductors*, http://public.itrs.net/, 2005.
2. M. Abadir and J. Zhu, "Transition test generation using replicate-and-reduce transform for scan-based designs," *Proc. VLSI Test Symp.*, pp. 22–27, 2003.
3. M. Abramovici, M. A. Breuer and A. D. Friedman, *Digital systems testing and testable design*, IEEE/Wiley Press, 1990.

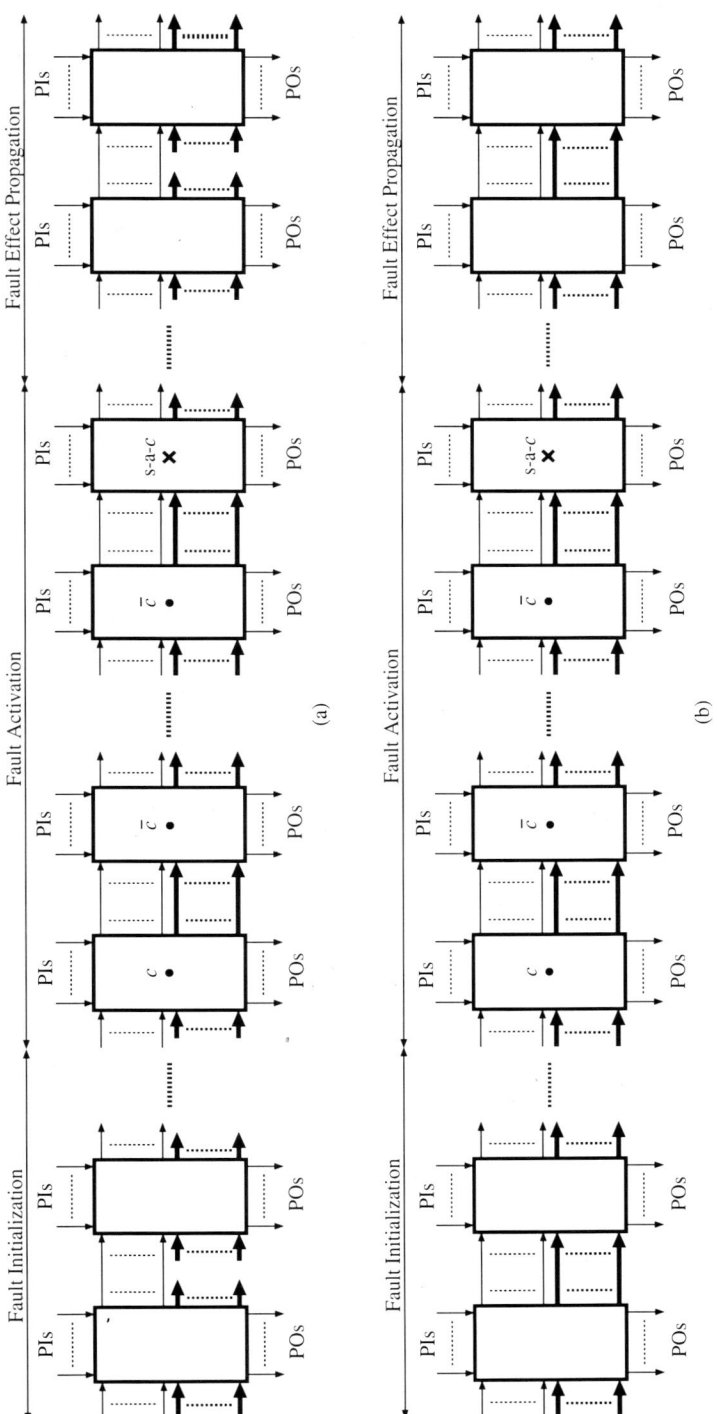

Fig. 9. Multiple activation: (a) scan-per-vector and (b) scan-per-test

4 S. T. Chakradhar, A. Balakrishnan and V. D. Agrawal, "An exact algorithm for selecting partial scan flip-flops," *Design Automation Conf.*, pp. 81–86, 1994.

5 K.-T. Cheng, "Transition fault testing for sequential circuits," *IEEE Trans. on CAD*, Vol. 12, No. 12, pp. 1971–1983, Dec. 1993.

6. K.-T. Cheng, S. Devadas and K. Keutzer, "Robust delay-fault test generation and synthesis for testability under a standard scan design methodology," *Proc. Design Automation Conf*, pp. 80–86, 1991

7. B. I. Dervisoglu and G. E. Stong, "Design for testability: using scanpath techniques for path-delay test and measurement," *Proc. Int. Test Conf.*, pp. 365–374, 1991.

8. T. Inoue, T. Hosokawa, T. Mihara and H. Fujiwara, "An optimal time expansion model based on combinational ATPG for RT level circuits," *Proc. Asian Test Symp.*, pp. 190–197, 1998.

9. T. Iwagaki, S. Ohtake and H. Fujiwara, "Acceleration of transition test generation for acyclic sequential circuits utilizing constrained combinational stuck-at test generation," *Proc. European Test Symp.*, pp. 48–53, 2005.

10. A. Krstić and K.-T. Cheng, *Delay fault testing for VLSI circuits*, Kluwer Academic Publishers, 1998.

11. I. Pomeranz and S. M. Reddy, "Theorems for identifying undetectable faults in partial-scan circuits," *IEEE Trans. on CAD*, Vol. 22, No. 8, pp. 1092–1097, Aug. 2003.

12. J. Rearick, "Too much delay fault coverage is a bad thing," *Proc. Int. Test Conf.*, pp. 624–633, 2001.

13. J. Savir and S. Patil, "Scan-based transition test," *IEEE Trans. on CAD*, Vol. 12, No. 8, pp. 1232–1241, Aug. 1993.

14. J. Savir and S. Patil, "Broad-side delay test," *IEEE Trans. on CAD*, Vol. 13, No. 8, pp. 1057–1064, Aug. 1994.

15. Y. Shao, I. Pomeranz and S. M. Reddy, "Path delay fault test generation for standard scan designs using state tuples," *Proc. Asia South Pacific Design Automation Conf. and Int. Conf. on VLSI Design*, pp. 767–772, 2002.

16. T. Takasaki, T. Inoue and H. Fujiwara, "A high-level synthesis approach to partial scan design," Proc. Asian Test Symp., pp. 309–314, 1999.

17. B. Underwood, W.-O. Law, S. Kang and H. Konuk, "Fastpath: a path-delay test generator for standard scan designs," *Proc. Int. Test Conf.*, pp. 154–163, 1994.

18. P. Varma, "On path-delay testing in a standard scan environment," *Proc. Int. Test Conf.*, pp. 164–173, 1994.

19. T. W. Williams and K. P. Parker, "Testing logic networks and designing for testability," Computer, Vol. 12, No. 10, pp. 9–21, Oct. 1979.

20. H. Wittmann and M. Henftling, "Path delay ATPG for standard scan design," *Proc. European Design Automation Conf.*, pp. 202–207, 1995.

21. Z. Zhang, S. M. Reddy, I. Pomeranz, X. Lin and J. Rajski, "Scan tests with multiple fault activation cycles for delay faults," *Proc. VLSI Test Symp.*, pp. 343–348, 2006.

Comparison of the Æthereal Network on Chip and Traditional Interconnects - Two Case Studies

Arno Moonen[1], Chris Bartels[1], Marco Bekooij[2], René van den Berg[2],
Harpreet Bhullar[2], Kees Goossens[2,3], Patrick Groeneveld[1], Jos Huisken[4], and
Jef van Meerbergen[1,5]

[1] Eindhoven University of Technology, The Netherlands
[2] NXP Semiconductors, The Netherlands
[3] Delft University of Technology, The Netherlands
[4] Silicon Hive, The Netherlands
[5] Philips Research, The Netherlands
a.j.m.moonen@tue.nl

Abstract. The growing complexity of multiprocessor systems on chip make the integration of Intellectual Property (IP) blocks into a working system a major challenge. Networks-on-Chip (NoCs) facilitate a modular design approach which addresses the hardware challenges in designing such a system. Guaranteed communication services, offered by the Æthereal NoC, address the software challenges by making the system more robust and easier to design.

This paper describes two existing bus-based reference designs and compares the original interconnects with an Æthereal NoC. We show through these two case study implementations that the area cost of the NoC, which is dominated by the number of network connections, is competitive with traditional interconnects. Furthermore, we show that the latency in the NoC-based design is still acceptable for our application.

1 Introduction

The integration of different types of cores like CPUs, DSPs, ASIPs and accelerators into a working system is a major challenge. The bottleneck in such multiprocessor architectures shifts from computation towards communication. Getting the right data at the right place at the right time will dominate the architecture. Currently busses and custom interconnects (point-to-point, crossbar switches) are often used, but with an increasing number of cores designed in technologies with decreasing dimension, they do not sufficiently address hardware problems (deep sub-micron VLSI design) and software problems (application programming). Networks-on-Chip (NoCs) tackle these problems and therefore are a better answer to the integration challenges.

First, *hardware problems*: NoCs help to answer some basic *deep sub-micron questions* because they structure the top level wires in a chip, and facilitate modular design [17]. Structured wiring results in predictable electrical parameters, such as crosstalk, etc. NoC interconnects are *segmented* and *multi-hop*. The advantage of segments is that only those segments are activated that are actually used in the communication. So only

Please use the following format when citing this chapter:

Moonen, A., Bartels, C., Bekooij, M., van den Berg, R., Bhullar, H., Goossens, K., Groeneveld, P., Huisken, J. and van Meerbergen, J., 2007, in IFIP International Federation for Information Processing, Volume 249, VLSI-SoC: Research Trends in VLSI and Systems on Chip, eds. De Micheli, G., Mir, S., Reis, R., (Boston: Springer), pp. 317–336.

those segments dissipate power. Multi-hop is needed because the transport delay from source to destination can become longer than the clock period.

Second, *software problems*: To reduce the programming effort proper transport level services have to be defined. In particular, networks on chip that offer *guaranteed communication services* (such as the Æthereal NoC [7] used in this paper) make systems on chip more robust, easier to design [8] and easier to program with a much lower non-recurring engineering cost. NoCs also provide concurrency, i.e. several transactions can be dealt with simultaneously.

NoCs are modular because they are built with only two parameterisable components (routers and network interfaces), that are combined in a scalable fashion to form the complete interconnect. New IP blocks can easily be added without changing the existing ones and guaranteed communication services assure that the performance of an IP block is not affected by the performance of other IP blocks. To guarantee bandwidth and latency, resources such as buffers and links must be allocated to connections [4], as we shall see later. The use of an automated tool chain that generates and verifies NoC hardware and software [6] is a key ingredient for successful deployment of NoCs.

Considering the analysis above the introduction of NoCs is unavoidable and the question becomes "What is the impact on area and performance?". This isn't easy to quantify. In [12] a general (artificial) design example is used. This paper follows a different approach. We start from two real-life applications and use two bus-based reference designs, one for an audio application and one for a video application. Audio and video applications have different demands in terms of communication, i.e. the required communication bandwidth and burst size for video are larger than for audio. The reference design for audio is NXPs in-car digital radio [18, 2]. The reference design for video is a programmable multi-standard Orthogonal Frequency-Division Multiplexing (OFDM) receiver [11, 9]. We compare the existing bus-based reference system-on-chips and compare it with several alternative NoC-based solutions.

The outline of this paper is as follows. Section 2 describes the in-car digital radio solution which is used as the reference design for our audio application. Section 3 describes the multi-standard OFDM demodulator and decoder which is used as the reference design for our video application. The NoC architecture is introduced in Section 4. Section 5 evaluates different NoC designs and compares these with the two reference designs. Finally, in Section 6 conclusions are drawn.

2   In-Car digital entertainment

In this section we introduce the reference design for our audio application and extract the application communication requirements. These requirements are used for dimensioning the NoCs, which eventually will be compared with the interconnect of the reference design.

The reference design is NXP's in-car digital radio chip SAF7780 [18, 2]. The SAF7780 is among others things capable of terrestrial reception, compressed audio playback and handsfree voice with acoustic echo cancellation, possibly in different use-cases like single versus dual media sound. Next to the audio application, the user application is

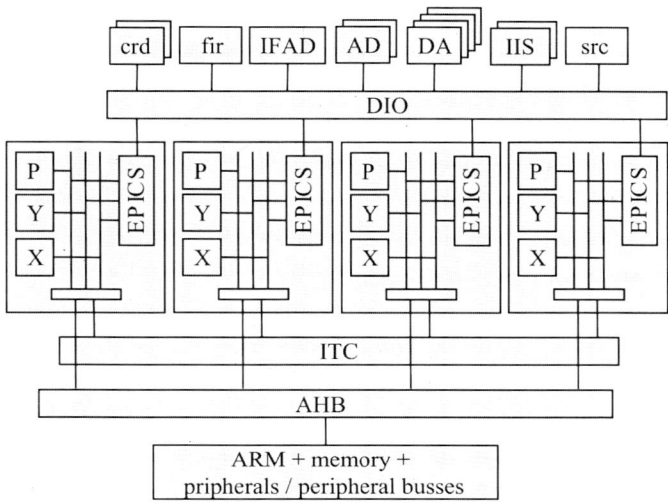

Fig. 1. The architecture of the SAF7780

executed on a programmable CPU which is integrated in the chip. In this paper our focus is on the audio application.

2.1 SAF7780 reference architecture

The SAF7780 reference architecture is shown in Fig. 1. It is a heterogeneous multiprocessor architecture combining a programmable CPU core (ARM), programmable DSP cores (EPICS), hardware accelerators (FIR, CRD) and peripherals. There are different interconnects in different parts of the architecture. The main interconnects are the Inter Tile Communication (ITC), the Digital In/Out (DIO) switch and a multilayer Amba High-speed Bus (AHB). An ARM subsystem, connected to this AHB, is used to configure and bootstrap the chip. Part of the user application is also executed on this ARM processor.

The four EPICs cores together with the ITC and DIO interconnects are the DSP subsystem where most of the signal processing takes place, e.g. audio processing. In our comparison we focus on this subsystem and replace the ITC and DIO interconnects with a NoC. The ITC channels and DIO switch are briefly described below:

ITC channels: an EPICS DSP core with its local memory (P, Y and X) is called a tile. An EPICS DPS core can write data in the memory of another tile, via an ITC channel. Reading from the memory of another tile is not implemented because it was not required by the application. There is an ITC channel from every tile to all other tiles. Tiles and the ITC interconnect are clocked at 125MHz.

ITC is based on address-based transactions. Each tile has its own address space. A specific region of this address space is mapped to an ITC channel. The ITC channel translates the address coming from the source tile to the address in the address space

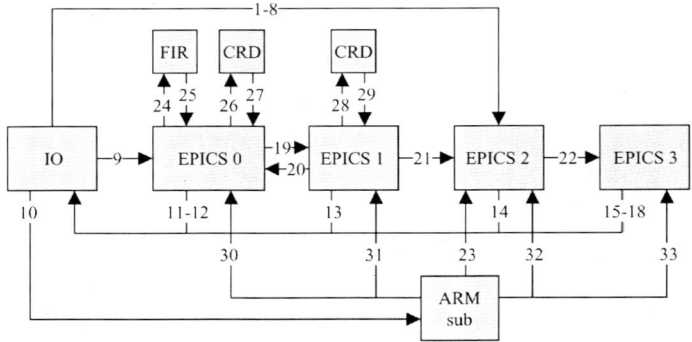

Fig. 2. Application requirements after mapping the audio application

of the destination tile. The address ranges and address translations are programmable at run time. Typically they are programmed only once per use case (mode).

DIO switch: The DIO switch connects four EPICS DSP cores to peripherals and application specific cores (hardware accelerators). Registers in peripherals and application specific cores are memory mapped in the address space of an EPICS DSP core. The peripherals, application specific cores, DIO switch and tiles are synchronous, clocked at 125MHz.

Each peripheral and application specific core is assigned to only one EPICS DSP core so that no arbitration is needed, therefore, the EPICS DSP core can access the data in one clock cycle. The assignment of peripherals and application specific cores to EPICS DSP cores is programmable at run time. This assignment is programmed only once per use-case (mode).

2.2   Communication requirements

The audio application processes streams of data and has real-time constraints. Such a streaming application can be presented by a graph that consists of tasks that communicate via channels, which are mapped onto the interconnect. The application communication requirements as a number of such connections is shown in Fig. 2.

Connections 1 through 18 have a peripheral as a source or destination. In Fig. 2 these peripherals are represented by the Input/Output (IO) box to keep the figure simple. In the NoC-based architectures each peripheral is connected to a network interface port.

The connections are data connections but there are also control connections. So there are two traffic classes:

- *Data connections (1-29)*: streaming connections represented by the edges in the task graph. Symbol sizes vary between 1 word and 512 words.
- *Programming connections (30-33)*: are used only at application start up to load the program memories and control registers of the various cores and IP blocks.

The Bandwidth and latency requirements are as follows:

Bandwidth requirements: the connection bandwidth requirements are derived from the overall symbol throughput and symbol sizes. The symbol throughput is 8 Khz for speech (telephone and navigation), between 40 and 48 Khz for audio and 325 KHz for the modulated radio signal. Symbol sizes vary from 1 word for a mono sample to 512 words for the input of the MP3 decoder.

The audio application has low average bandwidth requirements and most of the communication bursts are small. The average bandwidth requirements for connections 1-23 is between 40 KBytes/sec for the MP3 decoder input and 2 MByte/sec for the terrestrial radio demodulation input. The required average bandwidth between an EPICS DSP core and the Coordinate Rotation Digital computer (CRD) hardware accelerator is approximately 44 MByte/sec, which can be accommodated easily by the interconnect.

The amount of bandwidth assigned to programming connections effects only the stat up time of the application, which is not critical. Therefore, little bandwidth is given to these connections.

Latency requirements: Latency influences both (i) total time data takes to pass through the processing chain, and (ii) the throughput if the graph contains loops due to feedback or control. The loops cause a problem because they limit the possibility of pipelining and algorithmic transformations are needed to increase the performance.

The SAF7780 contains an adaptive filter with such a feedback loop. New filter coefficients are calculated and updated for every sample. The calculation of the filter coefficients is computed on the EPICS DSP core in cooperation with the CRD hardware accelerator. The round-trip latency, from the DSP to the accelerator and back, is composed of interconnect latency and computation latency. Backward compatibility of software is possible if the round-trip latency is not increased, after replacing the DIO switch with a NoC.

The SAF7780 is implemented in $0.18\ \mu$m technology. The EPICS DSP core and CRD hardware accelerator share the same clock with a clock frequency of 125 MHz. The EPICS accesses the input and output registers of the CRD in one clock cycle. The round-trip latency is determined by the computation latency of the task executed on the CRD, which is 36 clock cycles. Therefore, the round-trip latency is $36/(125 \cdot 10^6) = 288$ ns.

3   Digital Video Broadcasting - Terrestrial

In this section we introduce a second reference architecture which is a demonstrator and prototype of a fully programmable multi-standard OFDM demodulator and decoder using Silicon Hive cores. It is therefore a true software-defined radio design. In this paper we focus on the Digital Video Broadcasting - Terrestrial (DVB-T) [15, 3] application.

3.1   OFDM reference architecture

The OFDM reference architecture is shown in Fig. 3. It includes the processing cores (Bresca, Avispa1, Avispa2, Fec Inner, Fec Viterbi, Fec Outer) and peripherals with their

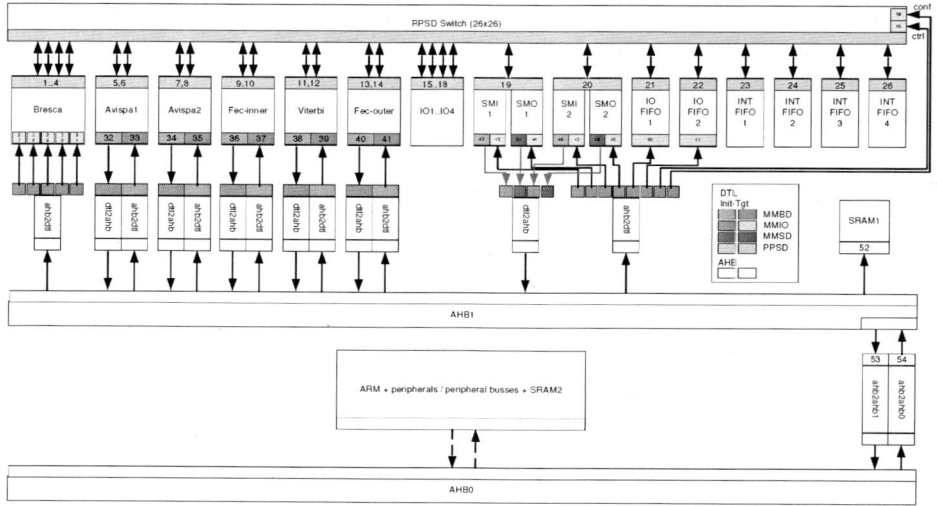

Fig. 3. Overview of the reference OFDM architecture. Note the AHB2DTL blocks that convert DTL to AHB and vice versa.

interconnects. The main interconnect structure is a bridged multilayer Amba High-speed Bus (AHB0 and AHB1) and a semi-static Peer-to-Peer Streaming Data (PPSD) switch. An ARM subsystem, connected to AHB0, is used to configure and bootstrap the processing cores. Most of the IP components use Philips's Device Transaction Level protocol (DTL) [14] as the interconnect-independent interface. The DTL is based on 4 profiles that support address-less streaming (PPSD) and single/burst/stream address-based transactions (MMIO/MMBD/MMSD, respectively). Adapters are used to convert from DTL to interconnect-specific protocols such as AHB and back. Notice that some adapter blocks also function as concentrators/distributors multiplexing bus traffic to/from multiple IP ports.

In our comparison we replaced AHB1 and the PPSD-switch with a NoC as these constitute the critical communication subsystem. The multilayer AHB and PPSD-switch are briefly described below:

Multilayer AHB: the Amba High-Speed Bus (AHB) [1] is a high-speed bus architecture. Multi-layer AHB (ML-AHB) and AHB-lite are super- and subsets, respectively, of this architecture. AHB-lite is a subset of the AHB bus protocol which only allows for one master, requiring no arbitration and saving some signals (request, grant, retry and split).

Multi-layer AHB (ML-AHB) is an interconnection architecture that extends the AHB bus architecture. It provides parallel accesses between multiple masters and slaves (Fig. 4) to increase the overall bus bandwidth and flexibility in the system architecture. The ML-AHB crossbar interconnection matrix has a higher area cost than standard AHB. The number of bus layers in one bus segment depends on performance and

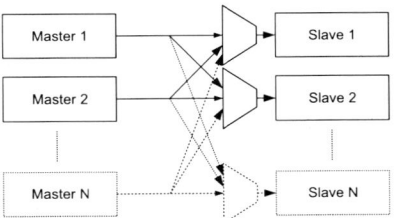

Fig. 4. Schematic of master-to-slave paths of a N-layer AHB(-lite) system.

clock-speed constraints (due to layout/placement). To achieve high clock speeds it may be desirable to split the bus.

The ML-AHB1 bus segment we take in our comparison is designed and verified for 80 MHz operation and contains 8 AHB-lite layers, each layer providing full connectivity to all slaves.

PPSD Switch: part of the interconnect structure is based on streaming point-to-point channels (DTL-PPSD). The PPSD switch allows connections to be programmed at run time. It consists of a single crossbar switch implemented using multiplexers and input/output FIFOs, and is clocked at 80 MHz. Connections are point to point and set up only once per use case (mode).

3.2 Communication requirements

Fig. 5 displays the DVB-T application communication requirements as a number of connections. Strictly speaking, the concept of connections does not exist for the original architecture because from the processor perspective, communication is address-based and the system is fully connected, i.e. each device can address any other device in the system.

The connections are categorised into four traffic classes:

- *Data connections (1-9)*: high-bandwidth streaming connections. Symbol sizes vary between 1 word and 8K words and are constant per connection.

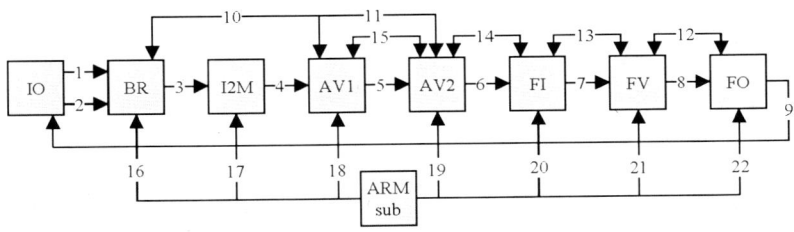

Fig. 5. Application requirements after mapping on cores.

– *Control connections (10-11)*: low-bandwidth streaming connections on which a control word is sent for every DVB-T symbol that a core has processed.
– *Token connections (12-15)*: low-bandwidth streaming connections used to send synchronisation tokens between processing cores. Synchronisation is based on available memory blocks.
– *Programming connections (16-22)*: are used only at application start up to load the program memories and control registers of the various cores and IP blocks.

The Bandwidth and latency requirements are as follows:

Bandwidth requirements: the connection bandwidth requirements are derived from system's overall required symbol throughput, symbol sizes and processor IO-rates. The DCT processing symbol sizes (8K words), and correspondingly the communication bursts, are large. Moreover, the processors operate a frequency higher than the bus frequency. As a result, the processors can saturate the bus. Hence the peak throughput of the high-bandwidth connections (520 MBytes/sec per connection) is limited by the AHB bus (320 MBytes/sec). The peak throughput is therefore spread out over time. This is allowed because the average throughput per connection ranges from 3 to 36 MWords/sec, which can be accommodated easily by the interconnect. The same reasoning applies to the programming connections. They are given little bandwidth because it affects only the start-up time of the application, which is not critical.

Latency requirements: the interconnect latency is negligible compared to the total processing time, making control and synchronization connections most latency critical. Cores send a few control tokens to the predecessor core halfway during the processing of a symbol, which should arrive before the next symbol is processed on the predecessor core. However, control loops are present only on cores that process relatively big symbols and, as a result, control latency requirements are low.

3.3   Area cost

Table 1 shows the area cost of the total interconnect. The busses and PPSD switch achieve 80 MHz after synthesis. The total interconnect area amounts to only a few percent of the total system-on-chip area.

In last two sections we introduced two reference designs and their application communication requirements. The NoC, which is described in next section, is dimensioned

Table 1. Interconnect area of the original design.

AHB routing logic	0.119 mm^2
AHB/DTL adapter	0.996 mm^2
PPSD switch	0.563 mm^2
Total interconnect area	1.68 mm^2

with the communication requirements and compared with the interconnects of the reference designs in Section 5.

4 Æthereal NoC

In this section we introduce the relevant characteristics of the Æthereal NoC [7], in particular the network interface (NI) [16].

4.1 NoC architecture

The NoC is composed of NIs and routers interconnected by links. NIs translate the IP protocols to NoC-internal packet-based protocols, offering two types of connections (or service classes): *guaranteed throughput* (GT), and *best effort* (BE). Data that is sent on BE connections is guaranteed to arrive at the destination, but without minimum bandwidth and maximum latency bounds. End-to-end flow control is used to ensure loss-less data transfer. GT connections use time-division multiple access (TDMA) to give hard (worst-case) guarantees on minimum bandwidth and maximum latency. Both GT and BE connections use source routing, i.e. the path to the destination is decided at the initiator NI. The initiator NI must be configured with this path, as we shall see later.

Data is sent from one NI to another using packets and is buffered using wormhole routing for low buffering costs. Every router contains GT input buffers consisting of one flit (3 words of 32 bits), and BE buffers of eight flits (24 words). TDMA router buffers require only one flit, as GT packets never stall in the router network. This is accomplished by globally scheduling packet injection from the NIs to the routers in such a way that packets never use the same link at the same time (thus avoiding contention). The pipelined virtual circuits that are implemented this way have a guaranteed minimum bandwidth (roughly, the number of slots reserved for the GT connection) and bounded latency (roughly, the waiting time until the appropriate slot, plus three cycles per router along the path). The TDMA slot allocation is an optimisation problem, per use case (or mode) of the NoC. We currently solve it at design time, resulting in a number of configurations. At run time these configurations are programmed (or loaded) in the NoC.

BE connections use slots that have not been reserved, or have not been used by GT packets. BE packets are scheduled dynamically at run time, and their behaviour (bandwidth, latency) is therefore not predictable.

4.2 Network interfaces

The network interface (Fig. 6) is split in a fixed *kernel* and variable *shells*. A NI shell converts transactions (e.g. read and write) of a particular IP protocols, such as DTL [14], to transport-layer messages. The NI kernel converts these generic messages into network-layer GT or BE packets. Shells are a modular layered approach: they confine protocol specific functionality; they can be composed to build complex protocols; and they allow multiple different IP ports to use a single NI [16].

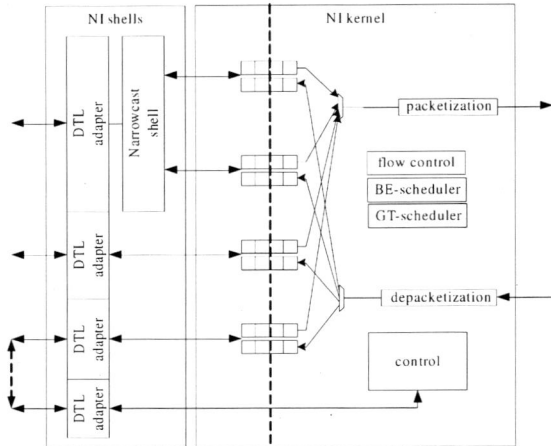

Fig. 6. Simplified network interface architecture.

The NI kernel contains FIFOs for three purposes. (i) They implement the clock boundary between IP blocks and the NoC. (ii) They decouple and isolate IP communication behaviour from the NoC behaviour. That is, data bursts from IP are buffered to fit the TDMA transmission schedule, and vice versa. (iii) They hide the round-trip latency of end-to-end flow control credits, which increases the buffer sizes by a few percent.

A connection uses two channels for every master-slave pair, with two buffers each. In order of use the connection buffers are: the initiator NI request buffer, the target NI request buffer, the target NI response buffer, and the initiator NI response buffer. As an example, Fig. 6 shows an initiator NI with three connections. The bottom two connections are simple connections of a master to a single slave, each using a request and a response buffer. The top connection is a narrowcast connection, in which a master communicates with multiple (in this case two) slaves using two channels. The connection uses four buffers in the initiator NI. In traditional address-based interconnects the processor can address each device. The narrowcast shells transparently implement the address-to-connection conversion for backward compatibility.

NIs must be programmed with the appropriate configuration at run time. This is performed using a memory-mapped IO (MMIO DTL profile [14]) configuration port on each NI. This configuration port is looped back to a target IP port (the bottom port in Fig. 6). The NoC is configured using itself, and no separate control interconnect is required [7].

4.3   NoC design flow

The NoC design flow we use consists of a number of tools for NoC generation, IP mapping, configuration, performance verification and simulation as shown in Fig. 7. Tools communicate using XML formats [6]. Note that our experiments do not include recent improvements to mapping, routing, and TDMA slot and buffer allocation [10].

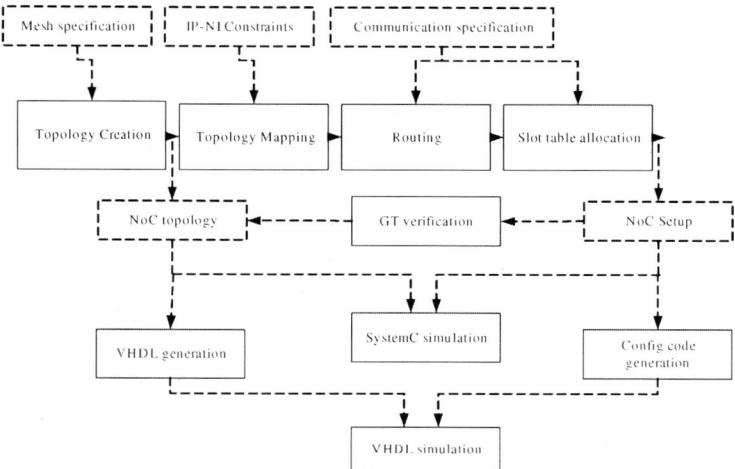

Fig. 7. NoC design flow overview.

The input to the NoC flow consists of the specification of the required communications (i.e. connections) for each use case (mode). For each connection the required protocol, bandwidth, latency, and burst size is specified. For all ports on all IP blocks the protocol and protocol related settings are also given. First, the NoC topology is selected, and the mapping of IP ports on the NI ports is determined. The topology XML file, with back-annotated buffer sizes, is used to generate RTL VHDL for gate-level synthesis.

The routes through the NoC of all connections, and the TDMA slots of all GT connections are then computed. The resulting XML file can be used to configure the NoC directly, or can be translated to C for compilation on embedded processors that configure the NoC.

The NoC description, the IP port mapping, and the configuration are used by the GT verification tool, which analytically verifies the guaranteed performance of GT connections, i.e. minimum bandwidth, maximum latency, and required buffer sizes [4].

4.4   NoC Area Cost

The NoC cell area is composed of router area and NI area. The router area depends on the number of routers, their degree (number of inputs and outputs) and the type of router (GT+BE, GT only). The GT and BE buffers in the router have a fixed size. The number of routers and their degree is determined by the topology. We select the smallest topology for which a successful mapping and configuration can be found from a set of templates (meshes in this case). There are two types of routers, GT+BE and GT-only. The GT+BE router contains GT and BE buffers whereas the GT-only router contains only GT buffers. Therefore, the area cost of a GT-only router is smaller than a GT+BE router. For example, a 6x6 GT+BE router occupies 0.175 mm^2, and a 6x6 GT-only router 0.033 mm^2 [7].

The NI area depends on the number of network interfaces, the number of connections and the NI buffers. As mentioned before, the buffers decouple the IP behaviour from the network behaviour and vice versa. A larger transaction burst size means more bursty traffic, and a larger buffer is required to decouple the IP and the network. The buffers must also hide the round-trip latency of end-to-end flow control credits. The size of NI buffers therefore depends on the connection's transaction burst size and round-trip latency, which in turn depend on the NoC topology, the mapping of IP ports to NI ports, the routing, the number of slots in the TDMA table, and the TDMA slot allocation. These parameters are mutually dependent.

The TDMA table size and slot allocation are determined by the usage of the NoC links. TDMA serves two purposes: to allocate and enforce different bandwidths to different connections, and to avoid contention (described before). Contention occurs within the router network, but also at the links between routers and NIs. Especially the latter depends very much on the mapping of IP ports to NI ports: if many connections use the same NI-router link a large TDMA table is required. The former depends mostly on the topology. A star topology, for example, funnels all connections to a single bottleneck, and requires a large TDMA table. A highly connected topology has less contention because links are less used, and because alternative paths may be available to route around congested areas.

Thus, the TDMA table size, and the slot allocation are determined by the quality of the mapping, routing, and TDMA slot allocation algorithms. We use XY routing, with an incremental slot allocation algorithm. IP port to NI port mapping balances IP port bandwidths over the NIs, clustering IP ports that communicate heavily on the same NI. It then minimises the distance (number of hops) between heavily communicating NIs, taking care not to overload any link. The improved UMARS algorithm [10] that reduces the TDMA table size, and improves the slot allocation for small buffers was not yet available at the time of our experiments for the ODFM case study. The improves UMARS algorithm is used for the SAF7780 case study.

Reducing the area of the NoC requires a trade off between minimising the number of routers and NIs, and minimising contention (which is easier in a larger NoC).

Assuming 500 MHz operation, testable, with worst-case military back-annotated lay-out timing, in Philips's 0.13 μm process technology, with GT+BE routers, [5] determined the following estimations for the router (Equation 1) and NI (Equation 2) area, respectively. In Equation 1 and Equation 2, p denotes the number of ports, c the number of connections per port, q the average buffer depth, and a the router degree. For the NI and router buffers, Equation 1 and Equation 2 count for hardware ripple-through FIFOs [19] which are faster and smaller than flip-flop-based FIFOs.

$$A_R(a) = (0.808a^2 + 23a) \cdot 10^{-3} \text{ mm}^2 \qquad (1)$$
$$A_{NI}(p, c, q) = (19.6pc + 0.72pcq + 4.8) \cdot 10^{-3} \text{ mm}^2 \qquad (2)$$

The designs we compare with in Section 5.1 are all based on minimal TDMA tables. Furthermore the number of NIs connected to the processing cores and their mapping was chosen largely equal to the traditional interconnect structure (e.g. each processing core has a single NI connected to it) as to facilitate the comparison. The tools assume operation of the NoC at 500 MHz.

4.5   Connection latency

Latency on a connection from source to destination is defined as the difference between the time at which the first word of a message has been offered to the network and the time the last word has been delivered to the destination. That is the so-called *total latency*. Total latency is composed of the *waiting latency* and *network latency*. If the NI has no space to accept data coming from the IP, the IP is stalled.

Waiting latency is defined as the difference in time at which the first word of a message has been written in the initiator NI request buffer and the time this word has been scheduled for packetization. Two factors contribute to this latency, (i) the first word of a message has to wait until it reaches the head of the buffer and (ii) it has to wait until it is scheduled. The latter depends on the distance between two allocated slots in the TDMA table.

Network latency is a consequence of latency in the NI shells, NI kernels, clock domain crossings, routers, arbitration and end-to-end flow control. The NI shell introduces two cycles latency in our DTL master shell (due to sequentialization, as part of packetization) and zero to two cycles latency in the narrowcast and multicast shells (depending on the NI instance). Between one and three cycles latency is introduced by the NI kernels (as data needs to be aligned to a three word flit boundary). The clock domain crossing, between source and destination, introduces two clock cycles latency at the destination clock. Three clock cycles latency is introduced per router. The TDMA arbiter causes additional delay. Per TDMA wheel rotation a predefined number of words can be send over the network. Therefore, the additional latency is a predefined number of TDMA wheel rotations, which are necessary for transferring the message. The round-trip latency of end-to-end flow control credits is hidden by sufficiently large NI buffers [4].

Note that a messages sent on a BE connection has an unbounded latency, because in the TDMA table no slots are allocated to this connection. From the messages sent on GT connections it is possible to compute an upper bound from a multi-rate dataflow model of the GT connection [13].

5   Comparison

In this section we compare the traditional interconnect of each of the two reference designs with replacement Æthereal NoCs. The area cost of the NoC is computed for various NoC configurations. The latency of the critical connections is computed for one NoC configuration.

5.1   Area comparison

In this subsection we assess the impact of (i) the NoC topology, (ii) the use of GT+BE versus GT-only routers, (iii) the number of connections in the design, and (iv) buffer depth optimisation.

(i) NoC topology: to explore the impact of the topology on the NoC area we implemented different NoCs without further optimisations. Table 2 contains the estimated

Table 2. Effects of topology scaling

Design	saf7780_1	saf7780_2	ofdm_1	ofdm_2	ofdm_3	ofdm_4
Mesh	1x1	1x2	1x1	1x2	2x2	3x3
# NIs	8	8	8	8	8	9
# TDMA slots	11	11	3	8	8	5
# buffers	148	148	132	132	132	134
Avg. buffer size (words)	6.23	6.52	8.81	9.30	9.42	9.19
FF-FIFO-based synth. (mm^2)	-	-	4.16	4.43	4.84	5.98
Opt.-FIFO-based synth. est. (mm^2)	-	-	1.79	1.84	1.95	2.33
Opt.-FIFO-based est. (mm^2)	2.39	2.45	1.99	2.04	2.20	2.66

area results. For synthesis of the NoC we use the Synopsys Ultra Design Compiler using Philips's $0.13\mu m$ technology and the same wire-load model as the OFDM reference design. Synthesis effort was set to medium with 200 MHz target clock speed.[6] For the NI and router buffers, the NoC designs used either synthesisable flip-flop-based FIFOs (FF-FIFO-based), or estimated area for faster and smaller hardware ripple-through FIFOs [19], referred to as "optimised FIFOs" (Opt.-FIFO-based). The rows labelled "FF-FIFO-based synth." and "Opt.-FIFO-based synth. est." contain the NoC area as obtained by synthesis of the entire NoC using FF-based FIFOs, and a synthesis based estimate using optimised FIFOs, respectively. The row labelled "Opt.-FIFO-based est." shows the estimate made by the automated tool chain, using Equation 1 and Equation 2.

The saf7780_1 and saf7780_2 designs are based on 29 GT and 4 BE connections. Most of the connections have low bandwidth requirements. BE connections are used for programming connections. All connections can be programmed at run time to be either GT or BE. The configuration processor uses only one connection to program the NoC. The buffer sizes of the GT connections were computed by the NoC design flow and for the BE connections buffers of sizes 8 words where used.

The ofdm_1 - ofdm_4 designs are based on 9 GT and 7 BE connections. They contain 13 additional zero-bandwidth (ZB) connections. The ZB connections are not used in the DVB-T application, but provide connectivity for other OFDM-based use cases. The NoC therefore offers the same connectivity as the traditional interconnect, for a correct comparison. GT connections are high bandwidth and used for the data flow of the application. BE connections are used for the low-bandwidth tokens, i.e. control and programming data. For the BE and ZB connections buffers of size 8 words were used, supporting only low bandwidth communication. The buffer sizes of the GT connections were computed by the NoC design flow.

saf7780_1-saf7780_2 and ofdm_1-ofdm_4 are all meshes, but of different sizes. Recall that the TDMA table size is affected by the number of connections sharing links between NIs and router (depending mainly on the mapping), and the contention on links (depending on slot allocation and routing). The 1x1 meshes only have NI-router contention, leading to a TDMA table with 3 and 11 slots for ofdm_1 and saf7780_1, re-

---

[6] An unoptimised narrowcast shell limited the NoC speed to 200 MHz, all other parts of the design reached higher clock speeds.

spectively. The 1x2 (ofdm_2, saf7780_2) and 2x2 (ofdm_3) meshes additionally suffer from contention in the NoC. The (heuristic) mapping, routing, and slot allocation cannot compensate for this, and 8 slots are required for the OFDM designs. In saf7780_2 the TDMA table contains still 11 slots because of the NI-router contentions. The 3x3 mesh (ofdm_4) offers more freedom to the algorithms, reducing the TDMA table to 5 slots. Although the TDMA table size impacts the NI buffering cost of high-bandwidth connections, the large number of low-bandwidth connections lowers the impact on the NI area. The difference in router area has the most impact on the total NoC area.

Table 3. Area estimation of GT-only optimisation

Design	saf7780_1gt	saf7780_2gt	ofdm_1gt	ofdm_2gt	ofdm_3gt	ofdm_4gt
Mesh	1x1	1x2	1x1	1x2	2x2	3x3
# NIs	8	8	8	8	8	9
# TDMA slots	12	11	9	17	17	11
# buffers	148	148	132	132	132	134
Avg. buffer size (words)	7.42	7.02	9.86	11.29	11.33	10.24
Opt.-FIFO-based est. (mm^2)	2.32	2.30	1.85	1.93	2.00	2.15
Difference with BE+GT	-2.9 %	-6.1 %	-7.0 %	-5.4 %	-9.1 %	-19 %

(ii) GT+BE versus GT-only routers: the area of the NoC can be reduced by using GT-only routers, because the 6x6 GT+BE router occupies 0.175 mm^2, and a 6x6 GT-only router 0.033 mm^2 [7]. The row labelled "Opt.-FIFO-based est." of Table 3 contains the estimated NoC area with optimised hardware FIFOs and smaller GT-only routers. All BE connections are converted to GT connections. As a result, the size of the TDMA table increases to accommodate the additional connections. All buffer sizes are now computed by the NoC design flow, formerly for BE connections buffer sizes of 8 words where used. The average channel buffer sizes grow for both designs. Of course, the former BE connections now have a guaranteed throughput.

The previous designs demonstrate that the NoC cost is mainly determined by the number of connections (i.e. number of buffers) and the TDMA contention in the NoC (affecting the TDMA table and the sizes of the buffers). We have illustrated how a larger NoC (more routers) reduces TDMA contention (and hence buffer cost), with ofdm_1-ofdm_4. Larger NoCs approximate a fully connected switch with least TDMA contention (i.e. one router, which is not scalable). We also illustrate that converting BE connections to GT connections reduces the router area at the cost of increased TDMA contention (ofdm_1-ofdm_3 versus ofdm_1gt-ofdm_3gt and saf7780_1-saf7780_2 versus saf7780_1gt-saf7780_2gt).

Table 4. Area results after connection optimisation

Design	saf7780_2	saf7780_2b	ofdm_3	ofdm_3b	ofdm_3c	ofdm_3d
Mesh	1x2	1x2	2x2	2x2	2x2	2x2
# NIs	8	8	8	8	8	8
# TDMA slots	11	8	8	8	6	6
# buffers	148	104	132	52	52	52
Avg. buffer size (words)	6.52	7.15	9.42	9.23	7.46	5.58
FF-FIFO-based synth. (mm^2)	-	-	4.84	2.50	2.30	2.12
Opt.-FIFO-based synth. est. (mm^2)	-	-	1.95	0.98	0.94	0.90
Opt.-FIFO-based est. (mm^2)	2.45	1.86	2.20	1.14	1.11	1.07

(iii) number of connections: the following designs use specific optimisations that are design dependent, unlike the previous trade offs that could all be automatically generated by the design flow.

In saf7780_2b the number of GT connections is reduced from 29 to 18 for determining the impact on the number of connections. This reduction of connections in the NoC is achieved by sharing the low-bandwidth connections from and to peripherals by means of combining the peripherals in one tile. The number of buffers is reduced from 148 to 104. Although, the number of TDMA slots are reduced (11 to 8) the average buffer depth is slightly increased (6.52 to 7.15) because we remove mainly small buffers. When comparing the area of saf7780_2b with the traditional interconnect, the increase is a few percent on the total chip area.

ofdm_3b is based on the 9 high-bandwidth GT connections only, and unused ports are removed. This resembles the application's main data flow only. ofdm_3b serves to assess the impact of the low-bandwidth connections on the NoC. We remove them from the NoC with the assumption that we can share the low-bandwidth connections (i.e. buffers & TDMA slots). The number of TDMA slots is not lower, but the number of buffers is more than halved (132 to 52). However, the average buffer depth does not change much (9.42 to 9.23). In other words, the low-bandwidth connections (using either GT or BE) use a significant number of buffers, but do not cause much contention.

ofdm_3c further reduces NI buffering by limiting the peak throughput from the application's maximum (520 MBytes/sec), for which the NoC was dimensioned, to the theoretical maximum of the traditional interconnect (320 MBytes/sec). This gives a fairer comparison with the reference interconnect. The size of the TDMA table is reduced (from 8 to 6), as is buffering (9.23 to 7.46 average buffer depth).

(iv) buffer depth optimisation: ofdm_3d takes the previous optimisation one (dangerous) step further. Rather than allocate the maximum (worst-case) throughput, it uses simulation to determine the required buffer sizes. This can be achieved by simulating the entire SoC with infinite buffers and recording their maximum fillings. This reduces the maximum buffer sizes of the high-bandwidth connections from ~50 to ~10. Of course, these maxima result from a limited number of simulations, and may not be large enough to guarantee bandwidth and latency, unlike the analytically computed buffer sizes.

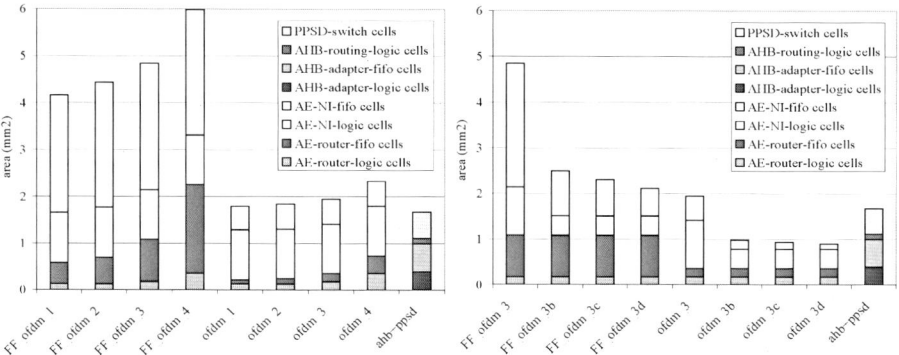

Fig. 8. (Left) Area comparison results for different mesh sizes. (Right) Area comparison results for connection-optimised Æthereal designs. The left-most four designs prefixed with FF are based on flip-flop FIFOs, the next four designs are based on optimised FIFOs. The right-most column contains the original interconnect area break down for the OFDM reference design.

The impact on the average buffer size and total area is small because only the request buffers of (write-only) high-bandwidth connections are reduced. The response buffers and programming connections are not changed. For the saf7780 designs the buffer sizes are already small, therefore, the impact on the average buffer size and total area is limited.

Fig. 8 contains a bar chart with the left-most eight bars showing the area of ofdm_1-ofdm_4 for FF-FIFO-based synth." and "Opt.-FIFO-based synth. est.", respectively. The right-most bar contains the original interconnect breakdown. We divided the area in logic for routing (bus or routers), logic for (bus or network) interfaces, and buffering cost (all buffers and state variables in the interconnect). This distinction can be easily obtained from gate-level synthesis. Buffering cost includes all flip-flops and FIFOs (RAMs are not used).

In this section we investigated the impact of the NoC topology, the use of GT+BE versus GT-only routers, the number of connections in the design, and buffer depth optimisations. Below we investigate the latency of the critical connections.

5.2   Latency comparison

In the OFDM designs control loops are present only on cores that process relatively big symbols and, as a result, there are no stringent latency requirements. However, in the SAF7780 design the audio application contains a control loop which lead to a tight latency constraint in the communication between the EPICS DSP core and the CRD hardware accelerator. This round-trip latency is 288ns with the current DIO switch running at 125 MHz, as described in Section 2.2. In a NoC-based architecture, the EPICS and CRD are attached to two different NIs. Clock domain crossing in the NI kernels enable the CRD to process at a higher clock frequency. The round-trip latency is composed of interconnect latency and computation latency. A higher clock frequency

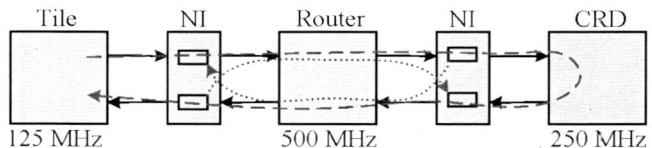

Fig. 9. Round-trip latency between the EPICS DSP core and CRD hardware accelerator

for the CRD results in a lower computation latency and more relaxed constraint for the interconnect latency.

The round-trip latency from the EPICS to the CRD and back to the EPICS is illustrated in Fig. 9 with the dashed arrow. For the purpose of analysing the round-trip latency we assume an implementation in 0.13 μm technology. The clock frequency of the EPICS is taken 125 MHz, which is the same as in the SAF7780. The network can run at a clock frequency of 500 MHz in 0.13 μm technology. In this technology it is expected that the CRD can run at a clock frequency of 250 MHz. There is a connection from the EPICS to CRD and a connection from the CRD to EPICS. Both connections are configured as GT connections. The end-to-end flow control credits of one connection are piggy-backed on messages send over the other connection, as illustrated with the dotted arrows in Fig. 9.

The task executed on the CRD has a computation latency of $36/(250 \cdot 10^6) = 144$ ns. The interconnect latency depends on the length of the message and allocation of slots in the TDMA table. The CRD reads four words from the input connection and writes two words to the output connection. In the case that address-less streaming communication (PPSD) is used, no extra data (e.g. control and address) is send in a message. Low latency results can be achieved by reserving many slots spread over the TDMA wheel, in such a way that the distance between reserved slots is small. For example, reserving one slot out of every two consecutive slots results in a 50% bandwidth allocation and an upper bound on the total NI latency of 108 ns. The latency introduced by the clock domain boundaries and router are 32 ns and 12 ns, respectively. Therefore, the total round-trip latency is 144+108+32+12=296 ns. This round-trip latency is 2.7 % higher than the round-trip latency in the SAF7780 (which is 288 ns) but is still acceptable.

In this section we investigated the impact of the NoC topology, the use of GT+BE versus GT-only routers, the number of connections in the design, and buffer depth optimisations. Finally, we computed the NoC latency for the critical connections in the SAF7780. Below we draw a number of conclusions.

6   Conclusions

In this paper we presented an interconnect comparison based on two existing software defined radio designs, one for in-car radio and one for DVB-T. For these two designs we conclude that it is feasible to replace the traditional interconnects by an Æthereal NoC and still meet the communication requirements (bandwidth and latency).

NoCs offer a structural and scalable approach for the integration of IP blocks into a working system. They help to master the deep sub-micron VLSI design problems by structuring the top level wires in a chip and facilitate modular design. On a software level the guaranteed communication services, offered by the Æthereal NoC [7], are a step forward in mastering the programming effort.

Based on two case-study implementations we conclude that the NoCs are competitive in terms of area with current dedicated interconnects. The NoC designs demonstrate that the NoC area cost is mainly determined by the number of connections (translating to a number of buffers) and the network topology (affecting the number of routers, the TDMA table and the sizes of the buffers).

We have illustrated how a larger NoC with more routers reduces TDMA contention and hence buffer cost. Larger NoCs approximate a fully connected switch with least TDMA contention (i.e. one router, which is not scalable).

The GT-BE trade off (using BE connections and GT+BE routers, or only GT connections and GT-only routers) is valuable, leading to an area reduction of 19% for a 3x3 mesh NoC. Converting BE to GT increases TDMA contention and hence buffer sizes, but this is offset by the lower cost of GT-only routers (0.033 mm^2 instead of 0.175 mm^2 for GT+BE routers).

The large number of low-bandwidth peripheral connections causes most problems. Either they use GT connections and increase TDMA contention (but not too much), or they result in the use of BE connections and (expensive) GT+BE routers. Essentially it is their number rather than their low-bandwidth that causes most cost.

The current Æthereal design flow already automatically finds the smallest regular topology (mesh, etc.), with the smallest TDMA table and optimised FIFO sizes. Optimised ripple-through hardware FIFOs are an essential component of the Æthereal NoC, leading to area reductions of around 60%. The experiments in this paper have shown that it is worthwhile to also automate the BE-GT trade off.

Furthermore, future work will include converting multiple BE connections to a single connection with shared buffers and shared TDMA bandwidth. Although the resulting connections are still BE, it reduces the number of connections and the number of buffers, as well as the TDMA contention and the depth of the buffers. The NoC can also use (inexpensive) GT-only routers.

References

1. ARM. Multi-layer AHB. overview., 2001.
2. H. Bhullar, R. van den Berg, J. Josten, and F. Zegers. Serving digital radio and audio processing requirements with sea-of-dsps for automotive applications the philips way. In *Proc. GSPx Conference*, 2004.

3. European Standard (EN) 300 744 V1.5.1. *Digital Video Broadcasting (DVB); Framing structure, channel coding and modulation for terrestrial television.*

4. O. P. Gangwal, A. Rădulescu, K. Goossens, S. González Pestana, and E. Rijpkema. Building predictable systems on chip: An analysis of guaranteed communication in the Æthereal network on chip. In P. van der Stok, editor, *Dynamic and Robust Streaming In And Between Connected Consumer-Electronics Devices*, volume 3 of *Philips Research Book Series*, chapter 1, pages 1–36. Springer, 2005.

5. S. González Pestana, E. Rijpkema, A. Rădulescu, K. Goossens, and O. P. Gangwal. Cost-performance trade-offs in networks on chip: A simulation-based approach. In *Proc. Design, Automation and Test in Europe Conference and Exhibition (DATE)*, pages 764–769, Washington, DC, USA, Feb. 2004. IEEE Computer Society.

6. K. Goossens, J. Dielissen, O. P. Gangwal, S. González Pestana, A. Rădulescu, and E. Rijpkema. A design flow for application-specific networks on chip with guaranteed performance to accelerate SOC design and verification. In *Proc. Design, Automation and Test in Europe Conference and Exhibition (DATE)*, pages 1182–1187, Washington, DC, USA, Mar. 2005. IEEE Computer Society.

7. K. Goossens, J. Dielissen, and A. Rădulescu. The Æthereal network on chip: Concepts, architectures, and implementations. *IEEE Design and Test of Computers*, 22(5):414–421, Sept-Oct 2005.

8. K. Goossens, J. Dielissen, J. van Meerbergen, P. Poplavko, A. Rădulescu, E. Rijpkema, E. Waterlander, and P. Wielage. Guaranteeing the quality of services in networks on chip. In A. Jantsch and H. Tenhunen, editors, *Networks on Chip*, chapter 4, pages 61–82. Kluwer Academic Publishers, Hingham, MA, USA, 2003.

9. P. Gruijters, K. Koch, and G. Burns. Flexible embedded processors for developing multi-standard broadcast receivers. In *Proc. GSPx Conference*, 2004.

10. A. Hansson, K. Goossens, and A. Rădulescu. A unified approach to constrained mapping and routing on network-on-chip architectures. In *Int'l Conf. on Hardware/Software Codesign and System Synthesis (CODES+ISSS)*, pages 75–80, Sept. 2005.

11. I. Held and B. Vandewiele. Avispa ch - embedded communications signal processor for multi-standard digital television. In *Proc. GSPx Conference*, 2006.

12. P. Martin. A comparison of network-on-chip and busses. Technical report, white paper downloadable from the Arteris website (www.arteris.com), 2005.

13. A. Moonen, M. Bekooij, and J. van Meerbergen. Timing analysis model for network based multiprocessor systems. In *Proc. ProRISC, 15th annual Workshop of Circuits, System and Signal Processing*, 2004.

14. Philips Semiconductors. *Device Transaction Level (DTL) Protocol Specification. Version 2.2*, July 2002.

15. U. Reimers. *DVB-The family of international standards for digital video broadcasting.* Springer-Verlag, 2nd edition, 2005.

16. A. Rădulescu, J. Dielissen, S. González Pestana, O. P. Gangwal, E. Rijpkema, P. Wielage, and K. Goossens. An efficient on-chip network interface offering guaranteed services, shared-memory abstraction, and flexible network programming. *IEEE Transactions on CAD of Integrated Circuits and Systems*, 24(1):4–17, Jan. 2005.

17. M. Sgroi, M. Sheets, A. Mihal, K. Keutzer, S. Malik, J. Rabaey, and A. Sangiovanni-Vincentelli. Addressing the system-on-a-chip interconnect woes through communication-based design. In *Proc. Design Automation Conference (DAC)*, pages 667–672, June 2001.

18. R. van den Berg and H. Bhullar. Next generation philips digital car radios, based on a sea-of-dsp concept. In *Proc. GSPx Conference*, 2004.

19. P. Wielage, E. J. Marinissen, and C. Wouters. Design and DFT of a high-speed area-efficient embedded asynchronous FIFO. In *Proc. Design, Automation and Test in Europe Conference and Exhibition (DATE)*, 2007.

Designing Routing and Message-Dependent Deadlock Free Networks on Chips

Srinivasan Murali[1], Paolo Meloni[2], Federico Angiolini[3], David Atienza[4],[6], Salvatore Carta[5], Luca Benini[3], Giovanni De Micheli[4], and Luigi Raffo[2]

[1] CSL, Stanford University, Stanford, USA, smurali@stanford.edu
[2] DIEE, University of Cagliari, Cagliari, Italy,
{paolo.meloni@diee.unica.it, luigi@diee.unica.it}
[3] DEIS, Univerity of Bologna, Bologna, Italy,
{fangiolini@deis.unibo.it, lbenini@deis.unibo.it}
[4] LSI, EPFL, Lausanne, Switzerland,
{david.atienza, giovanni.demicheli@epfl.ch}
[5] DMI, University of Cagliari, Cagliari, Italy, salvatore@unica.it
[6] DACYA, Complutense University of Madrid (UCM), Madrid, Spain.

Abstract. *Networks on Chip* (NoC) has emerged as the paradigm for designing scalable communication architecture for Systems on Chips (SoCs). Avoiding the conditions that can lead to deadlocks in the network is critical for using NoCs in real designs. Methods that can lead to deadlock-free operation with minimum power and area overhead are important for designing application-specific NoCs. The deadlocks that can occur in NoCs can be broadly categorized into two classes: *routing-dependent* deadlocks and *message-dependent* deadlocks. In this work, we present methods to design NoCs that avoid both types of deadlocks. The methods are integrated with the topology synthesis phase of the NoC design flow. We show that by considering the deadlock avoidance issue during topology synthesis, we can obtain a significantly better NoC design than traditional methods, where the deadlock avoidance issue is dealt with separately. Our experiments on several SoC benchmarks show that our proposed scheme provides large reduction in NoC power consumption (an average of 38.5%) and NoC area (an average of 30.7%) when compared to traditional approaches.

Keywords: Networks on Chips, Systems on Chips, Message-dependent deadlocks, routing-dependent deadlocks, topology, synthesis.

1 Introduction

Today's *Systems on Chips (SoCs)* consist of a large number of computing and storage cores that are interconnected by means of single or multiple layers of buses In order to cope with the large communication demands of such SoCs, a modular, scalable interconnect based on *Networks on Chips (NoCs)* is needed [1]-[6].

Please use the following format when citing this chapter:

Murali, S., Meloni, P., Angiolini, F., Atienza, D., Carta, S., Benini, L., De Micheli, G. and Raffo, L., 2007, in IFIP International Federation for Information Processing, Volume 249, VLSI-SoC: Research Trends in VLSI and Systems on Chip, eds. De Micheli, G., Mir, S., Reis, R., (Boston: Springer), pp. 337–355

Designing a custom-tailored interconnect that satisfies the performance and design constraints of the SoC is important to achieve efficient NoC designs [27]-[32]. A critical, but often neglected issue when designing NoCs is that they have to guarantee deadlock-free operation. If the NoC has no support to either avoid or recover from deadlocks, then correct functionality of the system cannot be guaranteed. This can lead to system crashes and unexpected system behavior, which is clearly unacceptable for SoCs. Designing efficient methods that avoid such a situation with minimum power and area overhead is an important research area in the NoC domain.

The deadlocks that can occur in NoCs can be broadly categorized into two classes: *routing-dependent* deadlocks and *message-dependent* deadlocks [33], [7]-[12]. Routing-dependent deadlocks occur when there is a cyclic dependency of resources created by the packets on the various paths in the network. For regular topologies (such as the mesh, torus), the use of restricted routing functions based on turn models is an effective way to avoid routing-dependent deadlocks [9], [10]. For custom application-specific NoCs, obtaining deadlock-free paths is a bigger challenge [12], [22], [32]. The major focus of this paper is to address this important issue of obtaining routing and message-dependent deadlock-free network operation.

Message-dependent deadlocks occur when interactions and dependencies are created between different message types at network endpoints, when they share resources in the network. Even when the underlying network is designed to be free from routing-dependent deadlocks, the message-level deadlocks can block the network indefinitely, thereby affecting the proper system operation. An example situation where a message-dependent deadlock occurs is presented in Figure 1(a). In this example, two of the cores are masters and two other cores are slaves. In this system, we assume two kinds of messages: *request* and *response*. Consider the following situation: Master 1 sends a request to Slave 1 (*Req 1*), Slave 1 is replying to a previously issued request to Master 1 (*Resp 1*) and at the same time, Slave 2 sends a response to Master 2 (*Resp 2*). When requests and responses share the same links, *Resp 2* is waiting for link 1 which is used by *Req 1* and *Resp 1* waits for link 4 used by *Resp 2*. Meanwhile, *Req1* is waiting for Slave 1, the operation of which has been stalled as *Resp 1* could not complete. Thus, none of the messages can move ahead, leading to a deadlock situation. An interesting point to note here is that message-level deadlocks can be avoided if the receivers have infinitely large buffering or if they have perfectly ideal operation (consuming all received data instantly), which would avoid queuing of the packets in the network. Obviously, such a solution is not feasible to obtain in practice.

In traditional multi-processor interconnection networks, the most common ways to avoid message-dependent deadlocks are the use of separate logical or physical networks for the different message types [13]-[21]. This would ensure that the different message types do not share the network components, thereby guaranteeing freedom from message-dependent deadlocks. The most common method to achieve separate logical networks is the of use of separate virtual channels for the different message types [13]. For the example design presented

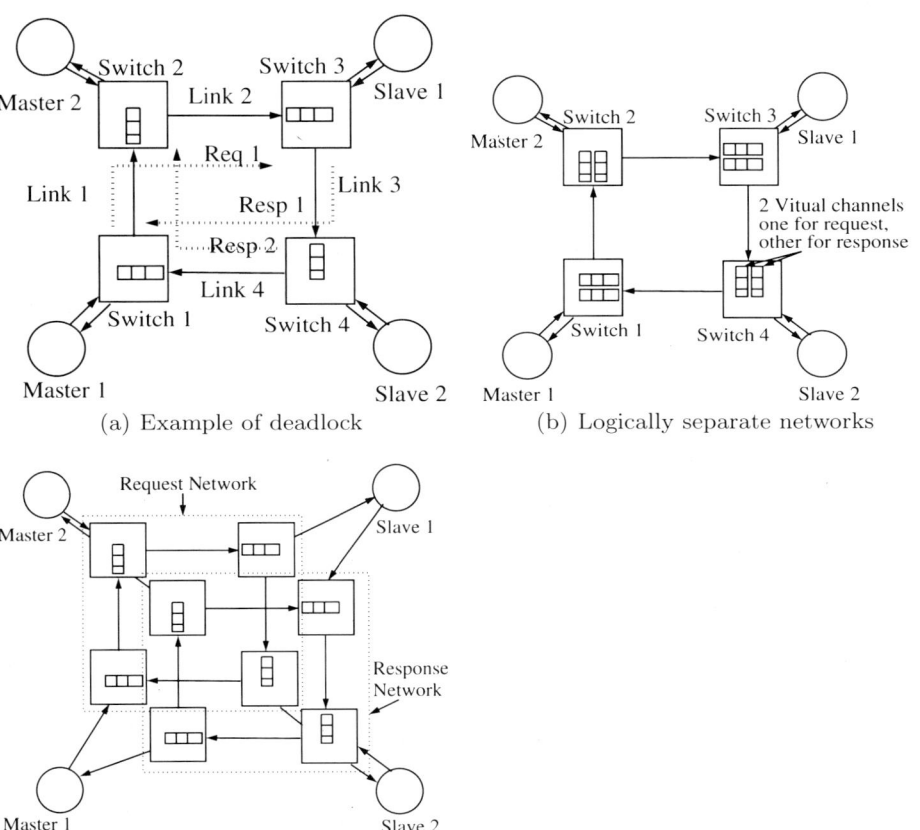

(a) Example of deadlock

(b) Logically separate networks

(c) Physically separate networks

Fig. 1. Example of a message-dependent deadlock and traditional methods available to remove it.

in Figure 1(a), each router input will need two virtual channels: one for the request messages and the other for the response messages (refer Figure 1(b)). This separation of message types is maintained at all the switches in the network. In the case of separate physical networks, the request network is built separately from the response network, an example of which is shown in Figure 1(c). This is the most commonly used solution in complex bus designs such as the STBus from STMicroelectronics [19] and several multi-processor designs [20], [21].

In this work, we show that by mapping the different message types onto different network resources during the topology mapping and synthesis phase, we can achieve much better NoC designs (in terms of power consumption and network area) than traditional approaches. We present a topology synthesis algorithm that specifically considers the message types and ensures the creation of a network that is free from message-dependent deadlocks. We also implement the common methods of deadlock avoidance: having separate virtual channels and having physically separate networks for the message types. For all the schemes, we make the underlying network operation free from routing-dependent deadlocks by integrating existing methods with our topology synthesis process [12]. We perform experiments on several SoC designs, which show that our proposed scheme provides large reduction in the NoC power consumption (an average of 38.5%) and area (an average of 30.7%) when compared to the traditional approaches.

2   Previous Work

The motivation for the use of NoCs has been established in several works [1]-[6]. The use of turn models to avoid deadlocks in mesh and torus networks has been presented in [10]. There has been a large body of work that have focused on developing routing-dependent deadlock-free operation for interconnection networks [9]-[12]. Several other works exist in the area of recovering from deadlocks in networks [7], [8].

The design of application specific NoCs has been explored in several works [24]-[30]. None of these works address the issue of message-level deadlock avoidance, which is critical for proper system operation. Avoiding routing-dependent deadlocks for mesh topologies has been considered in [24]. Avoiding routing deadlocks for custom NoC topologies have been presented in [22], [31].

The use of logically separated networks to avoid message-dependent deadlocks has been utilized in several industrial multi-processors, such as [14]-[17]. The use of physically separated networks to remove message-dependent deadlocks is used in many designs, such as [20], [21]. In [5], message-level deadlock freedom is achieved by a different mechanism than using logically or physically separated networks. It utilizes an end-to-end flow control scheme, which ensures that messages are sent from the sender only when the receiver has enough buffering resources to store them. This is coupled together with a network design that uses time division multiplexing to divide the network resources among the various communicating elements, providing guaranteed throughput to con-

nections. This leads to buffering free network for such connections and removal
from message-level deadlocks. The deadlock avoidance mechanism using their
protocol is presented in [23]. As we target general NoC designs that need not
support such end-to-end flow control mechanisms, we do not compare such a
scheme with our method presented here.

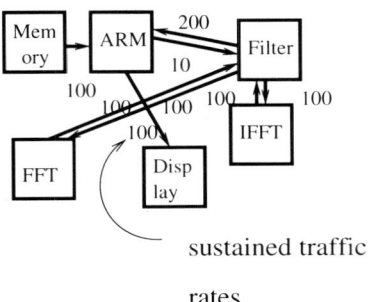

sustained traffic

rates

Fig. 2. Example filter application

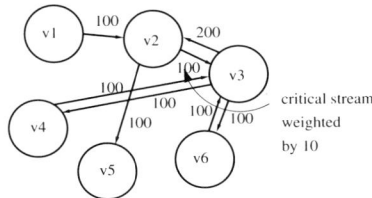

Fig. 3. Core graph with sustained rates
and critical streams

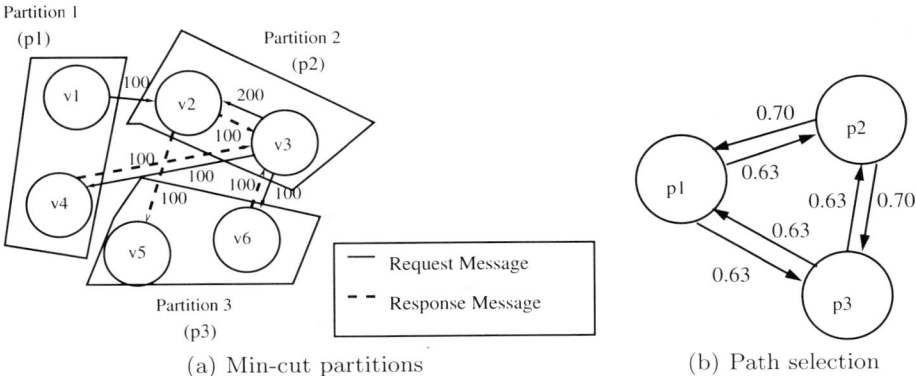

(a) Min-cut partitions

(b) Path selection

Fig. 4. Algorithm Examples

3 Topology Synthesis with Deadlock Freedom

We implement our routing and message-dependent deadlock-free path selection
routine as a plug-in to an established NoC topology synthesis flow.

3.1 Topology Synthesis Process

We assume that the application kernels are parallelized and mapped onto different processors and hardware cores using existing tools, as done in earlier works

[24]-[32]. The communication traffic flow between the various cores is represented by a core graph, which is taken as the input to the topology synthesis flow. The core graph for a small filter example (Figure 2) is shown in Figure 3. The edges of the core graph are annotated with the sustained rate of traffic flow, multiplied by the criticality level of the flow, as done in [26].

Before presenting the path selection routine, we first present the basic topology synthesis flow. In the topology synthesis procedure (Algorithm 1), we synthesize several topologies: starting from a topology where all the cores are connected to a single switch to a topology where each core is connected to a separate switch. For the chosen switch count, the input core graph is partitioned into those many min-cut partitions (refer to step 2 of Algorithm 1). At this point, the communication traffic flows within a partition has been resolved.

The reason for synthesizing these many topologies is that it cannot be predicted beforehand as to whether a design with few bigger switches can be more power efficient than a design with more smaller switches. A larger switch has more power consumption than a smaller switch to support the same traffic. This is because, a larger switch has a bigger crossbar and arbiter. On the other hand, in a design with many smaller switches, the packets may need to travel more hops to reach the destination. Thus the total switching activity will be higher than a design with fewer hops, which can lead to higher power consumption.

The partitioning is done in such a way that the edges of the graph that are cut between the partitions have lower weights than the edges that are within a partition and the number of vertices assigned to each partition is almost the same. Thus, those traffic flows with large bandwidth requirements or higher criticality level are assigned to the same partition and hence use the same switch for communication. Hence, the power consumption and the hop-delay for such flows will be smaller than the other flows that cross the partitions.

Now, we integrate in the main flow the core contribution of this work (in step 4 of Algorithm 1), i.e. an algorithm (*PATH_COMPUTE*) that maps the communication flows to physical paths while guaranteeing deadlock freedom. This algorithm is explained in detail in the following paragraphs. Once the paths for a topology are selected, Algorithm 1 resumes, where the design area, power consumption and wire-length for the topologies are obtained. Then, the topology that best optimizes the user objectives and satisfies all the design constraints is chosen. The topology synthesis flow, without considering freedom from message level deadlocks, has been presented by us in detail in [32].

In the following subsections, we explain the path selection mechanism (Algorithm 2) that guarantees routing and message-dependent deadlock-free operation of the NoC.

3.2   Avoiding Routing Dependent Deadlocks

When the *PATH_COMPUTE* procedure is invoked, the number of switches in the NoC and their connectivity with the cores has already been determined (cores in the same partition share the same switch). The procedure is used to

Algorithm 1 Topology Design Algorithm

1: Vary the number of switches in the design from 1 to the total number of cores in the design. Repeat steps 2 to 7 for each switch count.
2: For the chosen switch count, find that many min-cut partitions of the communication graph. Cores in each partition are attached to the same switch.
3: Check for bandwidth constraint violations when establishing the switches. The bandwidth of each link is the product of the NoC operating frequency and link width, which are inputs to the flow.
4: Find the connectivity between the switches using the function *PATH_COMPUTE* (presented in Algorithm 2).
5: Evaluate the switch power consumption and average hop-delay based on the selected paths.
6: Perform floorplan of the design. Obtain design area, wire-lengths. Check for timing violations on the wires and evaluate the power consumption on wires.
7: If solution minimizes objective, satisfies all constraints, note the design point and the topology.
8: Choose the best topology and design point based on the user objectives.

connect the different switches together and find paths for the traffic that flows across the partitions.

In the first step of the procedure, we build a complete graph, with each vertex in the graph representing a switch in the network.

In [12], the authors present a scheme for removing deadlocks in general networks. The approach is also utilized in [22] for removing routing-dependent deadlocks in NoCs. The approach removes routing-dependent deadlocks by prohibiting certain turns for the packets, thereby avoiding cycles in the network. In the next step of the *PATH_COMPUTE* algorithm, we invoke the *BLOCK_TURNS* procedure (Algorithm 3) to remove turns in the logical graph to avoid deadlocks. When we compute paths later in the *PATH_COMPUTE* procedure, we only use those turns that have not been blocked by the *BLOCK_TURNS* procedure.

3.3 Avoiding Message-Dependent Deadlocks

In the next step (step 3) of the *PATH_COMPUTE* procedure, the flows are ordered in decreasing rate requirements, such that bigger flows are assigned first. The heuristic of assigning bigger flows first has been shown to provide better results (such as lower power consumption and more easily satisfying bandwidth constraints) in several earlier works [25], [31]. Then, for each flow in order, we first evaluate the message type of the flow (step 4 of Algorithm 2). The message types can either be fed explicitly by the user, or can be implicitly considered by the tool. As an example for implicitly considering the type, in shared memory systems, all the traffic flows that originate from processors and terminate into memory devices are of request type. While those that originate from the memories and terminate in the processors are of response type. Note that in shared memory systems, all inter-processor communication occur through the memory devices. Note that, if the connection between any pair of cores constitutes multiple message types, then each message type needs to be treated as a separate traffic flow.

Algorithm 2 PATH_COMPUTE

1: Build a fully connected logical graph, with each vertex representing a switch in the NoC.
2: Invoke the *BLOCK_TURNS* procedure, to find the set of turns that are prohibited.
3: For each traffic flow in decreasing order of the bandwidth requirements, perform steps 4 to 8.
4: Find the message type supported by the chosen traffic flow.
5: For i1 from 1 to number of switches in the current design and j1 from 1 to number of switches in the current design, repeat steps 6 and 7.
6: If one or more physical links exists between the switches i1 and j1, evaluate whether any link exists that has already been supporting the current message type & has bandwidth to support the current flow. If so, find the marginal power consumption to re-use this existing link.
7: Else find the marginal power consumption for opening and using the link for this traffic flow.
8: Find the least cost path (path with least power consumption) across the switches. For any links that were newly established for this traffic flow, associate the message type of this flow to the links. When selecting paths, choose only those paths that have turns not prohibited for removing routing-dependent deadlocks (based on the method from [12]).
9: Return the chosen paths, new switch sizes, connectivity between switches and the type of message supported by each of the links.

Next, we evaluate the amount of power that will be dissipated across each of the switches, if the traffic for the chosen flow uses that switch. This power dissipation value on each switch depends on the size of the switch, the amount of traffic already routed on the switch and the frequency of operation. It also depends on how the switch is reached (from which other switch) and whether an already existing physical channel will be used to reach the switch or a new physical channel will have to be opened. The last information is needed, because opening a new physical channel increases the switch sizes and hence the power consumption of this flow and others that are routed through the switch.

In our NoC architecture, we permit the instantiation of multiple physical links between any two switches. When finding whether a switch is reachable from another switch for the current traffic flow, we evaluate whether any physical links between the switches have already been established. If so, we see the message type of the traffic flows that have already been routed on the links. From the set of established links, we choose a link that supports the same message type as the current traffic flow and has enough bandwidth available to support the current flow. If no such link is available between the switches, we evaluate the cost of opening up a physical link for the current traffic flow.

The process of evaluating the power consumption for the current traffic flow is repeated for all pairs of switches. Finally (in step 8 of Algorithm 2), the set of links from the source to destination of the flow that has the least power consumption is chosen. When choosing the paths, only those paths that do not use any turns blocked by the *BLOCK_TURNS* procedure is considered. Now physical connections are actually established on the chosen path and the message type of the current flow is assigned to the links that have been used for the flow.

Algorithm 3 BLOCK_TURNS

1: Select a node with minimum degree from the logical graph.
2: Mark all turns around this node as blocked, and allow all turns that start from this node.
3: Remove the node and all its adjacent edges from further consideration for this procedure.
4: Repeat all the above steps, until all the nodes have been considered.

Example 1. Let us consider the example from Figure 4(a). The input core graph has been partitioned into 4 partitions. We assume 2 different message types: request and response for the various traffic flows. Each partition p_i corresponds to the cores attached to the same switch. Let us consider routing the flow with a bandwidth value of 100 MB/S between the vertices $v1$ and $v2$, across the partitions $p1$ and $p2$. The traffic flow is of the message type request. Initially no physical paths have been established across any of the switches. If we have to route the flow across a link between any two switches, we have to first establish the link. The cost of routing the flow across any pair of switches is obtained. We annotate the edges between the switches by the cost (marginal increase in power consumption) of sending the traffic flow through the switches (Figure 4(b)). The cost on the edges from $p2$ are different from the others due to the difference in initial traffic rates within $p2$ when compared to the other switches. This is because, the switch $p2$ has to support flows between the vertices $v2$ and $v3$ within the partition. The least cost path for the flow, which is across switches $p1$ and $p2$ is chosen. Now we have actually established a physical path and a link between these switches. We associate the message type request for this particular link. This is considered when routing the other flows and only those traffic flows that are of request type can use this particular physical link. We also note the size and switching activity of these switches that have changed due to the routing of the current flow.

4 Experimental Platform

We have built accurate analytical models for calculating the power consumption, area and delay of the ×pipes network components [36].

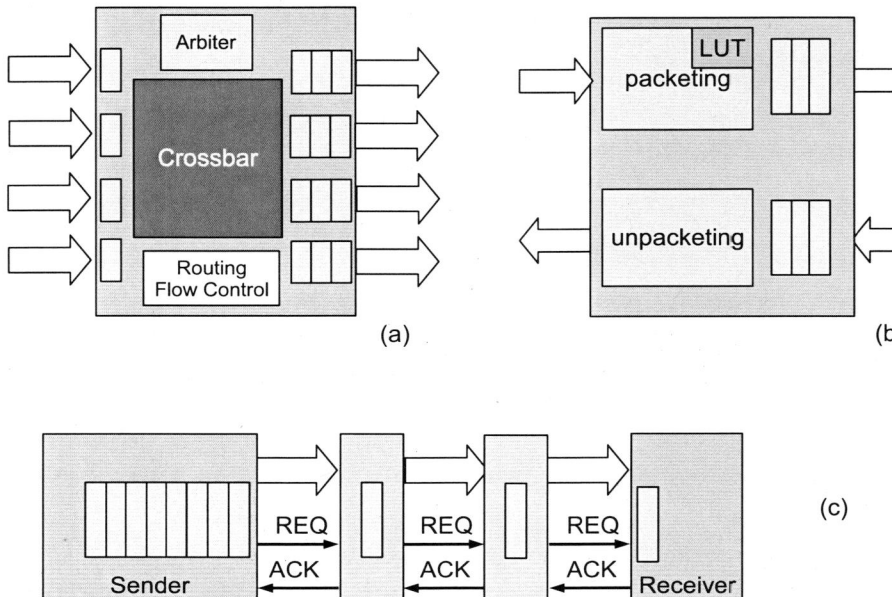

Fig. 5. ×pipes building blocks: (a) switch, (b) NI, (c) link

4.1 Architectural Overview

The ×pipes NoC [36], [37] library consists of a set of parameterizable soft-macros for the network components (Figure 5). The NoC is instantiated by deploying a set of components in an arbitrary topology and by configuring them. There are three main components in the ×pipes library: switches, *Network Interfaces (NIs)* and links.

The backbone of the NoC is composed of switches, whose main function is to route packets from sources to destinations. Arbitrary switch connectivity is possible, allowing for implementation of any topology. Switches provide buffering resources to lower congestion and improve performance. In ×pipes, output buffering scheme is utilized, where FIFOs are present on each output port of the switch. Switches also handle flow control [38] issues of the NoC, which resolves the conflicts among the packets that request access to the same physical links. The ×pipes architecture supports the use of different flow control strategies, such as the ACK/NACK and STALL/GO protocols. For the experiments performed in this paper, we use the ACK/NACK protocol for flow control.

An NI is needed to connect each IP core to the NoC. NIs convert transaction requests/responses into packets and vice versa. Packets are then split into a sequence of smaller units, referred to as *flits (FLow control unITS)*. In ×pipes, two separate NIs are defined, an *initiator* and a *target* one, respectively associated to system masters and system slaves. A master/slave device will require an NI of each type to be attached to it. The interface among IP cores and NIs is

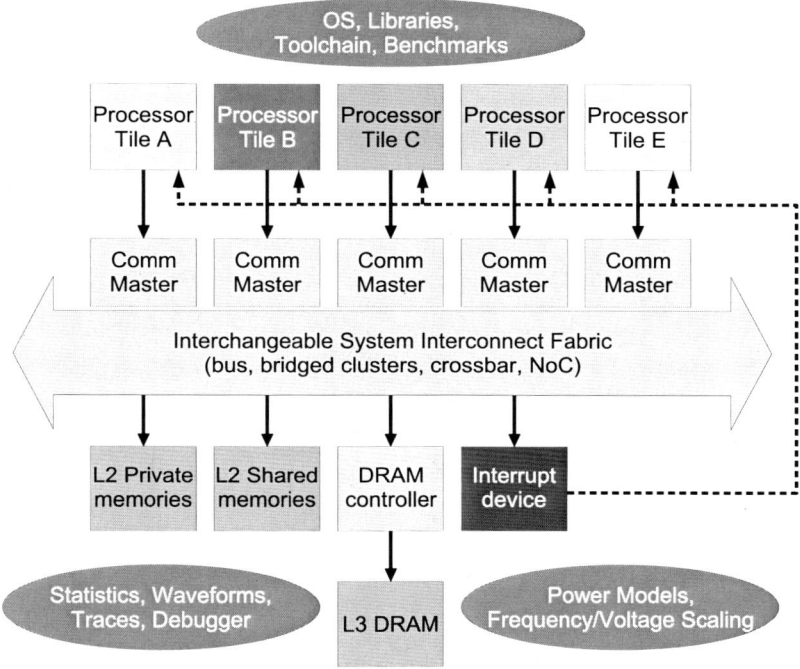

Fig. 6. The MPARM SystemC virtual platform

point-to-point and follows the OCP 2.0 [39] protocol, guaranteeing maximum re-usability. We use source based routing scheme, where each NI has a look-up table to specify the path that packets will follow in the network to reach their destination. Two different clock signals can be attached to NIs: one to drive the NI front-end (OCP interface), the other to drive the NI back-end (×pipes interface). The ×pipes clock frequency must be an integer multiple of the OCP one. This arrangement allows the NoC to run at a fast clock even though some or all of the attached IP cores are slower, which is crucial to keep transaction latency low. Since each IP core can run at a different divider of the ×pipes frequency, mixed-clock platforms are possible.

To get accurate simulation in a flexible environment, we integrate the NoC in MPARM (Figure 6). MPARM allows for accurate injection of functional traffic patterns as generated by real IP cores (processors, DMA engines, *etc.*) during a benchmark run. Further, it provides facilities for debugging, statistics collection and tracing.

4.2   Area and Power Models

To get an accurate estimate of these parameters, the place&route of the components is performed using Cadence SoC Encounter [35]. From the layout-level

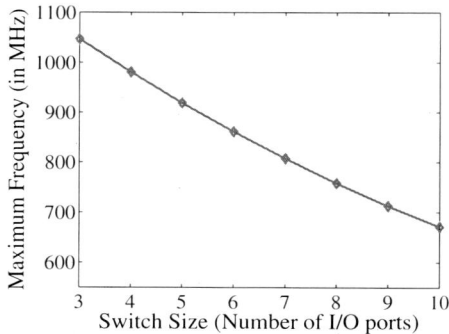

Fig. 7. Maximum frequency variation with switch size

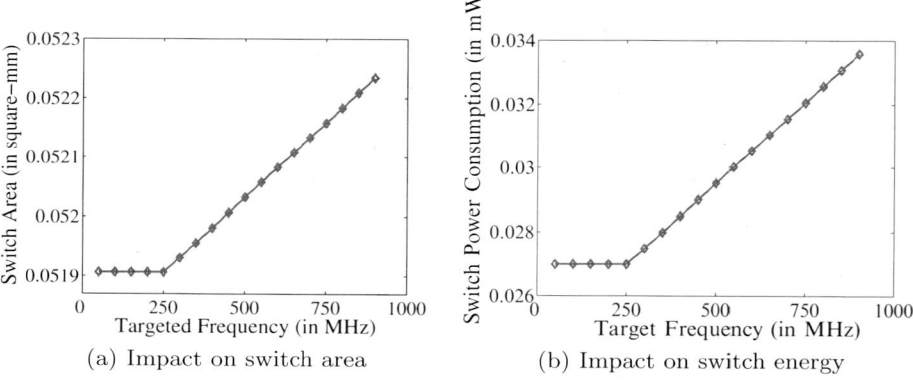

(a) Impact on switch area   (b) Impact on switch energy

Fig. 8. Impact of frequency on the area and energy of a 5×5 switch, for $0.13\mu m$ technology

implementations, the back-annotated accurate wire capacitances and resistances are obtained, with a $0.13\mu m$ technology library. The switching activity in the network components is varied by injecting functional traffic. The capacitance, resistance and the switching activity report are combined to estimate power consumption using Synopsys PrimePower [34].

A large number of implementation runs were performed, varying several parameters, such as the number of input, output ports, link-width and the amount of switching activity for the NoC switches. When the size of a NoC switch increases, the size of the arbiter and the crossbar matrix inside the switch also increases, thereby increasing the critical path of the switch. To have accurate delay estimates of the switches, we model the maximum frequency that can be supported by the switches, as a function of the switch size. An example set of values are presented in Figure 7.

We used linear regression to build analytical models for the area and power consumption of the components as a function of these parameters. Due to the intrinsic modularity and symmetry of NoC components, the models built are very accurate (with maximum and mean error of less than 7% and 5%, respectively) when compared to the actual values. In the ×pipes architecture, each core is connected to a separate NI [36]. Hence, we consider the power consumption of the NI to be part of the power consumption of the core.

The impact of the targeted frequency of operation on the area and energy consumption of an example 5×5 switch obtained from layout-level estimates is presented in Figure 8. Note that we plot the energy values (in power/MHz) instead of the total power. This is to avoid the inherent increase in power consumption due to increase in frequency of the network. When the targeted frequency of operation is below a certain frequency, referred to as the nominal operating frequency (around 250 MHz in the plots), the area and energy values for the switch remains the same. However, as the targeted frequency increases beyond the nominal frequency, the area and energy values start increasing linearly with frequency. This is because, the synthesis tool (such as Synopsys DC [34]) tries to match the desired high operating frequency by utilizing faster components that have large area and energy overhead. When performing the area, power estimates, we also model this impact of desired operating frequency on the switch area, power consumption.

5   Experimental Results

In this section, we present detailed experimental studies of our approach (which we further refer to as *INT-TOP* meaning message-dependent deadlock avoidance integrated with topology synthesis process) and compare it with traditional approaches:

(1) Using logically separate networks (*L-SEP*): In this scheme, we use separate buffers at each input, with as many buffers as the different message types, modeling the virtual channel based approach to remove message-dependent deadlocks.

(2) Using physically separate networks (*P-SEP*): In this scheme, we design physically different networks for each message type. For both these schemes we apply our topology synthesis procedure to obtain the network topologies.

(3) With a design that has no support to avoid message-dependent deadlocks (*ORIG*). Note that this base system cannot be employed in SoCs, as it cannot guarantee proper system operation. We present the experimental results for this scheme to only evaluate the overhead incurred in the other schemes to support deadlock-free operation.

5.1   Comparison on SoC designs

We apply the deadlock prevention methods to five different SoC designs: *Multimedia system (MULT 30 cores), IMage Processing application (IMP-27 cores), Video PROCessor (VPROC-42 cores), MPEG4 decoder (12 cores) and Video Object Plane Decoder (VOPD-12 cores).* The communication characteristics of some of these benchmarks is presented in [40]. There are two types of messages

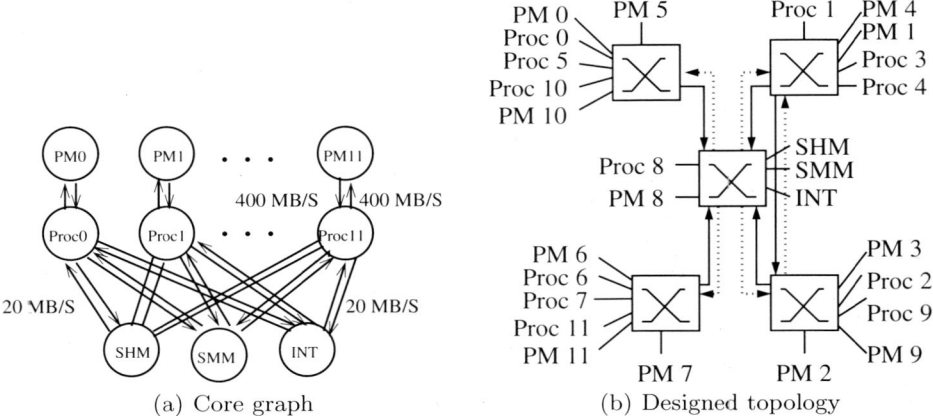

(a) Core graph   (b) Designed topology

Fig. 9. Core graph and designed topology for IMP

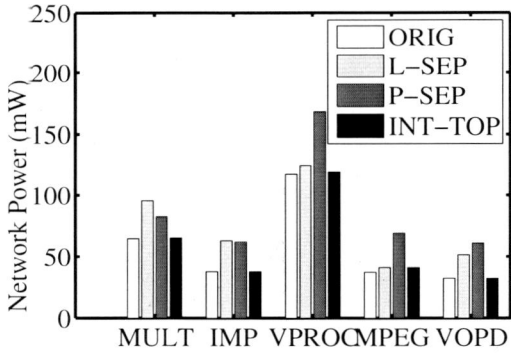

Fig. 10. Power consumption of different schemes

that are supported in each design: *request* and *response*. Each design consists of almost equal number of request and response traffic flows. This is because, every processor core communicates through the memory core, necessitating two-way communication (hence a request and response traffic flow) between the processors and memories. To make a fair comparison of the different schemes, we use the same synthesis approach and design constraints for synthesizing the topologies.

The communication pattern (core graph) for one of the applications (*IMP*) and the best synthesized topology for our proposed scheme (*INT-TOP*) are presented in Figures 9(a) and 9(b). The design consists of 12 processors (Proc 0 to Proc 11), a private memory for each processor (PM 0 to PM 11), a shared memory (SHM), a semaphore memory (SMM) and an interrupt device (INT). In the application, all communication from the processors are of request message type and communication to the processors are of response message type. In Figure 9(b), those links that support request message type are in bold and those links that support response message type are dashed.

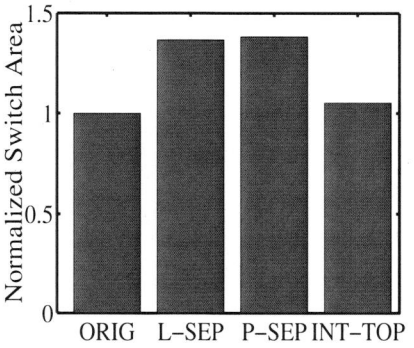

Fig. 11. Normalized switch area for the different schemes

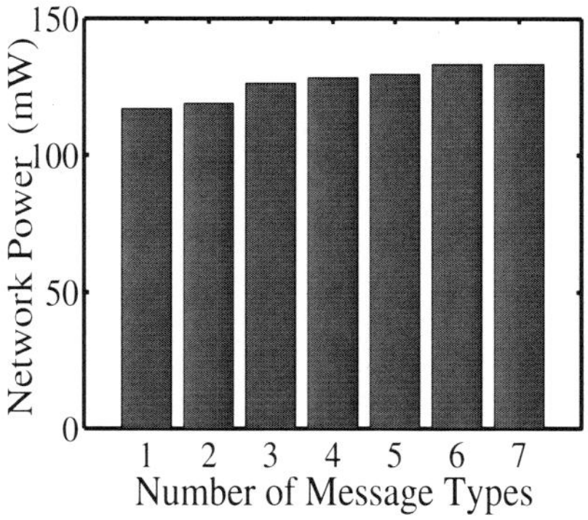

Fig. 12. Effect of number of message types

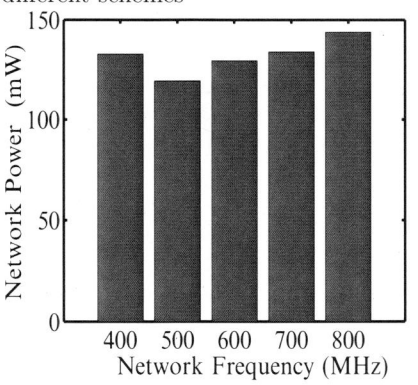

Fig. 13. Power consumption variation with frequency

The network power consumption, based on the functional traffic for the various designs using the different schemes is presented in Figure 10. As seen from this figure, the *INT-TOP* scheme presented in this work, outperforms the two conventional message-dependent deadlock avoidance schemes: *L-SEP* and *P-SEP*. Our proposed scheme leads to an average of 38.5% reduction in NoC power consumption when compared to the state-of-the-art deadlock avoidance schemes. When compared to our *INT-TOP* scheme, the *L-SEP* scheme requires large buffering requirements, as each virtual channel needs separate buffering resources. The *P-SEP* scheme requires more switches than the *INT-TOP* scheme, as the request and response messages utilize different networks. Interestingly, our proposed scheme incurs only a 2.5% increase in power consumption when compared to the *ORIG* scheme, where no message-dependent deadlock avoidance support is provided. This is mostly due to the efficient allocation of links to the different message types by our topology synthesis procedure. The switch area for the different schemes for the SoC designs, normalized with respect to the area of the base system (*ORIG*) is presented in Figure 11. The proposed method results in an average of 30.68% reduction in area when compared to the state-of-the-art schemes.

5.2   Effect of Different Number of Message Types

In this sub-section, we examine the power consumption of the proposed scheme, when the number of different message types is varied. The number of message types in a system depends on the underlying computation architecture. Cache coherent systems typically support several different message types. As an example, the S-1 multi-processor supports 4 different message types [18] and each type must be mapped onto different resources in the network. In [17], a more sophisticated protocol is used, which leads to seven different message types. To see the impact on the number of different message types, we created a synthetic benchmark having the traffic characteristics of the *VPROC* design. In this benchmark, around 80 different traffic flows exist, each one representing a message. We fixed the number of messages and varied the number of message types in the design from 1 to 7. The network power consumption for our proposed scheme, for the different number of message types is presented in Figure 12. This figure shows that our proposed scheme results in efficient designs, even for a large number of message types. Moreover, the rise in power consumption with an increasing number of message types saturates (designs with 6 and 7 message types have nearly the same power consumption), as most messages are already mapped onto unique links in the network.

5.3   Frequency Trade-offs

The algorithm presented here can also be used to perform frequency selection for a certain design. In this case, the frequency of operation of the NoC can be varied and the best topology can be synthesized for each frequency point. A higher operating frequency results in links having more bandwidth. Thus a smaller NoC can satisfy the design constraints. A trade-off curve for frequency vs power consumption of the network for the VPROC is presented in Figure 13. From such a curve, the most power-efficient operating frequency can be chosen for the design.

6   Conclusions

For *Networks on Chips (NoCs)* to be used in industrial designs, NoCs should guarantee proper system operation under all conditions. Achieving deadlock-free operation of the network with minimum power consumption and area overhead is critical for application-specific NoCs. In this work, we have focused on addressing the major issue of avoiding routing and message-dependent deadlocks during the network operation. We have shown that by mapping the different message types onto different network resources during the topology mapping and synthesis phase, we can achieve large reductions in network power consumption and network area when compared to the state-of-the-art approaches. In future work, we plan to compare deadlock recovery schemes with the proposed scheme for NoCs.

7 Acknowledgments

This work is supported by the US National Science Foundation (NSF, contract CCR-0305718) for Stanford University. It is also supported by the Swiss National Science Foundation (FNS, Grant 20021-109450/1) and the Spanish Government Research Grant TIN2005-5619. The work is also supported by a grant from Semiconductor Research Corporation (SRC project number 1188) and a grant by STMicroelectronics for DEIS.

References

1. L. Benini and G.De Micheli, "Networks on Chips: A New SoC Paradigm", IEEE Computers, pp. 70-78, Jan. 2002.
2. M. Sgroi et al., "Addressing the System-on-a-Chip Interconnect Woes Through Communication-Based Design", Proc. DAC, pp. 667-672, June 2001.
3. S. Kumar et al., "A Network on Chip Architecture and Design Methodology", Proc. ISVLSI, pp. 117-122, April 2002
4. P. Guerrier, A. Greiner, "A generic architecture for on-chip packet-switched interconnections", Proc. DATE, pp. 250-256, March 2000.
5. K. Goossens et al., "The Aethereal network on chip: Concepts, architectures, and implementations", IEEE Design and Test of Computers, Vol. 22(5), pp. 21-31, Sept-Oct 2005.
6. W. Dally, B. Towles, "Route Packets, not Wires: On-Chip Interconnection Networks", Proc. DAC, pp. 684-689, June 2001.
7. Y. H. Song, T. M. Pinkston, "A Progressive Approach to Handling Message-Dependent Deadlock in Parallel Computer Systems", IEEE TPDS, Vol. 14(3), pp. 259-275, March 2003.
8. Y. Choi, "Deadlock Recovery Based Router Architectures for High Performance Networks", PhD Dissertation, University of Southern California, June 2001.
9. G. Chiu, "The Odd-Even Turn Model for Adaptive Routing", IEEE TPDS, Vol. 11(7), pp. 729-738, July 2000.
10. C. Glass, L. Ni, "The turn model for adaptive routing", Proc. ISCA, pp. 278-287, 1992.
11. J. Duato, "A New Theory of Deadlock-Free Adaptive Routing in Wormhole Networks", IEEE TPDS, Vol. 8(8), pp. 790-802, Aug 1997.
12. D. Starobinksi et al., "Application of network calculus to general topologies using turn-prohibition", IEEE/ACM Transactions on Networking, Vol. 11, Issue 3, pp. 411-421, June 2003.

13. W. J. Dally, H. Aoki, "Deadlock-Free Adaptive Routing in Multi-computer Networks Using Virtual Channels", IEEE TPDS, Vol. 4(4), pp. 466-475, April 1993.
14. S. Scott, G. Thorson, Optimized Routing in the Cray T3D", Proc. Workshop Parallel Computer Routing and Comm., pp. 281-294, May 1994.
15. S. Scott, G. Thorson, "The Cray T3E Network: Adaptive Routing in a High Performance 3D Torus", Proc. Symp. Hot Interconnects IV, pp. 147-156, Aug. 1996.
16. J. Carbonaro, Cavallino, "The Teraflops Router and NIC", Proc. Symp. Hot Interconnects IV, pp. 157-160, Aug. 1996.
17. S.S. Mukherjee et al., "The Alpha 21364 Network Architecture", Proc. Symp. HOT Interconnects 9, pp. 113-117, Aug. 2001.
18. L. Widdoes, S. Correll, :The S-1 Project: Developing High Performance Computers", Proc. COMPCON, pp. 282-291, Spring 1980.
19. "http://www.st.com".
20. J. Laudon, D. Lenoski," The SGI Origin: A ccNUMA Highly Scalable Server", Proc. ISCA, pp. 241-251, June 1997.
21. D. Lenoski et al., " The Directory-Based Cache Coherence Protocol for the DASH Multiprocessor", Proc. ISCA, pp. 148-159, 1990.
22. A. Hansson, K. Goossens, A. Radulescu, "UMARS: A Unified Approach to Mapping and Routing on a Combined Guaranteed Service and Best-Effort Network-on-Chip Architecture", Technical Report 2005/00340, Philips Research, April 2005.
23. B. Gebremichael et al., "Deadlock Prevention in the Aethereal Protocol", Proc. Working Conference on Correct Hardware Design and Verification Methods (CHARME), Oct 2005.
24. J. Hu, R. Marculescu, 'Exploiting the Routing Flexibility for Energy/Performance Aware Mapping of Regular NoC Architectures', Proc. DATE, March 2003.
25. S. Murali, G. De Micheli, "SUNMAP: A Tool for Automatic Topology Selection and Generation for NoCs", Proc. DAC 2004.
26. S. Murali et al., "Mapping and Physical Planning of Networks on Chip Architectures with Quality-of-Service Guarantees", Proc. ASPDAC 2005.
27. A.Pinto et al., "Efficient Synthesis of Networks on Chip", ICCD 2003, pp. 146-150, Oct 2003.
28. W.H.Ho, T.M.Pinkston, "A Methodology for Designing Efficient On-Chip Interconnects on Well-Behaved Communication Patterns", HPCA 2003, pp. 377-388, Feb 2003.
29. T. Ahonen et al. "Topology Optimization for Application Specific Networks on Chip", Proc. SLIP 04.
30. K. Srinivasan et al., "An Automated Technique for Topology and Route Generation of Application Specific On-Chip Interconnection Networks", Proc. ICCAD '05.
31. A. Hansson et al., "A unified approach to constrained mapping and routing on network-on-chip architectures", pp. 75-80, Proc. ISSS 2005.
32. S. Murali et al., "Designing Application-Specific Networks on Chips using Floorplan Information", Proc. ICCAD 2006.
33. W. J. Dally, B. Towles, "Principles and Practices of Interconnection Networks", Morgan Kaufmann , Dec 2003.
34. "http://www.synopsys.com".
35. "http://www.cadence.com".
36. S. Stergiou et al., "×pipesLite: a Synthesis Oriented Design Library for Networks on Chips", pp. 1188-1193, Proc. DATE 2005.
37. F. Angiolini et al., "Contrasting a NoC and a Traditional Interconnect Fabric with Layout Awareness", pp. 124-129, Proc. DATE 2006.

38. A. Pullini, F. Angiolini, D. Bertozzi, L. Benini, "Fault Tolerance Overhead in Network-on-Chip Flow Control Schemes", Proc. SBCCI, pp. 224-229, 2005.
39. www.ocpip.org
40. D. Bertozzi et al., "NoC Synthesis Flow for Customized Domain Specific Multi-Processor Systems-on-Chip", IEEE Transactions on Parallel and Distributed Systems, Feb 2005.

Dynamic Reconfigurable Architecture Exploration based on Parameterized Reconfigurable Processor Model

Ittetsu Taniguchi, Keishi Sakanushi, Kyoko Ueda,
Yoshinori Takeuchi, and Masaharu Imai

Graduate School of Information Science and Technology
Osaka University

Abstract. In recent years, dynamic reconfigurable processor which can achieve reconfiguration with a few cycles is proposed. The fast reconfiguration makes run-time reconfiguration possible, and the run-time reconfiguration gives a new possibility to the dynamic reconfigurable processor, i.e. the dynamic reconfigurable processor can also execute partitioned independent subtasks with repeated reconfigurations and executions. However, to achieve an execution with the run-time reconfiguration, performance should be evaluated with various overheads: reconfiguration, memory accesses, etc. The overheads depend on reconfigurable architectures, and it is generally difficult to evaluate the overhead. As the overhead may critically affect the performance, designers should carefully explore design space for suitable architectures. In this paper, we propose a dynamic reconfigurable architecture exploration method based on Parameterized Reconfigurable Processor model (PRP-model) and task partitioning optimization algorithm for architecture exploration corresponding to proposed PRP-model. Experimental results showed that the proposed PRP-model and the task partitioning algorithm for PRP-model can fast evaluate various reconfigurable architectures, and designers can easily find suitable reconfigurable architectures by changing the PRP-model parameters.

1 Introduction

Portable information systems such as cellular phones and mobile MP3 players are widely spreading in our daily life. In general, there are various requirements for a design of portable information systems, e.g. low hardware cost and high performance. The requirements of low hardware cost lead to inexpensive and portable products, and the high performance is usually needed for media processing. Moreover, flexibility is required to adapt various coding standards using the same chips on board. To fulfill these requirements, system designers always design the products under the hard constraints to take account of design quality metrics: performance, area, and power. When designing embedded systems, it is essential to explore design solution space and choose a solution which they really need.

Please use the following format when citing this chapter:

Taniguchi, I., Sakanushi, K., Ueda, K., Takeuchi, Y. and Imai, M., 2007, in IFIP International Federation for Information Processing, Volume 249, VLSI-SoC: Research Trends in VLSI and Systems on Chip, eds. De Micheli, G., Mir, S., Reis, R., (Boston: Springer), pp. 357–376

To meet these requirements, Instruction Set Processors (ISPs) or ASICs have usually been used for their design. However they cannot completely satisfy these requirements. While ISPs can flexibly execute various functions by changing software, they cannot achieve high performance. While ASICs can realize high performance for specific applications, it is difficult to use them for other applications. Therefore, as a new approach to meet them, dynamic reconfigurable processors that feature both the flexibility of ISPs and the high performance of ASICs are focused [1, 2].

Dynamic reconfigurable processors usually have many coarse grain processing elements (PEs), and each PE is connected each other by flexible interconnections. Since many PEs are simultaneously executed, the tasks with high parallelism are effectively executed. The function of dynamic reconfigurable processor is defined by configuration data, e.g. setting information of interconnections and function of each PE, that is, according to the configuration data prepared before hand, the dynamic reconfigurable processor can reconfigure itself into various circuits. The reconfiguration speed depends on total amount of configuration data and the reconfigurable architecture specification, and the reconfiguration timing is decided by the reconfiguration speed.

In recent years, dynamic reconfigurable processor which can achieve a reconfiguration with a few cycles is proposed. The fast reconfiguration makes run-time reconfiguration possible, and the run-time reconfiguration gives a new possibility to the dynamic reconfigurable processor, i.e. the dynamic reconfigurable processor can also execute partitioned independent subtasks with repeated reconfigurations and executions. The run-time reconfiguration is a special feature that traditional programmable device does not have, and the authors pay attention to this feature and its potential.

However, to achieve an execution with the run-time reconfiguration, performance should be evaluated with various overheads: reconfiguration, memory accesses, etc. The overheads depend on reconfigurable architectures, and it is generally difficult to evaluate the overhead. As the overhead may critically affect the performance, designers should carefully explore design space for suitable architectures. The authors claim varieties of reconfigurable architectures and difficulty of architecture evaluation confuse designers to explore vast design space for the best solution and fast evaluation method for various reconfigurable architectures is needed.

In this paper, we propose a Parameterized Reconfigurable Processor model (PRP-model) and a task partitioning optimization algorithm for architecture exploration corresponding to proposed PRP-model. The task partitioning optimization algorithm divides tasks into subtasks to minimize execution cycles. The algorithm is applicable to various reconfigurable architectures and supports the evaluation of various architectures for specific applications by changing PRP-model parameters. To realize run-time reconfiguration, designers can easily find the suitable reconfigurable architectures.

This paper is structured as follows: section 2 gives an overview of related work and highlights our contribution. Sections 3, 4, and 5 present a Parameterized Reconfigurable Processor model, a port expansion DFG, and proposed algorithm, respectively. Experimental results are given in section 6. In section 7, we conclude this paper.

2 Related Work

Various reconfigurable processor architectures have been proposed, which are classified by reconfiguration granularity or reconfiguration time [1]. To use the architectures effectively, many design methods, algorithms, and applications have been studied. Especially, task partitioning problem is one of essential dilemmas to use dynamic reconfigurable processors, and so far many studies have been conducted. However, we have never seen a task partitioning problem for execution model of repeated reconfigurations and executions.

[3] proposed a HW/SW task partitioning that considered task assignment for Instruction Set Processors (ISPs). [4] proposed a task partitioning method that partitions tasks into two reconfigurable processors with different granularities. It cannot be applied to different reconfigurable processor architectures. [5] proposed a task partitioning algorithm considering reconfigurable overhead which means the number of CLBs of FPGA for the communication. [6] proposed behavior partitioning method which does high level synthesis of each task of the behavior simultaneously. It can get the optimal solution because the problem is come down to NLP, but it takes a long time even if the target application is small. [7] proposed a task partitioning method under the constraints of the number of memory ports, and can get the optimal solution. [8] considers a task partitioning limiting the number of partitioned subtasks. In this paper, we do not limit the number of partitioned subtasks. [9] proposed the task partitioning method for DRL architecture. [3–8] proposed methods for FPGA platforms. FPGA needs to store intermediate data to an external memory at a reconfiguration, because it cannot hold the data during reconfiguration. In recent years, many dynamic reconfigurable architectures with registers in the array are proposed, and the dynamic reconfigurable architectures can hold the data in the array during reconfiguration. In this paper, we propose a task partitioning method for not only FPGA but also the dynamic reconfiguration architectures which can hold the data in the array during reconfiguration.

[10, 11] proposed reconfigurable architecture exploration method. [10] proposed a design space exploration method using the task partitioning method proposed in [6]. [11] proposed ADRES architecture template, which is unique architecture consisting of VLIW processor and a PE array, and architecture exploration method only applicable to the template. In this paper, to realize the execution model of repeated reconfigurations and executions, we propose architecture exploration method which offers designers fast evaluation of various reconfigurable architectures by changing architecture parameters.

3   Parameterized Reconfigurable Processor Model

To evaluate many reconfigurable architectures in a short time, the reconfigurable processor model which covers many kinds of processing elements and memory architectures is needed. In this paper, we propose a Parameterized Reconfigurable Processor model (PRP-model) and task partitioning algorithm based on the PRP-model.

3.1   Processor Structure Model

Figure 1 illustrates the proposed PRP-model. PRP-model includes PE array arranged processing elements (PEs), internal memories with different capacity, and configuration memory to store configuration data.

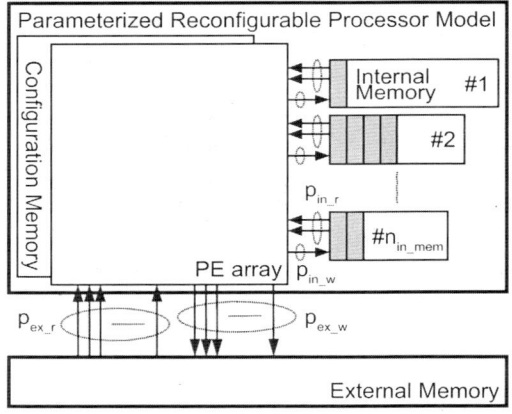

Fig. 1. Parameterized Reconfigurable Processor Model

PE Array A PE array is composed of three types of PEs: pPE, rPE, and prPE. pPE (Processing PE) has an ALU, and rPE (Register PE) has a register file, respectively. prPE (Processing and Register PE) has both an ALU and a register file. The numbers of pPEs, rPEs, and prPEs in a PRP-model are denoted as n_{pPE}, n_{rPE}, and n_{prPE}, respectively. pPEs can perform some operations, but pPEs cannot hold data at reconfiguration since it does not have any registers. rPEs can hold data at reconfiguration, but rPEs cannot perform any operations since it does not have any ALUs. Since prPEs have both ALUs and registers, it can operate and hold data. ALUs included in pPEs or prPEs have the same functionalities, and we assume that the application tasks are resolved into operations which pPEs or prPEs can execute on PEs. In this model, we treat the total amount of data as the number of data packet, and the numbers of data that rPEs and prPEs can hold are denoted as n_{reg_rPE} and n_{reg_prPE}, respectively. In this model, we consider the available number of all PEs and not the interconnection or placement of PEs because the interconnection and placement can be defined after the number of PE is decided.

Memory Structure Memory structure of PRP-model comprises three types of memories: external memory, internal memory, and register file of rPEs or prPEs. The number of internal memory is n_{in_mem}. Each memory has some read or write access ports, and can read and write as much data as access ports at reconfiguration.

External memory is large enough to hold any data, and it has p_{ex_r} read access ports and p_{ex_w} write access ports. One read access needs t_{ex_r} cycles, and one write access needs t_{ex_w} cycles. Thus, up to p_{ex_r} data are read from external memory at t_{ex_r} cycles, and up to p_{ex_w} data are written to external memory at t_{ex_w} cycles.

The i-th internal memory keeps $n_{in\_mem\_dat}(i)$ data. The numbers of read port and write port of internal memory are p_{in_r} and p_{in_w}, respectively, and t_{in_r} cycles or t_{in_w} cycles are needed to read or write data.

rPE and prPE have p_{reg_r} read ports and p_{reg_w} write ports, and t_{reg_r} cycles or t_{reg_w} cycles are needed to read or write data.

In PRP-model, memory access of data read or write can be done in parallel, and execution of calculation starts after all memory accesses finish.

Configuration Memory Configuration memory can store n_{config} configuration data. All configuration data are the same size, and reconfiguration always needs t_{config} cycles. After k-th reconfiguration, a new configuration data can be overwritten to k-th configuration data from external memory. It takes t_{cfg_r} cycles to store one configuration data to configuration memory.

Let $n_{subtask}$ be the number of partitioned subtasks to execute, that is, $n_{subtask}$ configurations are needed to execute the target task. When $n_{subtask}$ is not over n_{config}, all configuration data can be kept at configuration memory. On the other hand, when $n_{subtask}$ is greater than n_{config}, $n_{subtask} - n_{config}$ configuration data cannot be kept at configuration memory. Thus, at run-time, the configuration data which cannot be kept at configuration memory are read from external memory by its reconfiguration. The configuration data read and task execution can execute simultaneously.

Usually, memory specification is defined by bit width and memory depth. Let BW_{CM} and MD_{CM} be the bit width and the memory depth of the configuration memory, respectively. The total bit of configuration memory TB_{CM} is calculated as follows:

$$TB_{CM} = BW_{CM} \cdot MD_{CM}. \tag{1}$$

Let TB_{config} be the number of total bit of one configuration data, and TB_{config} is expressed by Eq.(2).

$$TB_{config} = Scale \cdot (n_{pPE} + n_{prPE} + n_{rPE}), \tag{2}$$

where *Scale* is a parameter of reconfigurable architecture complexity, and it is defined by each reconfigurable architecture, e.g. function of each PE, interconnection, etc. In this research, we assume the configuration data is increased linearly according to the number of PEs because the function of PE and the interconnection are not considered in PRP-model.

Then, n_{config} and t_{cfg_r} is expressed by Eq.(3) and Eq.(4).

$$n_{config} = \left\lfloor \frac{TB_{CM}}{TB_{config}} \right\rfloor. \tag{3}$$

$$t_{cfg\_r} = \left\lceil \frac{TB_{config}}{BW_{CM}} \right\rceil. \tag{4}$$

3.2   Memory Access Overhead

PRP-model has storage resources: external memory, internal memory, and register file of rPEs and prPEs. The input data of application are read from external memory and the output data of application are finally written to external memory. Thus, internal memories and register files in PEs are used to keep temporal data at reconfiguration. When the temporal data cannot be kept at internal memories or register file due to the limitation of the capacity, the data are stored to external memory.

The memory access cycles on PRP-model are calculated as follows. Let $rec_{in\_r}(i,j)$ be the number of requested data from the i-th internal memory at configuration j. The data read cycles at configuration j, $T_{in\_r}(j)$, are expressed by Eq.(5).

$$T_{in\_r}(j) = \max_{0 \le i < n_{in\_mem}} \{T_{in\_r}(i,j)\}, \tag{5}$$

where $T_{in\_r}(i,j)$, the data read cycles from i-th internal memory at configuration j, are as follows:

$$T_{in\_r}(i,j) = \left\lceil \frac{req_{in\_r}(i,j)}{p_{in\_r}} \right\rceil \cdot t_{in\_r}. \tag{6}$$

Let $req_{ex\_r}(j)$ be the number of requested data from external memory at configuration j. The data read cycles at configuration j, $T_{ex\_r}(j)$, are expressed by Eq.(7).

$$T_{ex\_r}(j) = \left\lceil \frac{req_{ex\_r}(j)}{p_{ex\_r}} \right\rceil \cdot t_{ex\_r}. \tag{7}$$

Let $req_{prPE\_r}(i,j)$ be the number of requested data from i-th prPE at configuration j. The data read cycles at configuration j, $T_{prPE\_r}(j)$, are expressed by Eq.(8).

$$T_{prPE\_r}(j) = \max_{0 \le i < n_{prPE}} \{T_{prPE\_r}(i,j)\}, \tag{8}$$

where $T_{prPE\_r}(i,j)$, the data read cycles from i-th prPE at configuration j, are as follows:

$$T_{prPE\_r}(i,j) = \left\lceil \frac{req_{prPE\_r}(i,j)}{p_{reg\_r}} \right\rceil \cdot t_{reg\_r}. \tag{9}$$

Let $req_{rPE\_r}(i,j)$ be the number of requested data from i-th rPE at configuration j. The data read cycles at configuration j, $T_{rPE\_r}(j)$, are expressed by Eq.(10).

$$T_{rPE\_r}(j) = \max_{0 \le i < n_{rPE}} \{T_{rPE\_r}(i,j)\}, \tag{10}$$

where $T_{rPE\_r}(i,j)$, the data read cycles from i-th rPE at configuration j, are as follows:

$$T_{rPE\_r}(i,j) = \left\lceil \frac{req_{rPE\_r}(i,j)}{p_{reg\_r}} \right\rceil \cdot t_{reg\_r}. \tag{11}$$

The data read cycles ("Memory Access (Data Read)" in figure 2) of configuration j are expressed by Eq.(12).

$$T_r(j) = \max \{T_{ex\_r}(j), T_{in\_r}(j), T_{PE\_r}(j)\}, \tag{12}$$

where $T_{PE\_r}(j)$, the data read cycles from prPE and rPE at configuration j, are as follows:

$$T_{PE\_r}(j) = \max \{T_{prPE\_r}(j), T_{rPE\_r}(j)\}. \tag{13}$$

The data write cycles at configuration j, $T_w(j)$, are expressed the same way.

3.3   Processing and Reconfiguration

Figure 2 shows processing flow of PRP-model.

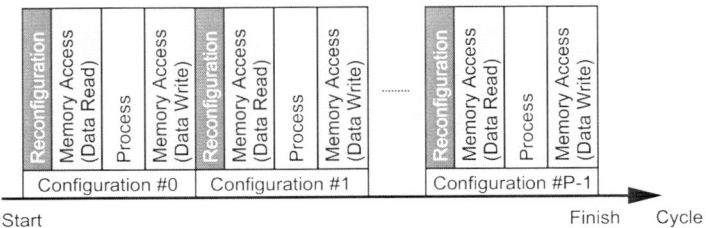

Fig. 2. Processing Flow

PRP-model is configured to the target circuit according to the configuration data, and input data or intermediate data are read from external memory, internal memory, and register file of rPEs and prPEs. Then the partitioned sub-tasks are processed after reading data. Finally, output data or intermediate data are stored to storage resources. PRP-model processes this processing flow repeatedly. However, as the limited configuration memory, the processing flow of PRP-model may be stalled according to the reading cycles of the configuration data needed to the reconfiguration. The waiting cycles for configuration data read are calculated as follows.

Let $End_{exec}(i)$ and $End_{cfg\_r}(i)$ be the end time of the execution of configuration i and the end time of i-th configuration data read from external memory, respectively. The end time of i-th reconfiguration is expressed by Eq.(14).

$$End_{reconf}(i) = Start_{reconf}(i) + t_{config}, \tag{14}$$

where $Start_{reconf}(i)$ is as follows:

$$Start_{reconf}(i) = \max\left\{End_{exec}(i-1), End_{cfg\_r}(i)\right\}. \tag{15}$$

After i-th reconfiguration, configuration i is executed. Let $T_{proc}(i)$ be the processing cycles of i-th subtask without memory access cycles, and the end time of execution of configuration i are expressed by Eq.(16).

$$End_{exec}(i) = Start_{exec}(i) + T_r(i) + T_{proc}(i) + T_w(i), \tag{16}$$

where $Start_{exec}(i)$ is as follows:

$$Start_{exec}(i) = End_{reconf}(i). \tag{17}$$

PRP-model can simultaneously read the configuration data and execute the task. The start time of configuration data read is expressed by Eq.(18).

$$Start_{cfg\_r}(i) = \max\left\{End_{cfg\_r}(i-1), End_{reconf}(i - n_{config})\right\}, \tag{18}$$

where $End_{cfg\_r}(i)$ is as follows:

$$End_{cfg\_r}(i) = Start_{cfg\_r}(i) + t_{cfg\_r}. \tag{19}$$

Therefore, the waiting cycles according to the stalled processing flow are expressed by Eq.(20).

$$T_{wait} = \sum_{i=0}^{P-1}\left\{Start_{reconf}(i+1) - End_{exec}(i)\right\}. \tag{20}$$

Total execution cycles are evaluated including T_{wait}.

4   Port Expansion DFG

We define port expansion DFG to evaluate reconfiguration overhead in terms of exactly calculating the number of memory accesses. Data flow graph (DFG) is one method of task representation with execution dependency. Traditional DFG is composed of nodes and edges. Node represents an operation, and the edge connected to a pair of nodes means the data flow between them. Incoming edges of a node mean the input data used for the corresponding operation, and outgoing edges of a node mean the results from corresponding operation.

When the data are used at some nodes, data read sometimes occurs but data write occurs only one time. The traditional DFG cannot precisely represent the number of memory accesses because the differences of the data represented by outgoing edges cannot be distinguished. Thus, we label the input/output of node of traditional DFG as "port", which represents the input or output data. We call this DFG as port expansion DFG. Each node of the port expansion DFG has ports, and each edge connects to a pair of ports.

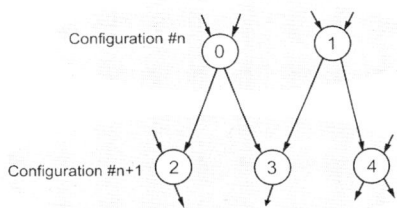

Fig. 3. Traditional DFG

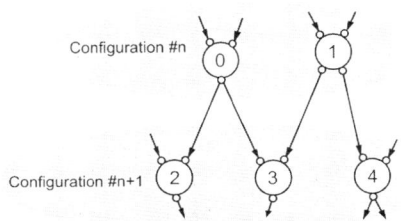

Fig. 4. Port Expansion DFG

Figure 4 shows an example of a port expansion DFG based on a traditional DFG showed in figure 3. In figure 3, it is difficult to calculate the number of memory accesses when the configuration n is reconfigured into the configuration $n + 1$. However, using a port expansion DFG, we can recognize that node 0 clearly outputs data used at nodes 2 and 3, and node 1 outputs two data that are respectively used at nodes 3 and 4. Thus, it can be obtained that three data writes are required after configuration n.

5   Task Partitioning Algorithm

5.1   Task Partitioning Problem

In this section, we define the task partitioning problem for PRP-model.

– TASK PARTITIONING PROBLEM –

For given a port expansion DFG of the target application and a PRP-model, to find a task partition and storage resource assignment whose execution cycles are minimum keeping the execution order defined by port expansion DFG.

□

5.2   Outline of Task Partitioning Algorithm

In this section, we propose a task partitioning algorithm using Simulated Annealing (SA) for PRP-model that consists of configuration and storage resource assignments. The following is the outline of task partitioning algorithm.

1. Initial solution decision.
2. The following steps are repeatedly processed until SA's final condition is satisfied:
   (a) Configuration assignment by MOVE operation.
   (b) Storage resource assignment.
   (c) Execution cycles estimation for the current solution.
   (d) Optimal solution update.

   In configuration assignment, the task is partitioned into several subtasks, and each subtask is assigned to a configuration according to an execution order. Then in storage resource assignment, the intermediate data between configurations are assigned to storage resources.

5.3   Configuration Assignment

In this section, we explain how to make neighbor solution (configuration assignment) using MOVE operation. The MOVE operation in SA is the movement of a randomly selected node to an adjacent configuration. We define two MOVE operations: $MOVE_{BWD}$ and $MOVE_{FWD}$. $MOVE_{BWD}$ is the movement of the node to the previous configuration, and $MOVE_{FWD}$ is the movement of the node to the next configuration (figure 5).

   Let $Child(x)$ and $Parent(x)$ be a set of child nodes of node x and a set of parent nodes of node x, respectively. Let $Cfg(s, x)$ and $Vacant(s, c)$ be the number of configuration assigned to node x of the solution s and the number of vacancies of configuration c of the solution s, respectively. We call configuration c *empty* when there is no nodes in the configuration c.

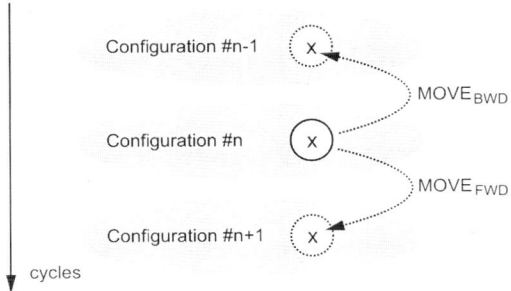

Fig. 5. $MOVE_{BWD}$ and $MOVE_{FWD}$ operations

When the node n in the solution s satisfies Eq.(21) and Eq.(22), $MOVE_{BWD}$ can apply to node n.

$$\forall x \in Parent(n); Cfg(s,x) < Cfg(s,n) \tag{21}$$

$$Vacant(s, Cfg(s,n) - 1) > 0 \tag{22}$$

Similarly, when the node n in the solution s satisfies Eq.(23) and Eq.(24), $MOVE_{FWD}$ can apply to node n.

$$\forall x \in Child(n); Cfg(s,x) > Cfg(s,n) \tag{23}$$

$$Vacant(s, Cfg(s,n) + 1) > 0 \tag{24}$$

$MOVE_{BWD}$ and $MOVE_{FWD}$ operations occur sometimes an *empty* configuration. In figure 6(a), there is *empty* configuration c in the configuration sequence because of $MOVE_{BWD}$ for the node k. In such case, $MOVE_{BWD}$ removes the *empty* configuration c after moving node k (figure 6(b)). We call this operation "Packing". By the packing, the number of the configuration is not always greater than the number of nodes of port expansion DFG.

5.4 Storage Resource Assignment

In this section, we define storage resource assignment.

"Data1" and "data2", illustrated in figure 7, should be saved as storage resources because these data cannot be kept at reconfiguration. Thus, the storage resource assignment algorithm assigns these interconfiguration data to storage resource according to the following policy after configuration assignment:

– Data are assigned to the highest priority resource with space.

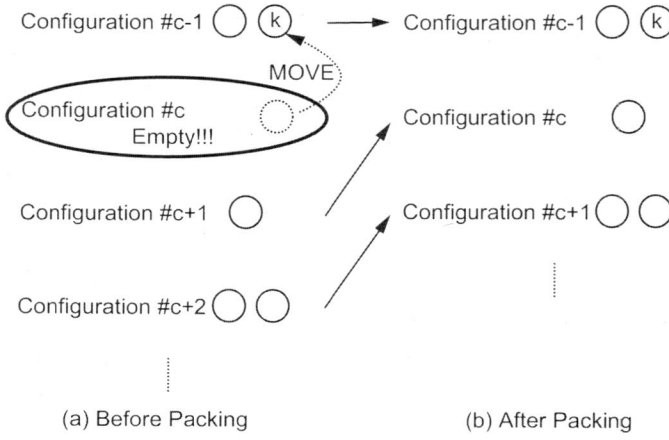

(a) Before Packing   (b) After Packing

Fig. 6. Packing for *empty* configuration

- The first priority resource is rPE, the second is prPE, the third is internal memory, and last is external memory.

Figure 8 shows an example of storage resource assignment, and "data1" and "data2" are stored external memory and internal memory at reconfiguration, respectively. Note that memory accesses to read or write are needed at reconfiguration. When the assigned data become unnecessary, other data can be overwritten.

Since the number of interconfiguration data is changed when the configuration assignment is changed by the MOVE operation, the storage resource assignment is recalculated after the MOVE operation.

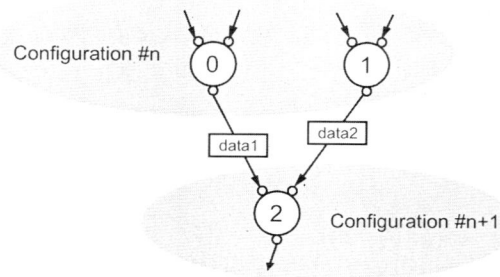

Fig. 7. The Data Crossed Border of Two Configurations

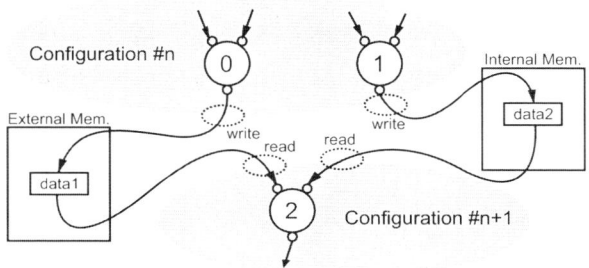

Fig. 8. Example of Storage Resource Assignment

6 Experiment

To demonstrate the efficiency of our proposed method, we show two experimental results: evaluation of task partitioning algorithm for PRP-model and reconfigurable architecture exploration. The experimental environment includes PentiumD 2.8 GHz, 2 GB memory, and Fedora 4. The Simulated Annealing (SA) parameters were as follows: initial temperature was 10, final temperature was 0.01, and the cooling ratio of the temperature was 0.98. Solutions obtained by SA are according to SA's random seed. Thus, we evaluate the average execution cycles and CPU times of 10 solutions obtained by different random seeds.

6.1 Evaluation of Task Partitioning Algorithm

To demonstrate the efficiency of the proposed algorithm, we applied it to two applications: DCT8 and CHENG. DCT8 is a one-dimensional eight point discrete cosine transform (DCT), implemented completely in parallel. CHENG is Cheng's DCT algorithm implemented completely in parallel, too. The number of nodes in DCT8's port expansion DFG is 50, which is 13 in CHENG. We evaluate the proposed task partitioning algorithm by focusing on the quality of solutions and CPU time. We evaluated 8 architectures under the variety of n_{pPE}, n_{prPE}, and n_{config}. The other architecture parameters were fixed and described in Table 1.

Table 1. Fixed Parameters

Parameter	Value	Parameter	Value
t_{ex_r}	2	t_{ex_w}	3
t_{in_r}	1	t_{in_w}	2
t_{reg_r}	1	t_{reg_w}	1
p_{ex_r}	4	p_{ex_w}	4
p_{in_r}	1	p_{in_w}	1
p_{reg_r}	1	p_{reg_w}	1
n_{in_mem}	0	n_{rPE}	0
n_{reg_prPE}	1	t_{config}	1
t_{cfg_r}	16		

Table 2 shows comparison results of the execution cycles of target application CHENG. "Proposed" in Table 2 is execution cycles obtained by the proposed task partitioning algorithm, and "Opt." is the optimal execution cycles obtained by branch and bound strategies. Experimental results showed that for all architectures the proposed task partitioning algorithm can obtain the same execution cycles as the optimal solutions. Notice that execution cycles decreased in Table 2 when the number of PEs changed from four to eight, because the more operations that are executable simultaneously, the fewer the reconfigurations. Using prPEs with internal register files decreases the execution cycles more than using pPE. Reconfiguration overhead is decreased because prPEs have register files that pPEs do not have at storage resource assignment.

Table 2. Comparison of Execution Cycles

No.	Parameters n_{pPE}	n_{prPE}	n_{config}	Architecture		Proposed	Opt.
1	0	4	1	4 prPEs	with Config. Mem. (1 Config.)	57	57
2	0	4	2		with Config. Mem. (2 Config.)	39	39
3	0	8	1	8 prPEs	with Config. Mem. (1 Config.)	25	25
4	0	8	2		with Config. Mem. (2 Config.)	16	16
5	4	0	1	4 pPEs	with Config. Mem. (1 Config.)	58	58
6	4	0	2		with Config. Mem. (2 Config.)	40	40
7	8	0	1	8 pPEs	with Config. Mem. (1 Config.)	27	27
8	8	0	2		with Config. Mem. (2 Config.)	20	20

Next, we compared CPU time under the same conditions as previous and target applications, CHENG and DCT8. Table 3 shows CPU time comparisons. "Proposed" is CPU time by the proposed algorithm, and "Opt." is CPU time to get an optimal solution by branch and bound strategies. Table 3 shows that the proposed algorithm can obtain CHENG's solution in 30.7 seconds in the worst case when the optimal solution is obtained in about 57 minutes (3391 sec.). Furthermore, when DCT8 is the target application, the optimal solution cannot be obtained in practical time. However the proposed algorithm can obtain solutions in about 150 sec. in the worst case. Thus, the proposed algorithm can obtain solutions for various architectures in practical time.

Table 3. CPU Time

No.	Architecture		CHENG		DCT8	
			Proposed [sec]	Opt. [sec]	Proposed [sec]	Opt. [sec]
1	4 prPEs	with Config. Mem. (1 Config.)	30.7	3391	153.8	NA
2		with Config. Mem. (2 Config.)	30.5	3398	153.0	NA
3	8 prPEs	with Config. Mem. (1 Config.)	29.1	4595	140.1	NA
4		with Config. Mem. (2 Config.)	29.3	4597	139.9	NA
5	4 pPEs	with Config. Mem. (1 Config.)	27.1	3202	140.2	NA
6		with Config. Mem. (2 Config.)	27.1	3201	140.3	NA
7	8 pPEs	with Config. Mem. (1 Config.)	26.0	4220	120.8	NA
8		with Config. Mem. (2 Config.)	25.9	4222	120.8	NA

Table 4. CPU Time for Complex Example [sec]

	$n_{pPE} = 8$	$n_{pPE} = 64$	$n_{pPE} = 256$
$N = 100$	310.9	242.5	242.2
$N = 300$	1390.7	915.7	811.4
$N = 500$	3160.3	1690.8	1427.5

Table 4 shows CPU time for more complex port expansion DFGs under the same conditions as previous. The number of port expansion DFG's nodes N equals 100, 300, and 500, and the number of pPE n_{pPE} equals 8, 64, and 256. In table 4, the worst search time is about 50 minutes (3160.3 sec.). Thus proposed algorithm can get the solution for complex input in the practical time with various architecture parameters.

However, from solution's quality perspective, proposed algorithm cannot always obtain a good solution. In case of $N = 500$ and $n_{pPE} = 256$, the number of configurations results in four, and this solution is not feasible. Analyzing this result, some configurations can be merged, and the execution cycles can be drastically reduced by decreasing number of configurations. When the number of PEs is $n_{pPE} = 256$, the same phenomenon occurs. The reason why proposed algorithm cannot obtain the good solution is discontinuity of solution space. The discontinuity of solution space prevents proposed algorithm from efficient search for a good solution. To solve bigger problem more than 128 PEs, the proposed algorithm should be modified.

6.2   Reconfigurable Architecture Exploration

In this section, we demonstrate reconfigurable architecture exploration using PRP-model under the variety of the number of PEs and configuration memories. Target application is the sample DFG, whose number of nodes equals to 500, used previous experiment. Experimental environment is the same conditions as previous. The above-mentioned experimental results show that proposed algorithm cannot always obtain a good solution in case of more than 128 PEs. Therefore, we demonstrate reconfigurable architecture exploration under the number of PEs from 16 to 128.

Table 5 shows fixed parameters of explored reconfigurable architectures, and table 6 shows specifications of configuration memories. We assume the parameter of reconfigurable architecture complexity *Scale* in Eq.(2) equals 128. Then, n_{config} and t_{cfg_r} are shown in Table 7. Table 7 shows that n_{config} is decreased half when the number of PEs is increased two times.

Figure 9 shows the execution cycles of each architecture with configuration memory A, B, and C, and the number of $prPE$ equals zero. In Figure 9, the execution cycles simply increase/decrease in the case of configuration memory A/C according to the increase of the number of PEs. On the other hand, in the case of configuration memory B, the execution cycles progressively decrease according to the increase of the number of PEs. However, when the number of

Table 5. Fixed Parameters for Reconfigurable Architecture Exploration

Parameter	Value	Parameter	Value
t_{ex_r}	2	t_{ex_w}	3
t_{in_r}	1	t_{in_w}	2
t_{reg_r}	1	t_{reg_w}	1
p_{ex_r}	4	p_{ex_w}	4
p_{in_r}	1	p_{in_w}	1
p_{reg_r}	1	p_{reg_w}	1
n_{in_mem}	0	n_{rPE}	0
n_{reg_prPE}	1	t_{config}	1

Table 6. Configuration Memory Specification

Config. Mem. Type	Total Bit TB_{CM} [bit]	Memory Depth MD_{CM}	Bit Width BW_{CM} [bit]
A	32k	2k	16
B	64k	4k	16
C	128k	8k	16
D	32k	1k	32
E	64k	2k	32
F	128k	4k	32

Table 7. n_{config} and t_{cfg_r} $(Scale = 128)$

#PEs	Configuration Memory											
	Type A		Type B		Type C		Type D		Type E		Type F	
	n_{config}	t_{cfg_r}	n_{config}	t_{cfg_r}	n_{config}	t_{cfg_r}	n_{config}	t_{cfg_r}	n_{config}	t_{cfg_r}	n_{config}	t_{cfg_r}
16	16	128	32	128	64	128	16	64	32	64	64	64
32	8	256	16	256	32	256	8	128	16	128	32	128
64	4	512	8	512	16	512	4	256	8	256	16	256
128	2	1024	4	1024	8	1024	2	512	4	512	8	512

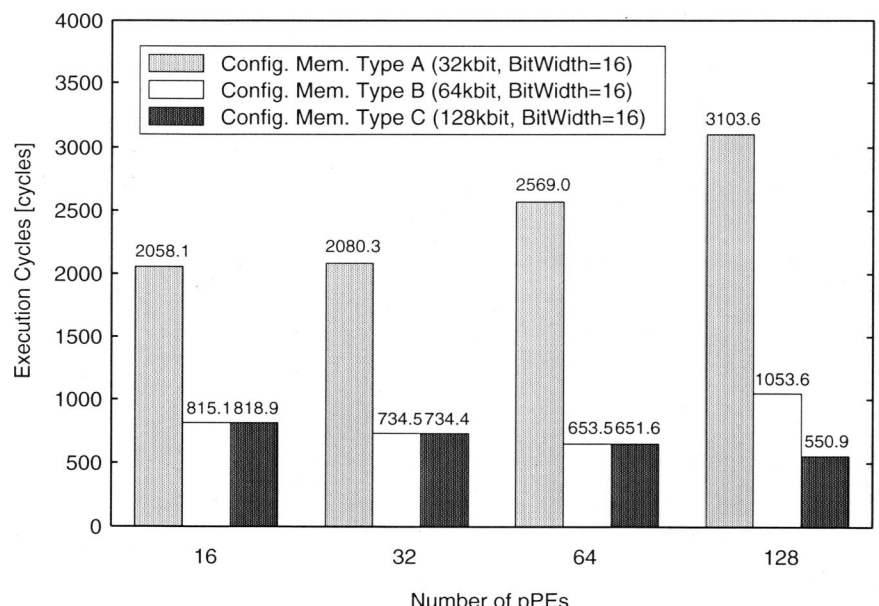

Fig. 9. Execution Cycles $(Scale = 128$, Config. Mem. Type is A, B, and C.)

PEs equals 128, the execution cycles increase greater than the execution cycles when the number of PEs equals 16.

Figure 10 shows the execution cycles of each architecture with configuration memory D, E, and F. We can also see that the execution cycles simply increase/decrease in the case of configuration memory D/F according to the increase of the number of PEs. The increasing rate of execution cycles is more slowly than Figure 9. At the same time, the execution cycles always decrease according to the increase of the number of PEs in the case of configuration memory E, whose total bit equals configuration memory B's one.

When the total bit of configuration memory is the same, the bit width of configuration memory affects the increase of the execution cycles. The increase of configuration memory bit width leads to the decrease of configuration data read cycles, i.e. the overhead of configuration data read decreases. For less overhead, designers should choose the configuration memory which is as wide bit width as possible.

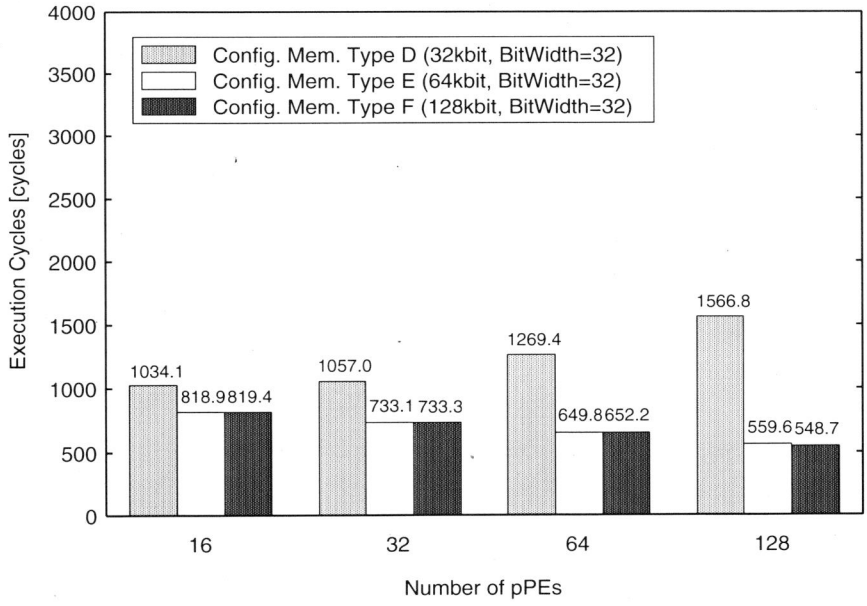

Fig. 10. Execution Cycles (*Scale* = 128, Config. Mem. Type is D, E, and F.)

Table 8 shows the execution cycles of each architecture and the ratio of waiting cycles included the execution cycles. In the increasing case of the execution cycles, we can see that waiting cycles account for a large percentage of the execution cycles. On the other hand, in the decreasing case, the ratio of waiting cycles always equals zero.

Table 8. Total Execution Cycles and Waiting Ratio (pPE Case)

#PEs	Configuration Memory											
	Type A		Type B		Type C		Type D		Type E		Type F	
	Total	Wait[%]	Total	Wait[%]	Total	Wait[%]	Total	Wait[%]	Total	Wait[%]	Total	Wait[%]
16	2058.1	58.5	815.1	0.0	818.9	0.0	1034.1	17.6	818.9	0.0	819.4	0.0
32	2080.3	63.1	734.5	0.0	734.4	0.0	1057.0	27.0	733.1	0.0	733.3	0.0
64	2569.0	73.2	653.5	0.0	651.6	0.0	1269.4	46.2	649.8	0.0	652.2	0.0
128	3103.6	81.2	1053.6	44.2	550.9	0.0	1566.8	62.7	559.6	0.3	548.7	0.0

The above-mentioned experimental results are obtained when $prPE$ is not used. Table 9 shows the execution cycles and the ratio of waiting cycles when $prPE$ is only used. Experimental results in Table 9 show the execution cycles are 150-300 cycles less than Table 8 in the case of the ratio of waiting cycles equals zero. However, the execution cycles are almost the same as Table 8 when the ratio of waiting cycles is not zero.

Table 9. Total Execution Cycles and Waiting Ratio ($prPE$ Case)

#PEs	Configuration Memory											
	Type A		Type B		Type C		Type D		Type E		Type F	
	Total	Wait[%]	Total	Wait[%]	Total	Wait[%]	Total	Wait[%]	Total	Wait[%]	Total	Wait[%]
16	2056.2	62.6	675.1	0.0	668.7	0.0	1032.1	25.1	677.0	0.0	668.5	0.0
32	2073.0	69.6	553.4	0.0	552.6	0.0	1048.9	40.3	560.9	0.0	545.7	0.0
64	2567.1	81.3	519.1	6.2	426.9	0.0	1287.4	63.2	427.4	0.0	424.1	0.0
128	3091.0	90.7	1043.0	72.8	234.4	0.0	1554.5	81.5	530.2	45.7	233.6	0.0

Because of the limitation of configuration memory, n_{config} simply decrease according to the increase of the number of PEs. In contrast, t_{cfg_r} simply increase according to the increase of the number of PEs because we assume the configuration data linearly increase according to the number of PEs. Therefore, the overhead of configuration data read drastically increases according to the increase of the number of PEs, and critically affects the execution cycles.

When the reconfigurable architecture which includes small amount of configuration memory is used, experimental results show that the overhead of configuration data read is dominant, and the overhead critically affects the execution cycles. To use the reconfigurable processor effectively, designers should carefully design the configuration memory and its parameters, i.e. t_{config}, n_{config}, and t_{cfg_r}.

Table 10. n_{config} and t_{cfg_r} ($Scale = 64$)

#PEs	Configuration Memory											
	Type A		Type B		Type C		Type D		Type E		Type F	
	n_{config}	t_{cfg_r}	n_{config}	t_{cfg_r}	n_{config}	t_{cfg_r}	n_{config}	t_{cfg_r}	n_{config}	t_{cfg_r}	n_{config}	t_{cfg_r}
16	32	64	64	64	128	64	32	32	64	32	128	32
32	16	128	32	128	64	128	16	64	32	64	64	64
64	8	256	16	256	32	256	8	128	16	128	32	128
128	4	512	8	512	16	512	4	256	8	256	16	256

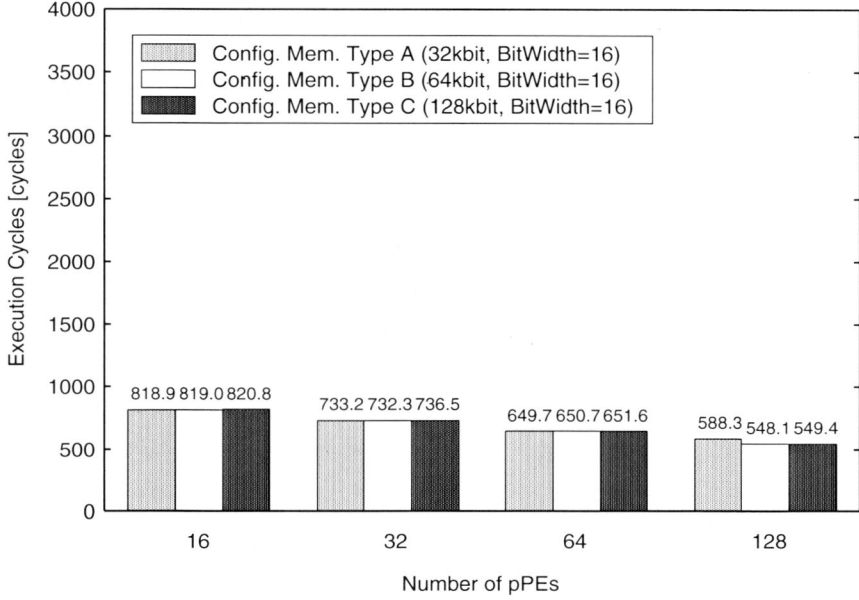

Fig. 11. Execution Cycles (*Scale* = 64, Config. Mem. Type is A, B, and C.)

Next, we assume the parameter of reconfigurable architecture complexity *Scale* in Eq.(2) is decreased to 64, i.e. designers use simpler reconfigurable architecture. Table 10 shows n_{config} and t_{cfg_r} in case of *Scale* = 64. Figure 11 shows the execution cycles of each architecture, whose *Scale* equals to 64, with configuration memory A, B, and C. In Figure 11, compared to Figure 9, the execution cycles always decrease according to the increase of the number of PEs because the decrease of *Scale* makes the configuration data half. To decrease overhead, the reduction of reconfigurable architecture complexity *Scale* also effectively affects.

When the performance is not enough, designers often add more PEs to execute effectively with rich HW. However, to use the execution with run-time reconfiguration, addition of PEs does not always improve the performance because of effect of the complex overhead. Considering the effect of the overhead carefully, the best reconfigurable architecture which satisfies the requirement should be chosen. It is the first step for the effective execution with run-time reconfiguration to analyze the details of reconfigurable architecture carefully.

7   Conclusion

In this paper, we proposed dynamic reconfigurable architecture exploration method based on Parameterized Reconfigurable Processor model and task partitioning optimization algorithm for reconfigurable architecture exploration corresponding to proposed PRP-model. Using proposed method, designers can fast evaluate various reconfigurable architectures, and easily find suitable reconfigurable architectures by changing PRP-model parameters. Future work includes the modification of task partitioning optimization algorithm corresponding to large size architectures, the establishment of reconfigurable architecture exploration that considers area and power, and the processing element function decision method according to the total configuration data constraint.

References

1 Reiner Hartenstein. A Decade of Reconfigurable Computing: a Visionary Restrospective. In *Proc. of DATE'01*, pages 642–649, 2001.
2 Andreu Mas-Corell, Michael Winston, and Jerry Green. Introduction to Reconfigurable Hardware. In *System Level Design of Reconfigurable System-on-Chips*, pages 15–26. Springer, 2005.
3 S. Banerjee, E. Bozorgzadeh, and N. Dutt. Physically-Aware HW-SW Partitioning for Reconfigurable Architectures with Partial Dynamic Reconfiguration. In *Proc. of DAC'05*, pages 335–340, 2005.
4 M. D. Galanis, A. Milidonis, G. Theodoridis, D. Soudris, and C. E. Goutis. A Methodology for Partitioning DSP Applications in Hybrid Reconfigurable Systems. In *Proc. of ISCAS'05*, pages 1206–1209, 2005.
5 Karthikeya M. Gajjala Purna and Dinesh Bhatia. Temporal Partitioning and Scheduling Data Flow Graphs for Reconfigurable Computers. *IEEE Trans. on Computers*, 48(6):579–590, June 1999.
6 M. Kaul and R. Vemuri. Optimal Temporal Partitioning and Synthesis for Reconfigurable Architecture. In *Proc. of DATE'98*, pages 389–396, 1998.
7 B. Ouni, A. Mtibaa, and M. Abid. Synthesis and Time Partitioning for Reconfigurable Systems. In *Design Automation for Embedded Systems*, volume 9, pages 177–191. Springer Netherlands, 2004.
8 V. Srinivasan, S. Govindarajan, and R. Vemuri. Fine-Grained and Coarse-Grained Behavioral Partitioning with Effective Utilization of Memory and Design Space Exploration for Multi-FPGA Architectures. *IEEE Trans. on VLSI Systems*, 9(1):140–158, 2001.
9 M. Meribout and M. Motomura. A Combined Approach to High-Level Synthesis for Dynamically Reconfigurable Systems. *IEEE Trans. on Computers*, 53(12):1508–1522, 2004.
10 M. Kaul and R. Vemuri. Temporal Partitioning combined with Design Space Exploration for Latency Minimization of Run-Time Reconfigured Designs. In *Proc. of DATE'99*, pages 202–209, 1999.
11 B. Mei, A. Lambrechts, D. Verkest, Jean-Yves Mignolet, and R. Lauwereins. Architecture Exploration for a Reconfigurable Architecture Template. *IEEE Design & Test of Computers*, pages 90–101, March-April 2005.

HUMAN++: Emerging Technology for Body Area Networks

Julien Penders[1], Bert Gyselinckx[1], Ruud Vullers[1], Olivier Rousseaux[1], Mladen Berekovic[1], Michael De Nil[3], Chris Van Hoof[2], Julien Ryckaert[2], Refet Firat Yazicioglu[2], Paolo Fiorini[2], Vladimir Leonov[2]

[1]IMEC Nl/Holst Centre, High Tech Campus 48, 5656AE Eindhoven, The Netherlands.
[2]IMEC, Kapeldreef 75, B-3001 Leuven, Belgium.
[3]Technische Universiteit Eindhoven, Den Dolech 2, 5612 AZ Eindhoven, The Netherlands.

Abstract. This paper gives an overview of results of the Human++ research program [1]. This research aims to achieve highly miniaturized and nearly autonomous sensor systems that assist our health and comfort. It combines expertise in wireless ultra-low power communications, packaging and 3D integration technologies, MEMS energy scavenging techniques and low-power design techniques.

Key words. BAN, microsystem, wireless, autonomous, integration

1. Introduction

Many national health services struggle in the face of financial resource constraints and shortages of skilled labor. The cost of healthcare delivery is steadily on an upward trend. A recent survey shows that by 2020, healthcare spending is projected to triple in dollars, consuming 21% of GDP in the US and 16% of GDP in other OECD countries. As a result, the pressure on health systems to step up efforts in cost containment and efficiency improvement keeps growing. Consensus about the main determinants of expenditure is not complete but revolves generally around cost drivers such as rising income and patient expectations; demographic change, in particular the ageing of population; and new technologies.

In wealthier nations consumer demand increases, leading to a higher spend on healthcare. Statistics show that the cost lowering effect of technology and automation is more than offset by the impact of an ageing society, consumerism, biotechnology and medical breakthroughs. This results in an overall increase in cost between 2-3% per year. As a result, alternative ways of increasing efficiency, productivity and usability while controlling cost are being sought. One strategy that is gaining major attention consists of offloading healthcare institutions by shifting the health management outside the expensive formal medical institutions. Other strategies seek to improve the appropriateness of treatment or emphasize preventive care rather than treatment. For example, the field of chronic diseases is a vast domain in which the provision of

Please use the following format when citing this chapter:

Penders, J., Gyselinckx, B., Vullers, R., Rousseaux, O., Berekovic, M., De Nil, M., Van Hoof, C., Ryckaert, J., Yazicioglu, R.F., Fiorini, P. and Leonov, V., 2007, in IFIP International Federation for Information Processing, Volume 249, VLSI-SoC: Research Trends in VLSI and Systems on Chip, eds. De Micheli, G., Mir, S., Reis, R., (Boston: Springer), pp. 377–397

real-time data from and to the patient anywhere and at any point in time may hold significant potential for cost reduction.

The supporting role of an adequate technology platform is critical here. E-health technology, enabling wireless and mobile based healthcare services, is increasingly co ned as the revolutionizing enabler for the next decades to come. As defined by the WHO, e-health refers to the use in the health sector, of digital data transmitted, stored and retrieved electronically for clinical, educational and administrative purposes, both at the local site and at a distance. There are three strong drivers for an e-Health technology platform. The first, demographic, is the evolution toward an ageing society, active ageing and independent living. This calls for radical changes in how care will be provided for the elderly and how technology may assist. Second is the epidemiologic transition from episodic to chronic healthcare needs. Future healthcare systems should thus focus on prevention, on effective provision of continuous treatment, on integrating lifestyle parameters, and should be customized to individual needs of each patient. Finally, patients expect ever more from health services, reinforcing the existing concept of a 'patient centric view' of healthcare which emphasizes the patient's experience and journey through a system that provides continuity of care to a proactive patient.

E-health is claimed to offer the potential to reduce costs, enable personalized healthcare, deliver remote health services and increase the delivery efficiency in real-time. However, at this early stage today it is only through the many pilot projects on e-health ongoing in different countries around the world that evidence will be gathered to determine the economic viability, and answer the question how e-health can enhance the healthcare system efficiency and resolve the associated cost burden.

2.   An Enabling Technology: Body Area Networks

In this text we analyze one component of e-health, the personal body area network (BAN) [2] that provides medical, lifestyle, assisted living, sports or entertainment functions for the user. This network comprises a series of miniature sensor/actuator nodes each of which has its own energy supply, consisting of storage and energy scavenging devices. Each node has enough intelligence to carry out its task. Each node is able to communicate with other sensor nodes or with a central node worn on the body. The central node communicates with the outside world using a standard telecommunication infrastructure such as a wireless local area or cellular phone network. Experts might then provide services to the individual wearing the BAN, such as management of chronic disease, medical diagnostic, home monitoring, biometrics, and sport and fitness tracking. Next generation of BAN will include feedback loops for disease management or drug and treatment delivery within so-called closed-loop systems, and will provide feedback to the individual about her lifestyle and health status, eventually leading to human-in-the-loop systems.

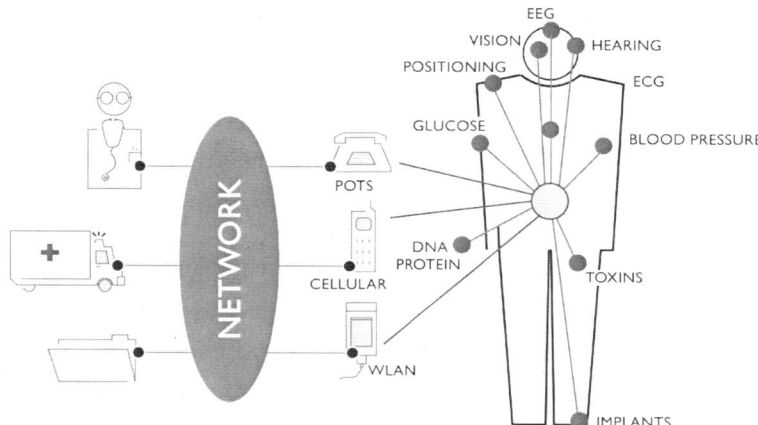

Fig. 1. The technology vision for the year 2010: people will be carrying their personal body area network and be connected with service providers regarding medical, lifestyle, assisted living, sports and entertainment functions.

The successful realization of this vision requires innovative solutions to remove the critical technological obstacles. First, the overall size should be compatible with the required form factor. This requires new integration and packaging technologies. Second, the energy autonomy of current battery-powered devices is limited and must be extended. Further, interaction between sensors and actuators should be enlarged to enable new applications such as multi-parameter biometrics or closed loop disease management systems. Next, intelligence should be added to the device at the node level so that each node is capable of storing, processing and transferring data continuously or on an event-triggered basis. Intelligence should also be introduced at the network level to deal with issues such as network management, data integration and data interpretation. Finally, the energy consumption of all building blocks needs to be drastically reduced to allow energy autonomy.

The Human++ program is looking into all of these generic BAN challenges. In the following sections we will have a closer look at the technologies under development. We will start with an overview of the test case, then we will discuss the enabling technologies such as wireless communication, micropower generation, digital signal processing and sensors. Finally, we show how advanced integration technology can bring together all the heterogeneous subcomponents in a compact form factor.

3. Ambulatory Multi-Parameter Monitoring as Test Case

We selected ambulatory multi-parameter monitoring as a driving application for Human++. The target of such a monitoring system is to acquire process, store, and visualize a number of physiological parameters in an unobtrusive way. In one case, we focused on the simultaneous acquisition of EEG/ECG/EMG biopotential signals. Traditionally, such signals are either captured in a clinical setting for immediate interpretation or they are recorded in an ambulant setting for post factum analysis via a Holter monitor. With a wireless ambulatory monitoring system we want to combine the real time features of the clinical system with the benefits of ambulatory monitoring from a Holter monitor. Fig. 2 shows a schematic drawing of typical set-up. The set-up consists of:

- 1 EEG sensor node that can acquire, process and transmit 1 to 24 EEG signals;
- 1 ECG sensor node that can acquire, process and transmit ECG signals;
- 1 EMG sensor node that can acquire, process and transmit EMG signals;
- 1 base-station that collects the information from the 3 sensor nodes.

All the sensors have very similar functionality. First, the incoming signals are amplified and filtered. The resulting signals are sampled at 1024Hz with a 12-bit resolution. If required, the bio-signals are then processed locally to extract relevant features, e.g. heart rate, energy expenditure or force. Finally, the digital signals are transmitted over a wireless link operating in the 2.4GHz ISM band. Because the sensors are very similar, they can be realized with the same programmable hardware, illustrated on Fig. 3.

The base-station acts as a data collector. The collected data are passed on to a PC or PDA through a USB interface. Further, the base-station also acts as a master node for the network, which manages the data-flow through the network. BSN are typical star topology networks, for which time division multiple access (TDMA) schemes are well suited.

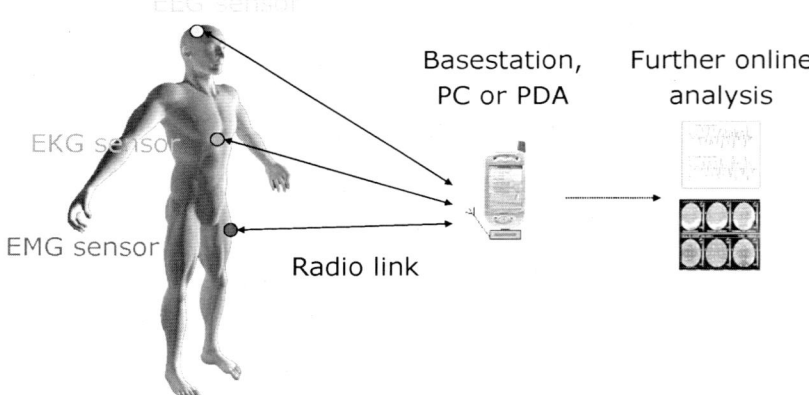

Fig. 2. Schematic overview of the BAN set up, consisting of EEG/ECG/EMG sensors wirelessly connected to a PC or PDA.

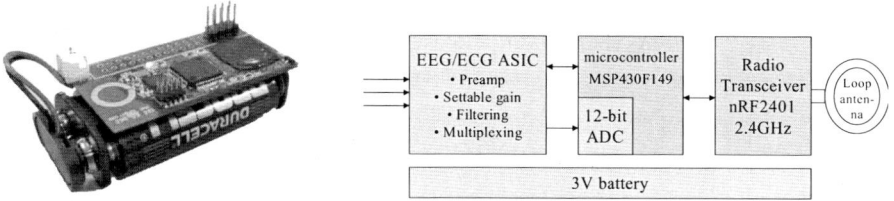

Fig. 3. Close up of the sensor node. On the left a picture and on the right a functional diagram.

A key design criterion for such system is the power of the sensor nodes because this will directly determine the size and the operational lifetime of the system. Analysis of the operation of the sensors shows that they are alternating between four different modes of operation:

1. Listen: the sensors receive their parameters from the base station;
2. Processing: the biopotential signals are monitored and processed;
3. Transmit: the sensors send their data to the base station;
4. Sleep: power save mode – most of the electronics are switched off.

The time spent in each of these modes is very much application dependent. As an example, Fig. 4(a) shows the relative time spent in each of the modes for a particular EMG measurement. In this particular case, signal processing consists of RMS computation. It is clear that for this sensor the idle time is very important and that the system needs to have a very low stand by power consumption.

Each of these modes has its own power consumption. Fig. 4(b) shows that the current consumption in listen and transmit mode is much higher than in processing or sleep mode. This is a direct consequence of the radio which is switched on in these modes and which consumes about 90% of the power when it is active. Bringing all of these data together we get to the total average power consumption for the sensor: with the current system consisting mainly of off the shelf components a prototype can be designed that consumes less than 1mW of power if the measurement interval is longer than 1s. If we assume that we use two AA batteries in series with a capacity of 2500mAh, the battery lifetime becomes approximately 3 months.

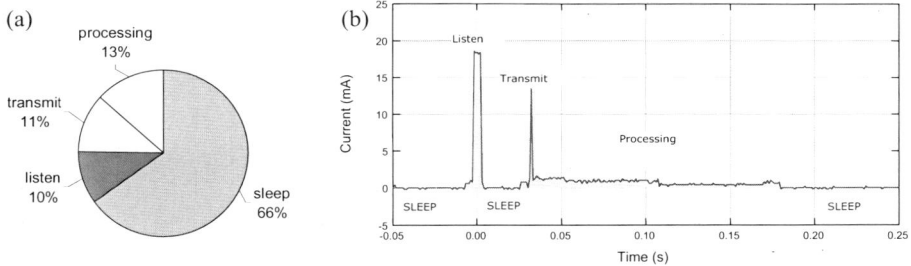

Fig. 4. (a) Relative time spent in different operation modes for a EMG sensor node performing RMS computing locally; (b) Current consumption of sensor node in different operating modes.

This clearly shows that with today's technology, first realistic demonstrators with a reasonable lifetime can be manufactured. However a couple of major challenges still have to be solved in order to come to a widespread deployment of BANs.

- Miniaturization: in most of current systems, batteries are the single largest contributor to the sensor node size. AA batteries are good for demonstration, but one would like to work with a coin or planar type of battery. These batteries have roughly 100 times less capacity than the AA cells. To keep the same battery lifetime the power of the electronics has to be reduced by a factor 100. Furthermore, development of advanced integration technology is required to achieve compact, flat and flexible sensor nodes, for embodiment in textile or clothing accessories. Making BSN invisible and non-intrusive is likely to determine their acceptance as a support for personalized healthcare and independent living.
- Autonomous systems: the system we demonstrated can run for months. However, to come to a truly autonomous system it should be able to operate over its full lifetime without maintenance. At a given battery capacity, lifetime can be increased by reducing the power of electronics. Alternatively, one can scavenge energy from the environment during the operation of the system. If the average scavenged energy is larger than the average consumed energy, the system can run eternally with the battery or a super capacitor acting only as a temporary energy buffer. A combination of these technologies appears as the optimal solution for achieving autonomous BSN.
- Integration of novel sensor and actuator concepts: the quality of the information resulting from a BAN is only as good as what you measure. Today, often only simple physical properties are measured such as bio-potentials, temperature and movement. Sensors that can measure these parameters in an ambulatory setting are much needed. Motion artifacts are often a major source of data corruption. Implantable sensors face additional issues such as biocompatibility. Besides the monitoring of vital signs and biomedical variables, additional sensors are required for sensing the environment, thus making the system "context aware". This set of heterogeneous sensors need to be secure and reliable. Moreover interference issues need to be addressed, especially for actuating systems where interferences might affect not only the monitoring but also the active part of the system.
- Providing more functionality: most of today's devices act as simple gateways, passing on the information to a central hub where the data is converted into actionable information. By adding intelligence to the sensors they can take decisions locally and the signaling overhead in terms of data and latencies can be reduced. Additionally, intelligence is required for the system to make decisions depending on the status of the environment, thus enabling context-awareness. Finally, embedded intelligence opens the door to closed-loop systems providing action or feedback to the user.
- Become manufacturable at low cost: today's systems cost around 100+ euros. A major reason for this is the low volume of the market so far, but another more technical reason is that there are no commercially available packaging technologies that can efficiently integrate such heterogeneous components as batteries, MEMS, processors, and radios in a single package.

In the remainder of this chapter we will show how advances in wireless communication, energy scavenging, sensors and system integration can enable such systems in the near future. The hunt for sensor systems that are 1000x more power efficient, that have ample intelligence to make decisions, that cost less than 1€ and that are unobtrusive is open.

4.   Wireless Communication

Commercially available low power radios, which typically rely on Bluetooth, Zigbee [3] or other proprietary narrowband communication schemes (e.g. Nordic [21]) cannot meet the stringent Wireless BAN power requirements that we are looking for. Typical chipsets for these radios consume in the order of 10 to 100mW for data rates of 100 to 1000kbps, leading to a power efficiency of roughly 100 to 1000mW/Mbps or nJ/bit. Some recent research prototypes or commercial chipsets with limited functionalities have shown efficiencies of 20 nJ/bit or slightly below [6][7][8]. As outlined in the previous section, we need a radio which is 1 to 2 orders of magnitude more power efficient.

A classical approach to reduce power consumption consists in *data-level* duty-cycling: sensor data are collected in a buffer and organized in packets periodically transmitted by the radio in short communication bursts. If the radio is switched off between communication bursts, this in turn reduces radio duty cycle and therefore reduces power consumption. However, none of the off-the-shelf low-power radio, even duty-cycled, have proven the ability to reduce the power consumption by the required orders of magnitude while offering the necessary communication performances.

Novel air interfaces relying on impulse radio ultra wideband (UWB-IR) radio signals show a strong potential for further reduction of the wireless communications power budget and are currently attracting a lot of attention for the development of next-generation ultra low power radios. In the rest of this section, we highlight how the use of UWB-IR signals allows reducing the power consumption of the radio and then giving an overview of recent achievements in the development and implementation of such systems.

4.1. UWB air interface: promising candidate for ultra low power wireless communications

UWB signals are formally defined as radio signals that have a bandwidth larger than 500 MHz. The Federal Communications Commission (FCC) has recently authorized UWB communications between 3.1GHz and 10.6GHz. The regulations on UWB radiation define a power spectral density (PSD) limit of -41dBm/MHz, but there are no specific regulations on the definition of the time-domain waveform and there are various ways of obtaining an UWB waveform. In UWB-IR systems, the transmitted signal consists in pulses of short duration (~1-2 ns) that are separated by longer silent periods (~20 ns or more) as illustrated in Fig. 5.

(a)

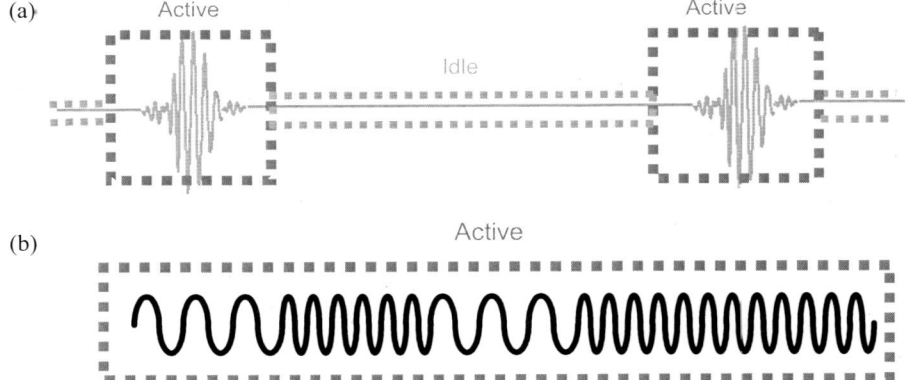

(b)

Fig. 5. (a) Schematic representation of a UWB-IR signal allowing signal-level duty cycling; (b) schematic representation of a narrowband radio signal.

Power consumption of UWB-IR systems can be reduced using *signal duty cycling*, i.e. duty cycling of the radio front-end *during* data transmission. Indeed, the structure of UWB-IR signals allows to switch off the radio front-end during the silent time intervals between UWB pulses, and to re-activate it only when a pulse needs to be generated, leading to a duty cycle ranging from 1% to 10% of the actual data transmission time. Combining duty cycling at signal and data level duty cycling potentially enables effective global duty cycles ranging from 0.1% to 0.01%. Such a signal duty-cycling approach is not an option in narrowband systems where a continuous waveform must be generated to transmit the data, requiring the front-end to stay in an active mode throughout the actual data transmission interval, as illustrated in Fig. 5.

4.2. Low Power UWB pulse generator

In order to achieve ultra-low power consumption, the circuits used for the generation of UWB pulses should have very short start-up times (i.e. a few nanoseconds) to enable efficient signal duty cycling. Furthermore, although no specific waveform is mentioned in international regulations, various UWB standard proposals ([5], [11]) have subdivided the entire UWB spectrum in 500MHz sub-bands as a solution to compensate for strong interferers, to improve multiple accesses and to compose with the different regulations on UWB emissions worldwide. The emerging 802.15.4a low-power UWB-IR standard [11] even imposes time-domain specifications for the generated pulses. In order to comply with these regulations and standards, the generated pulses of UWB impulse-radio (UWB-IR) approaches must fulfill stringent spectral masks that can feature such low bandwidths. This poses a serious challenge for the pulse generation of UWB-IR transmitters.

A low-power and low-complexity implementation of a pulse generator complying with these specifications is proposed in [13]. The proposed pulser architecture generates a triangular baseband pulse shape which is then multiplied with an RF carrier for up-conversion, an approach which is often referred to as *carrier-based UWB impulse radio*. The triangular waveform is in line with the standard recommendations both in terms of time-domain and spectral specifications with a side-lobe rejection of more than 20dB. The pulse generator architecture is presented in Fig. 6. A triangular pulse generator and a ring oscillator are activated simultaneously. The choice of a ring type of oscillator is motivated by its fast startup time that enables heavy signal duty-cycling. A gating circuit (ring activation circuit in Fig. 6) manages the signal duty cycling and activates the ring oscillator when a pulse must be transmitted, avoiding wasting power between the pulses.

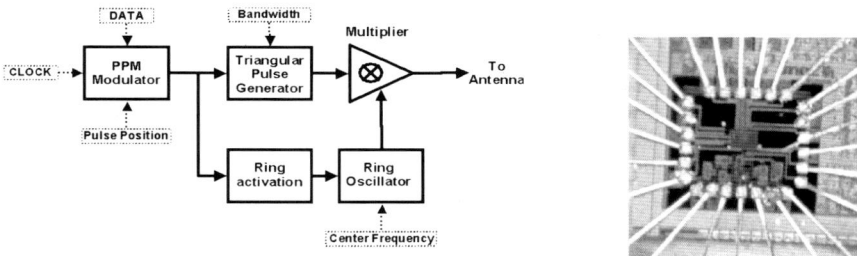

Fig. 6. Architecture of the pulse generator (left) and micrograph of the pulser die (right).

The pulser circuit has been implemented in a logic 0.18µm CMOS technology [14]. The system can deliver a pulse rate up to 40MHz with a measured energy consumption of 2mW or 50pJ per pulse at a 40MHz pulse repetition rate. In a typical scenario where one bit of information is encoded over 20 pulses, this represents a power efficiency of roughly 1nJ/bit. This is exactly the type of breakthrough low power consumption we were hoping to find.

4.3. UWB analog receiver

Processing wideband analog signals in the digital domain requires an extremely fast sampling ADC with a wide input bandwidth. Such solutions exist [15] but usually exhibit high power consumption. In order to minimize the overall sampling rate and total power consumption, analog preprocessing is an interesting alternative. Fig. 7 illustrates an analog-based correlation receiver architecture, which corresponds to a direct down-conversion architecture with an analog quadrature correlator. UWB signals are first down-converted in quadrature baseband, then correlated with a rectangular pulse template. The output of the analog baseband is then sampled at the pulse repetition frequency. The choice of such a simple receive template is driven by the simplicity of its implementation and the consequent low power consumption.

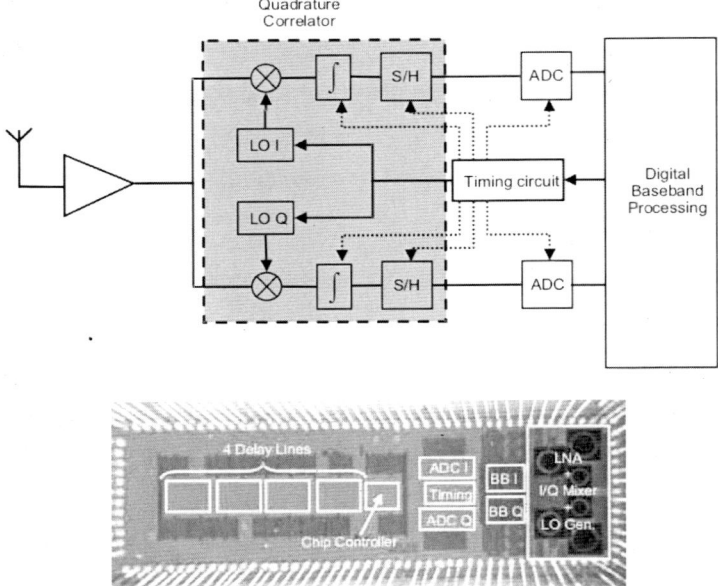

Fig. 7. UWB analog receiver: schematics and chip micrograph.

This receiver has been implemented on a 0.18μm CMOS technology [16]. The total power consumption of the chip is 30mW at 20MHz pulse repetition frequency. This power consumption is more than one order of magnitude higher than for the pulser, and corresponds to an energy efficiency of approx. 10 nJ/bit. The reason behind this increased power consumption is that signal duty cycling could not be applied in the receiver, since the startup time of some receiver components is larger than the average inter-pulse interval.

The proposed transmitter outperforms low-power radios by more than one order of magnitude, while the receiver has an energy efficiency that compares to narrowband receivers. This asymmetric power consumption is adequate for BAN applications since the most power-constrained elements are the sensors, while the receiver in the central station has slightly more relaxed power budget.

5.   Micropower Generation and Storage

Today, the batteries that are needed to power wireless autonomous transducer systems seriously limit the possibilities of this emerging technology. Modern electronic components become smaller and smaller, while the scaling of electrochemical batteries faces technological restrictions. As a consequence, either large batteries are used that give a longer autonomy but make the system bigger, or small batteries are used that make the system less autonomous. For this reason, a worldwide effort is ongoing to replace batteries with more efficient, miniaturized power sources. We aim at generating and storing power at the micro scale to improve the autonomy or reduce the size of wireless autonomous transducer systems. The envisaged solution takes its energy – thermal or mechanical – from the human body and converts it into electrical energy, stored in a micro-battery or (super)capacitor. The choice of scavenging principle depends on the application and the environment in which it is used. In this section, we will present thermal scavengers because they are well suited to convert the thermal body heat into electricity.

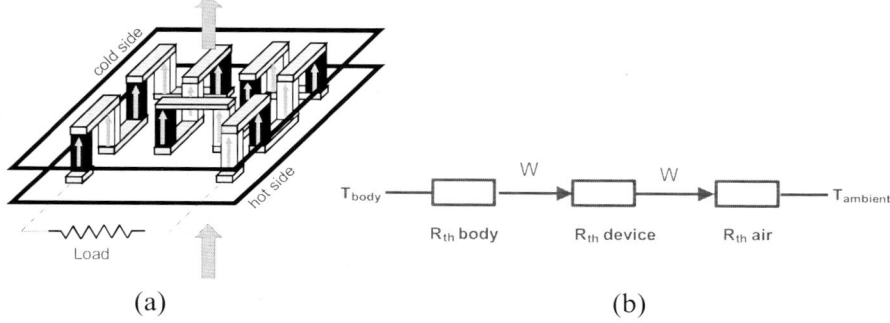

$$T_{body} \rightarrow R_{th}\ body \xrightarrow{W} R_{th}\ device \xrightarrow{W} R_{th}\ air \rightarrow T_{ambient}$$

(a)   (b)

Fig. 8. (a) Schematic of a thermoelectric generator. (b) The schematic thermal circuit representing the generator and its environment.

Thermal energy scavengers are thermoelectric generators which exploit the Seebeck effect to transform the temperature difference between the environment and the human body into electrical energy. A thermoelectric generator is made of thermopiles, which are in turn made of a large number of thermocouples connected thermally in parallel and electrically in series, as shown schematically in Fig. 8(a). The black and white pillars represent the two types of thermoelectric materials, and the metal interconnects are drawn in grey. They are sandwiched between two plates, the cold and hot sides of the device. Heat flows from the hot side to the cold side, through the pillars (indicated by the arrows). The maximum electrical power is generated when the two following conditions are met: (i) the load is matched to the electrical resistance of the generator and (ii) the thermal conductance of the thermocouples equals the one of the air between the plates. Under this condition, power increases when increasing the height of the pillars.

5.1. Thermoelectric generators and systems with commercial thermopiles

A number of TEGs has been built during the last few years. First generation TEGs were introduced to measure the heat flows on man and animals. Although they offer a good power output (0.1 mW), they usually produce less than 0.5 V output on the matched load. The second generation of TEGs is characterized by a voltage of more than 0.7 V which is enough for its effective utilization for powering electronic modules [22]. In 2005, the size of TEGs has been reduced to the one of a man watch thereby significantly improving their acceptance by the wearer [23]. These third generation TEGS are intended for use at a temperature of about 22°C, then providing a useful power of 0.2-0.3 mW on office workers and about 0.1 mW at night or e.g. on immobilized patients in hospitals (low metabolic rate), as represented on Fig. 9(a). Therefore, taking into account adverse illumination conditions at home, on transport and at night, the TEGs turn out to be much more powerful than the best solar cells [25].

Integration of thermopiles within sensor modules requires advanced power management schemes to optimize harvested power efficiency. Typically, the TEG continuously charges a battery or a supercapacitor, which then provides power to electronic modules. Voltage up-converters are usually added to match the need for higher voltage power-supply of different electronic components. However, since such converters significantly decrease the overall power efficiency, attention should be paid to match input-voltage of electronic components with the output-voltage of the battery/supercapacitor. In systems exploiting the last generation of TEGs, super-capacitors were preferred to batteries because they can start storing energy at a lower voltage. At a given temperature, this means more energy can eventually be stored; at a given power need, this means the device could work at higher air-temperature.

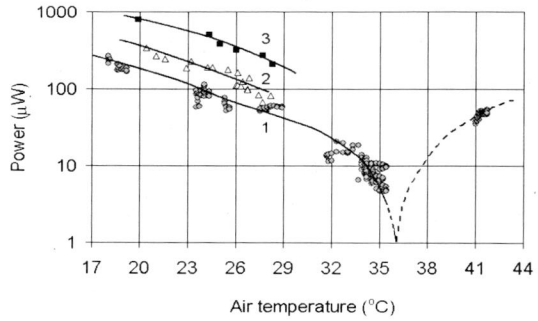

Fig. 9. Performance characteristics of the watch-like TEG. Power vs. air temperature produced at different metabolic rate levels: on a walking man (3), when working in office (2), and with no physical activity for prolonged period (1).

In 2006, we achieved the development of the first prototype of a non-trivial biomedical sensor fully powered by the patient's body heat: a wireless pulse oximeter for non-invasive measurement of pulse and blood oxygen [24], shown in Fig. 10. The device consists of the watch-size wrist TEG, sensor electronics, processor and radio, and a commercial transmission pulse oximetry finger sensor. The watch-size wrist TEG is connected to a supercapacitor as described above. For digital signal processing and radio communication, the system uses the wireless sensor platform presented in the test case. The sensor locally performs all signal processing on 3.9s blocks of measurements, and transmits the resulting pulse and oxygen saturation values (Fig. 10). This leads to a sustainable average power consumption of about 100 µW.

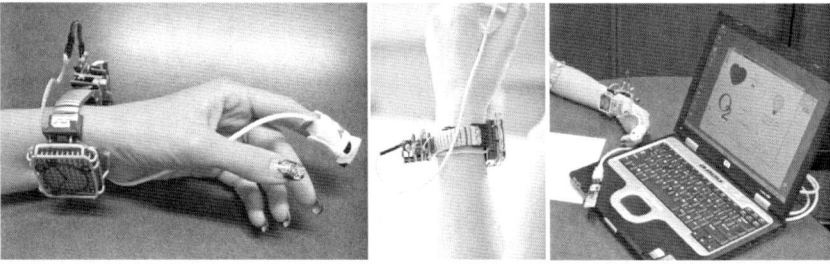

Fig. 10. Wireless body-powered pulse oximeter (left), a prototype of the battery-less electronics (middle) and the application running on a laptop (2006).

5.2. Micromachined thermopiles

The use of commercial thermopiles has proven that human heat can be used to power a sensor node. Nevertheless, the solution is non-optimal for two reasons: (i) it does not offer the possibility of optimizing the power and the voltage at the same time, and (ii) it is a very expensive one, because thermopiles fabrication techniques cannot be easily automated. A possible solution could be the use of micromachined thermopiles. These have already been presented in the scientific literature, and are used in miniaturized commercial thermoelectric coolers. Micromachining has the potential advantage of reducing the lateral size of the thermocouple. This means that a much larger number of thermocouples can be fabricated per unit area, thus allowing to matching of thermal conductance of thermopile to the one of the air. Such approach would allow combining a large voltage and a large power. Unfortunately, micromachined thermocouples have a height of a few microns only, which drastically reduces the thermal resistance of the generator and, hence, the generated power.

In order to overcome this difficulty, we have developed a special design of a micromachined thermoelectric generator for application on humans, which combines a large thermal resistance of the device with a large number of thermopiles. The schematic is shown in Fig. 11(a). Several thousands of thermocouples are micromachined on a silicon rim. The function of this rim is to increase the parasitic plate-to-plate thermal resistance of the generator. If Bi_2Te_3 is used as thermoelectric material an optimized device fabricated according to this scheme and positioned on the human wrist can generate up to 30 $\mu W/cm^2$ at a voltage exceeding 4 volts. Fig. 11(c) shows a realization of such a device based on SiGe thermocouples. Because of the inferior thermoelectric properties of this material with respect to Bi_2Te_3, an optimized device is expected to generate about 5 $\mu W/cm^2$ at a voltage of 1.5 Volts.

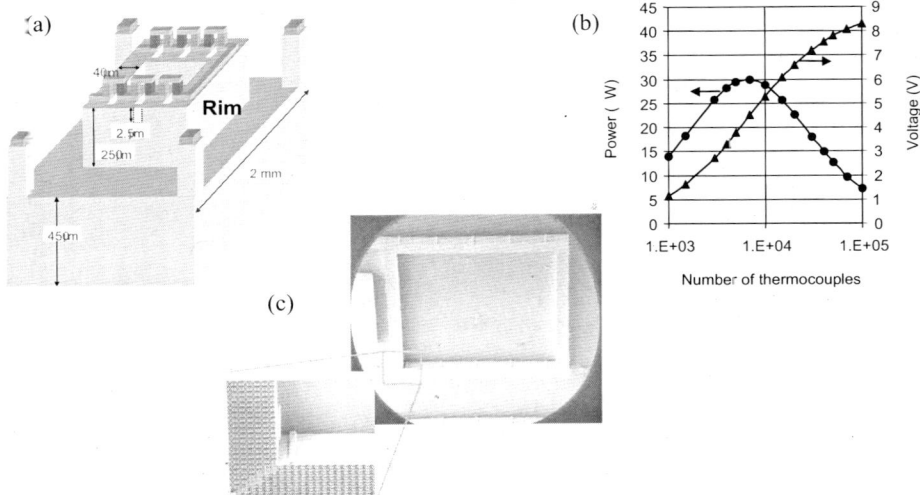

Fig. 11. (a) Schematic of the TEG capable of combining large power and large voltage (rim and thermopiles are not scaled to overall device dimensions). (b) Simulated performance of the TEG shown in (a). (c) Photograph of the fabricated thermoelectric generator. Thermocouples are made of SiGe.

6.   Digital Signal Processing

The ambulatory monitoring test-case has shown that the large majority of power is spent on the wireless transmission of the data (see Fig. 4(a)). As discussed, UWB communication is foreseen as an efficient way to reduce power consumption due to wireless communication. For most applications, further reduction can be obtained by performing local signal processing on the node, hence avoiding the transmission of the entire raw data (bandwidth reduction). Technologies to enable local ultra-low-power signal processing should thus be considered. Among these, Digital Signal Processors (DSPs) are particularly well-suited for tight power budget, while still offering a decent level of flexibility. Here we discuss the development of Ultra Low Power (ULP) Digital Signal Processors for BSN, which shall be capable of running on scavenged energy only.

As the intensity of requested local processing is highly dependent on the application, we will illustrate our developments on the particular case of ECG monitoring, part of the ambulatory multi-parameter monitoring test-case. An extensive overview of algorithms available for ECG analysis and QRS detection can be found in [26]. We illustrate our developments on an algorithm for R-peak detection described in [28]. Since only R-peak information is then transmitted through the radio, this typically reduces the data-rate by a factor 1000 (from 10+ kbps to 10+ bps), thus resulting in a better balance between computation and communication.

The typical power breakdown of a (DSP) microprocessor is given by:

- Leakage Power (P_{leak}): static dissipation while the processor is powered on.
- Active Power (P_{active}): dynamic dissipation consumed while the processor is active.
- Idle Power (P_{idle}): dynamic dissipation consumed while the processor is in stand-by mode without processing data.

Total power consumption is thus given by:

$$P_{Total} = P_{Leakage} + f_{sample} * ((P_{Active} \cdot t_{Active\_avg}) + (P_{Idle} \cdot t_{Idle\_avg})) \, ,$$

where f_{sample} is the sample rate of the incoming data, $t_{active,avg}$ the average time needed for processing one sample and $t_{idle,avg}$ the average time the processor is idle after processing a sample. Both $t_{active,avg}$ and $t_{idle,avg}$ depend on the processor clock speed with $1/f_{sample} = (t_{active,avg} + t_{idle,avg})$. Also P_{active} depends on the processor clock speed with $(P_{active} * t_{active,avg}) = $ energy/sample, constant for every active operation.

Simplistic approaches to achieving low power performance consist in reducing processor clock frequency to its minimal value, f_{sample} * max(cycles), where max(cycles) is the maximum number of cycles required to process a data sample. This technique can be applied to any processors and results in a reduction in idle time and active power while, however, increasing processor response time. Alternatively, more advanced optimization techniques can be used to achieve ultra low power performances. They fall under three main categories:

- Idle power is reduced by implementing a top level clock gating mechanism, on top of the existing fine-grain clock-gating, which eliminates any internal switching activity, thus reducing internal idle switching energy to zero.
- Active power reduction is achieved by decreasing the amount of execution cycles and optimizing the signal processing algorithm using embedded computing techniques including conversion from floating- to fixed-points, and implementation of approximations for divisions, trigonometric and other complex mathematical functions.
- Leakage power consumption is cut down by using high threshold voltage (High-V_T) libraries for standard cells and memories.
- Scaling down program memory from 128 to 64 bits, reducing data memory sizes and decreasing processor complexity.

We emphasize the impact of these optimizations on a benchmarking example, in which we mapped the R-peak detection algorithm onto a processor using the coarse-grained reconfigurable technology of Silicon Hive [27]. The algorithm executes on the Silicon Hive processor at 100MHz with a sample rate of 200Hz. The empirical results obtained before optimization suggested a duty cycle of 0.25%, showing that most of the power is consumed when the processor is idle. Dynamic and static optimization of power consumption led to the following results:

- Dynamic idle power consumption is reduced to $\sim 0\mu W$ (clock consumption).
- Cycle count is reduced by a factor of five, therefore driving active time down to only 0.05%.
- Leakage power is decreased by a factor 14, mainly thanks to the use of high threshold voltage (factor 7).

Power analysis after a gate-level simulation, with wire-capacitances back-annotated from layout, reports a power consumption of $7.81\mu W$ ($5.45\mu W$ leakage, $2.36\mu W$ active). Table 1 summarizes the power optimization achieved for our benchmarking case, compared to an initial Silicon Hive (generic purpose RISC) processor running at 100MHz and at 1.25MHz, minimal clock frequency in this case. Further reduction of leakage power during idle time could be achieved by dynamic power switching or dynamic voltage scaling. Since the processor spends most of its time in idle mode (Table 1), this should lead to a leakage power below 1 μW.

This discussion has shown that Digital Signal Processors, combining both flexibility and efficiency, are suitable for BSN applications powered by energy scavenging technology. Future research should focus on additional architectural explorations, application analysis and advanced power optimizations e.g. power gating and data path enhancement.

Table 1. Power breakdown for benchmarking example (see text).

	Original @ 100Mhz	Original @ 1.25Mhz	Optimized @ 100Mhz
Active	6.52 mW * 0.25%	81.55 µW * 20.00%	4.71 mW * 0.05%
Idle	0.75 mW * 99.75%	9.41 µW * 80.00%	0.00 mW * 99.95%
Leakage	100.04µW	100.04µW	5.45 µW
TOTAL P	867.47µW	123.84µW	7.81 µW

Acknowledgments. Results presented in this section were obtained in collaboration with Silicon Hive, the TU Eindhoven and Philips Research. The authors would like to thank Frank Bouwens, Lennart Yseboodt, Jos Huisken and Jef van Meerbergen for their contribution.

7.   Sensors and Actuators

Our multi-parameter monitoring system needs to acquire biopotential signals in a very power efficient way. The different biopotentials have different amplitude and frequency characteristics as shown in Fig. 12. For this purpose, we developed a low-power 25 channel biopotential ASIC [30]. The ASIC preprocesses typical biopotentials such as ECG and EEG signals. It can be configured in different operational modes thanks to its variable bandwidth and gain settings.

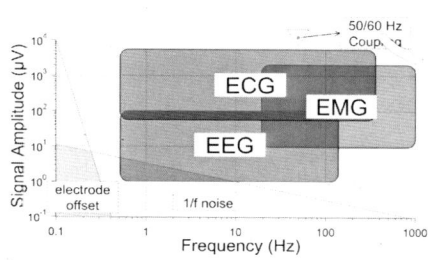

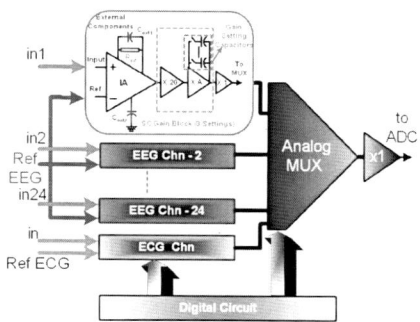

Fig. 12. Amplitude and frequency characteristics of different biopotential signals.

Fig. 13. 25 channel biopotential ASIC.

The mixed-signal ASIC consists of 25 channels (Fig. 13). In a typical configuration, 24 channels are configured for EEG measurements and 1 channel is configured as ECG channel. Each channel of the ASIC consists of a high CMRR instrumentation amplifier, followed by a variable gain amplifier. There are 8 different gain modes ranging from 200 to 10000 for the EEG channels and from 20 to 1000 for the ECG channel. The front end instrumentation amplifier has bandpass filter characteristics, where in-band gain and the cut-off frequencies are settable with external components. With an external capacitor of 1 μF, a bandwidth of 0.5 – 80Hz is selected. The CMRR is larger than 90dB at 50mV electrode offset. The total input referred voltage noise of each channel is less than 1 μVrms in the 0.5 Hz – 80 Hz bandwidth. These features allow suppressing the input common mode voltages coupled to the human body, while amplifying the microvolt level biopotential signals. The mixed signal ASIC is designed and fabricated in 0.5μm CMOS process. The ASIC can operate from a voltage supply ranging from 2.7 V – 3.3 V while dissipating less than 10.5 mW. All the channels of the ASIC are multiplexed with a frequency of 1 kHz per channel and buffered at the output. Therefore, a single ADC with a maximum input capacitance of 50 pF can sample all the channels of the ASIC.

In a test setup, two channels of the ASIC are connected to Ag/AgCl electrodes for reading the brain activity at the occipital cortex (backside of the head). A microcontroller with integrated ADC, is directly connected to the ASIC. Operation and gain settings of the ASIC are controlled from the microcontroller. When the patient closes his eyes, the typical alpha rhythm becomes clearly visible at the output (Fig. 14).

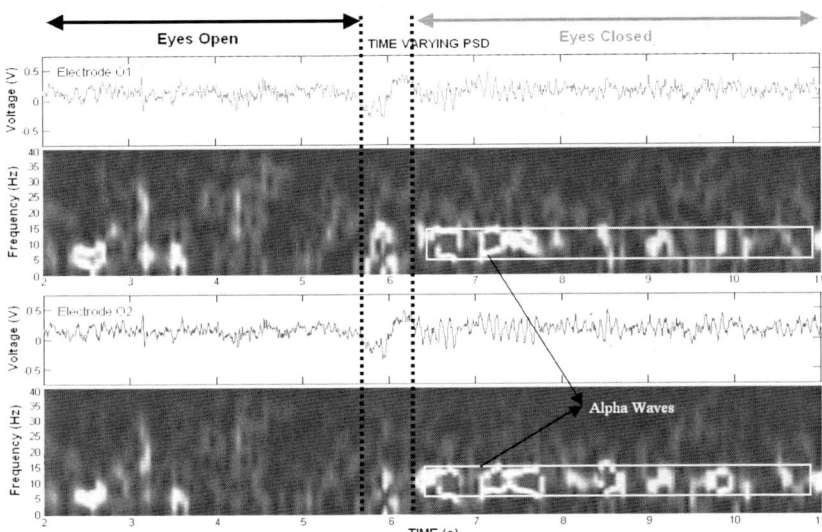

Fig. 14. Alpha activity from the two electrodes at occipital cortex, and their Short-Time Fourier Transform.

More recently we have also fabricated an alternative instrumentation amplifier architecture with a power consumption of 0.06mW per channel while maintaining a CMRR of 120dB [31]. This ASIC provides an additional factor of 20 in power savings per channel compared to the 24 channel version.

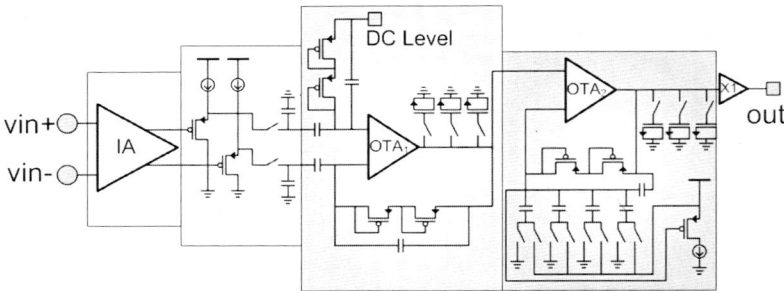

Fig. 15. Architecture of single channel biopotential ASIC.

8.   Integration Technology

One form factor suitable for many sensor applications is a small cubic sensor node. To this end, a prototype wireless sensor node has been integrated in a cubic centimeter (Fig. 16). In this so-called three-dimensional system-in-a-package approach (3D SIP) [19], the different functional components are designed on separate boards and afterwards stacked on top of each other through a dual row of 0.7mm solder balls with a pitch of 1.27mm. This system has the following advantages:

(i) modules can be tested separately, (ii) functional layers can be added or exchanged depending on the application, (iii) each layer can be developed in the most appropriate technology. The first generation 3D stack offers a complete System-in-a-Package (SiP) solution for low power intelligent wireless communication. The first generation 3D stack offers a complete System-in-a-Package (SiP) solution for low power intelligent wireless communication. The integrated stack includes a commercial low power 8 MIPS microcontroller [20] and 2.4GHz transceiver [21], crystals and all necessary passives, as well as a matched dipole antenna custom-designed on the top layer laminate substrate. The bottom layer has a BGA footprint, allowing standard techniques for module mounting. This sensor module has been integrated with the watch-like thermal scavenger and consists in a basis for sensor networks [29], which, unlike most of their predecessors, are fully energy autonomous.

9. Conclusion

This text gave an overview of the Human++ research program, which is targeted at developing key technologies and components for future wireless body area networks for health monitoring applications. Several enabling technologies and integrated modules were discussed.

Over the next years, we will see more BAN technologies being developed all over the world. Step by step these will bring us closer to the end goal: an unobtrusive portable system that keeps track of our health and fitness level at an affordable cost.

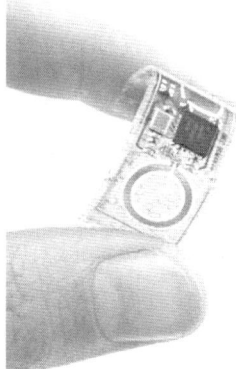

Fig. 16. 3D SiP Wireless autonomous sensor node. **Fig. 17.** Sensor in a flexible band aid.

Parallel research was started to implement the same technology on a flexible carrier. The ultimate target is to create a small and smart band-aid containing all the necessary technology for sensing and communication with a base station. It will provide a generic platform for various types of applications (wound healing, EEG, ECG, EMG...). The first prototype (Fig. 17) is 10 times smaller than a credit card

(12 x 35 mm) and about as thin as a compact disc (1-2 mm). The flexible 25 μm polyimide carrier contains a microprocessor and a wireless communication module (2.4 GHz radio). It enables us to optimize the antenna for its activity on human skin. Current focus lies on adding the necessary sensors and energy equipment (rechargeable battery, energy scavenger and advanced electronics to keep energy consumption as low as possible). We target an ultimate device thickness of approximately 100 μm.

The biggest challenges in developing this kind of modules are the extreme miniaturization and its effects on the functionality of the used components. Some of the many problems to tackle are the use of naked chips, chip scaling, assembly processes like wire bonding and flip-chip on a flexible substrate, application of thin-film batteries and solar cells and integration of the entire technology in a biocompatible package.

References

[1]   http://www.imec.be/ovinter/static_research/human++.shtml
[2]   R. Schmidt et al., "Body Area Network BAN, a key infrastructure element for patient-centered medical applications", Biomed Tech (Berl). 2002;47 suppl 1 pt 1:365–8
[3]   http://www.bluetooth.com, http://www.zigbee.org
[4]   http://www.bluetooth.com/Bluetooth/Learn/Technology/Compare/Technical/
[5]   IEEE 802.15.4a: http://www.ieee802.org/15/pub/tg4a.html
[6]   http://www.rfm.com/products/data/tr1001.pdf
[7]   B.W. Cook et al., "An ultra-low power 900 MHz RF Transceiver for wireless sensor networks in 0.13 um cmos with 400 mV supply and an integrated passive RX front-end", ISSCC, 2006, pp. 1460–1461
[8]   Y.H. Chee, A.M. Niknejad, and J.M. Rabaey, "An ultra-low power MEMS-based two-channel transceiver for wireless sensor networks," Symposium on VLSI Circuits, June 2004, pp. 20–23
[9]   M.Z. Win and R.A. Scholtz. Impulse radio: how it works. In IEEE Communications Letters, pp. 36–38, February 1998
[10] Federal Communications Commission (FCC). Revision of part 15 regarding ultra-wideband transmission systems. First Report and Order, ET Docket, 98–153, FCC 02–48, adopted Feb. 2002, released Apr. 2002, available at http://www.fcc.gov
[11] http://www.ecma-international.org/publications/standards/Ecma-368.htm
[12] standard draft proposal. IEEE 802.15.4a available at http://www.ieee802.org/15/pub/TG4a.html
[13] Y.H. Choi. Gated UWB pulse signal generation. In IEEE Joint International Workshop of UWBST and IWUWBS, pp. 122–124, May 2004
[14] J. Ryckaert, M. Badaroglu, C. Desset, V. de Heyn, G. Van der Plas, P. Wambacq, B. Van Poucke, and S. Donnay. Carrier-based uwb impulse radio: Simplicity, flexibility, and pulser implementation in 0.18μm cmos. International Conference on Ultrawideband, ICU 2005, 2005
[15] R. Blazquez, F.S. Lee, D.D. Wentzloff, B. Ginsburg, J. Powell, and A.P. Chandrakasan. Direct conversion pulsed uwb transceiver architecture. Proc. of Design, Automation and Test in Europe, March 2005
[16] J. Ryckaert, M. Badaroglu, V. De Heyn, G. Van der Plas, P. Nuzzo, A. Baschirotto, S. D'Amico, C. Desset, H. Suys, M. Libois, B. Van Poucke, P. Wambacq, and B. Gyselinckx. A 16mA UWB 3-to-5Ghz 20Mpulses/s quadrature analog correlation

receiver in 0.18μm CMOS. In ISSCC Digest of Technical Papers, pp. 114–115, February 2006

[17] J. Ryckaert, P. De Doncker, R. Meys, A. de Le Hoye, and S. Donnay, "Channel model for wireless communication around the human body", Electronics Letters, Vol. 40, Nr. 9, pp. 543–544, April 2004

[18] M. Verhelst, W. Vereecken, M. Steyaert, and W. Dehaene. Architectures for low power ultra-wideband radio receivers in the 3.1–5GHz band for data rates < 10Mbps. In ISLPED'04, International Symposium on Low Power Electronics and Design, pp. 280–285, August 2004

[19] S. Stoukatch, M. Ho, K. Vaesen, T. Webers, G. Carchon, W. De Raedt, E. Beyne, and J. De Baets, "Miniaturization using 3-D stack structure for SIP application", Proc. SMTA (Surface Mount Technology Association) International Conference, 21–29 September 2003; Chicago.

[20] TI MSP430F149, www.ti.com

[21] Nordic nRF2401, www.nvlsi.com

[22] V. Leonov, P. Fiorini, S. Sedky, T. Torfs, and C. Van Hoof, "Thermoelectric MEMS generators as a power supply for a body area network". Proc. 13-th Int. Conf. Solid-State Sensors, Actuators and Microsystems (Transducers'05), Seoul, Korea, June 5–9, 2005, pp. 291–294

[23] V. Leonov, T. Torfs, P. Fiorini, and C. Van Hoof, "Thermoelectric converters of human warmth for self-powered wireless sensor nodes". IEEE Sensors Journal, 2007 (in press)

[24] T. Torfs, V. Leonov, B. Gyselinckx, and C. Van Hoof, "Body-Heat Powered Autonomous Pulse Oximeter", proceedings, IEEE Sensors, 2006, Korea, in press

[25] S.J. Roundy, "Energy Scavenging for Wireless Sensor Nodes with a Focus on Vibration to Electricity Conversion", PhD Thesis, UC Berkeley, 2003

[26] B. Köhler, H. Carsten, and R. Orglmeister: *The Principles of Software QRS Detection*, Feb. 2002, IEEE Engineering in Medicine and Biology, 42–57

[27] T.R. Halfhill: *Silicon Hive Breaks Out*, Dec. 1st 2003, Microprocessor Report, www.MPRonline.com

[28] J. Pan and W.J. Tompkins: *A Real-Time QRS Detection Algorithm*, 1985, IEEE Transactions Biomedical Engineering, BME-32(3): 230–236

[29] M. Tubasihat and S. Madria, "Sensor networks: an overview", IEEE Potentials, 2003, Vol. 22(2), 20–23

[30] R.F. Yazicioglu, P. Merken, and C. Van Hoof, "Integrated Low-Power 24-Channel EEG Front-End", IEE Electronics Letters, Vol. 41, iss. 8, pp. 457–458, Apr. 2005

[31] R.F. Yazicioglu, P. Merken, R. Puers, and C. Van Hoof, "A 60 μW 60nV/√Hz Readout Front-End for Portable Biopotential Acquisition Systems," IEEE J. Solid-State Circuits, 2007

[32] A.A. Ahmed, "A Survey on Network Protocols for Wireless Sensor Networks", Information Technology: Research and Education, 2003 proceedings, pp. 301–305

[33] M.A.M. Vieira et al., "Survey on wireless sensor network devices", IEEE Emerging Technologies and Factory Automation 2003 proceedings, Vol. 1, pp. 537–544

[34] B.A. Warneke et al., "An autonomous 16 mm³ solarpowered node for distributed wireless sensor networks", IEEE Sensors 2002 proceedings, Vol. 2, pp. 1510–1515

[35] T. Torfs, S. Sanders, C. Winters, S. Brebels, and C. Van Hoof, "Wireless network of autonomous environmental sensors", Proceedings of IEEE Sensors 2004, Vienna, 24–27 October 2004.

Printed in the United States of America